# KEY NOTES ON ENTOMOLOGY, PLANT PATHOLOGY AND MICROBIOLOGY

*For Ready Reference to the*

**STUDENTS, TEACHERS, RESEARCHERS & ASPIRANTS OF COMPETITIVE EXAMINATIONS**

## THE EDITORS

**Dr. U.D. Chavan** obtained his M.Sc. (Agri. in Biochemistry) degree from Mahatma Phule Krishi Vidyapeeth, Rahuri. He received his Ph.D. degree in Food Science from Memorial University of Newfoundland St. John's Canada in 1999. He has done International Training on "Global Nutrition 2002" at Uppsala University Uppasala, Sweden in 2002. Dr. Chavan worked as Senior Research Assistant in the Department of Biochemistry & Food Science and Technology at MPKV Rahuri from 1988 to 2000. During his Ph.D., he worked as Technician/Research Associate at Atlantic Cool Climate Crop Research Center and Agriculture and Agri-Food Canada. He received D.Sc. degree in 2006 from USA.

Dr. Chavan is presently working as a Senior Cereal Food Technologist in the Department of Food Science & Technology at Mahatma Phule Krishi Vidyapeeth, Rahuri.

**Dr. J.V. Patil** obtained his M.Sc. (Agri.) from, MPKV, Rahuri. He completed his course work for Ph.D. at CCSHAU, Hisar and research at MPKV, Rahuri in 1992. He rendered his research and teaching services at MPKV Rahuri as Geneticist, Associate Professor, Plant Breeder and Professor of Genetics & Plant Breeding and Head, Genetics and Plant Breeding Department, MPKV, Rahuri. He also delivered many administrative responsibilities in the University. Dr. Patil joined as the Director, Directorate of Sorghum Research, Hyderabad in August 2010.

## THE CONTRIBUTORS

**Dr. B.P. Chavan** is an Assistant Professor in the Department of Agricultural Entomology at Mahatma Phule Krishi Vidyapeeth, Rahuri.

**Dr. A.G. Chandele** is a Head of the Department of Agriculture Entomology at Mahatma Phule Krishi Vidyapeeth, Rahuri.

**Dr. R.B. Sonawane**, Former Faculty in the Department of Plant Pathology and Microbiology at Mahatma Phule Krishi Vidyapeeth, Rahuri.

**Dr. G.P. Deshmukh** is an Assistant Professor in the Department of Plant Pathology and Microbiology at Mahatma Phule Krishi Vidyapeeth, Rahuri.

# KEY NOTES ON ENTOMOLOGY, PLANT PATHOLOGY AND MICROBIOLOGY

*For Ready Reference to the*

STUDENTS, TEACHERS, RESEARCHERS & ASPIRANTS OF COMPETITIVE EXAMINATIONS

*Editors*

## U.D. CHAVAN
&
## J.V. PATIL

*– Contributors –*

B.P. CHAVAN
A.G. CHANDELE
R.B. SONAWANE
G.P. DESHMUKH

**2015**

# Daya Publishing House®
*A Division of*
# Astral International (P) Ltd
New Delhi 110 002

*Published by* : **Daya Publishing House®**
*A Division of*
**Astral International Pvt. Ltd.**
– ISO 9001:2008 Certified Company –
4760-61/23, Ansari Road, Darya Ganj
New Delhi-110 002
Ph. 011-43549197, 23278134
E-mail: info@astralint.com
Website: www.astralint.com

*Laser Typesetting* : **Twinkle Graphics, Delhi**

*Printed at* : **Thomson Press India Limited**

PRINTED IN INDIA

# PREFACE

India is an agricultural country. The Indian economy is basically agarian. Inspite of economic and industrialization, agriculture is the backbone of the Indian economy. As Mahatma Gandhi said "India's lives in villages and agriculture is the soul of Indian economy". Agriculture is a vast subject and encompasses at least 20 major and minor subjects in it. New developments have lead to entirely a new face of agriculture. Study of agriculture has always been intrigued with a mosaic of interwove concepts, subjects, facts and figures. There are number of books and large literature on Entomology, Plant Pathology and Microbiology but the Key Notes type of book have not been compiled in a readable manner.

The present book *"Key Notes on Entomology, Plant Pathology and Microbiology"* has been designed to fulfill this long felt need of students, teachers, researchers and aspirants of competitive examinations. It is designed in such a way that give rapid, easy access to the core materials in a short format which facilitates easily learning and rapid revision. The book carries fundamentals of Entomology, Plant Pathology and Microbiology. The book is divided in two parts. The Part A of the book is Key Notes on Entomology and the Part B of the book is on Key Notes on Plant Pathology and Microbiology. The most recent information is provided along with a detailed list of references for further reading.

Hope this book would be highly useful for graduate and post-graduate students of agriculture, teachers and researches. This book will also useful for the aspirants of various competitive examinations such as Agricultural Research Service (ARS), ICAR- National Eligibility Test (NET), State Eligibility Test (SET), Junior Research Fellowship (JRF), Senior Research Fellowship (SRF), Civil Services, Allied Agricultural Examinations and Extension Workers for reference and easy answers of many complicated questions. Thus it is expected that this book will adequately meet the need of wider circle of students and readers for preparing their professional career.

We acknowledge the references that are used in this manuscript. Authors are also thankful to all scientists and friends who have helped directly or indirectly while preparing this manuscript. The editors of grateful to all the contributors for their cooperation, support and timely submission of their manuscripts for bringing out this publication. We would have like to acknowledge the patience and support of our families whilst we have spent many hours with drafts of manuscripts rather than with them. Lastly, our sincere thanks to publisher Astral International Pvt. Ltd., New Delhi who provides an opportunity to publish this book.

To all readers we extend an invitation to report that no doubts have escaped our attention and to offer suggestion for improvements that can be incorporated in future editions.

**U.D.Chavan and J.V. Patil**

**Editors**

# CONTENTS

# PART A

# KEY NOTES ON ENTOMOLOGY

# 1

# DISCOVERIES

**Entomology :** The work 'Entomology' is derived from two Greek words Entomon–Insects and Logous–discourage.

It can be defined as a branch of zoology which deals with the study with insects and various other aspects such as the taxonomy, anatomy, morphology, physiology and control measures of insects.

**Branches of Zoology :** Anthropology, Anthrozoology, Apiology, Arachnology, Arthropodology, Cetology, Conchology, Entomology, Ethology, Helminthology, Herpetology, Ichthyology, Malacology, Mammalogy, Myrmecology, Nematology, Neuroethology, Ornithology, Paleozoology, Planktology, Rodentology, etc.

**Branches of Applied Entomology :** Agricultural Entomology, Horticultural Entomology, Forest Entomology, Medical and Veternary, Entomology, etc.

**Agricultural entomology :** It is the study of insects in relation to agriculture.

**Branches of Agricultural Entomology :** Insect Morphology, Insect Anatomy and Physiology, Insect Ecology, Insect Taxonomy, Insecticide Toxicology, Insect Parasitology, Insect Pathology.

Man originated about a million years ago, and insects at least 500 million years ago. At some 1.3 million described species, insects account for more than two-thirds of all known organisms. Insects are omnipresent and each crop we cultivate is being attacked by at least a dozen of insect species called as pests. Apart from the pest insects there are several productive and useful insects. Insects are considered as one of the major constrain in increasing agricultural productivity. Hence it is important to understand about the insects, their biology, classification and management.

Our earliest knowledge about insects dates back to 6000 years as our Indian ancestors were well versed in the art of rearing silk worms and weaving silk cloth. Even during 3870 BC an Indian king sent various silken materials as presents to a Persian king. There are lot of mentions about insects in our mythological epics, the Ramayana (2550-2150 BC) and Mahabharata (1424-1366 BC). Mentions about silk, honey and lac were found in these epics. The first detailed classification of insects was done by Umaswati (0-100 AD). Classification of bees by the Indian physician Charaka (1200-1000 BC) and classification of ants, flies and mosquitoes by the surgeon Sushruta (100-200 AD) are the evidences for our earliest knowledge about insects.

Entomology in modern India must have taken its place in Natural History some time after the 16<sup>th</sup> century.

| Year | Discovery |
|------|-----------|
| 1758 | The beginning of the modern era of Indian Entomology came with the publication of the tenth edition of Carl Linnaeus's "Systema Naturae". This contained the earliest record of 28 species of Indian insects. |
| | The first entomologist who made any extensive study of Indian Insects was J.C. Fabricius. He was a Danish Professor of political economy turned in to a systematist and classified the insects in to 13 orders based on type of mouth parts. |
| 1767-79 | J.G. Koenig, a medical Officer from Denmark, a student of Carl Linnaeus and friend of Fabricius, for the first time during the 18th century collected number of insects from Coromandel area and Southern Peninsular India and his collections were studied and named by Professor Linnaeus himself. He also published a special account of the termites of Thanjavur District. Fabricius, made Koenig's name remembered for ever by naming the well known and destructive red cotton bug of this country as Dysdercus koenigi. |
| 1782 | Dr. Kerr published an account of lac insect. |
| 1785 | Asiatic Society of Bengal started in Calcutta. |
| 1791 | Dr. J. Anderson issued a monograph on Cochineal scale insects. |
| 1799 | Dr. Horsfield, an American doctor and first Keeper of the East India Museum published his famous book "A Catalogue of the Lepidopterous Insects in the Museum of the Honourable East India Company, 2 vols. (1857, 1858-59). |
| 1800 | Buchanan (Traveller studying the natural wealth of India) wrote on the cultivation of lac in India and on sericulture in some parts of South India. |
| | Edward Denovan published an illustrated book entitled "An epitome of the natural history of insects of India and the Islands in the Indian seas" which was the first pictorial documentation on the insects of Asia and was revised in 1842 by West Wood. |
| 1875- | Foundation of the "Indian Museum" at Calcutta. |
| 1883 | "Bombay Natural History Society" was started. Numerous contributions of Indian insects were published in the Journal of the Bombay Natural History. |
| | Commencement of "Fauna of British India" series under the editorship of W.T. Blandford. |

*Contd...*

| Year | Discovery |
|------|-----------|
| 1892 | Entomological part of the "Fauna of British India" (now Fauna of India) series started with Sir George Hampson contributed first of the four volumes on the moths of India. |
| 1893 | Rothney published on Indian Ants (earliest record of biological pest control in India), i.e., white ants attack on stationary items was kept free by red ants. |
| 1889-1903 | Indian Museum, Calcutta published the "Indian Museum Notes" in five volumes, which contributed much on economic entomology and applied entomology in India. |
| 1901 | Lionel de Nicevelle was posted as the first entomologist to the Government of India. |
| 1903 | Professor Maxwell Lefroy succeeded Nicevelle as Government Entomologist. |
| 1905 | Establishment of Imperial Agricultural Research Institute at Pusa, Bihar and Professor Lefroy became the first Imperial Entomologist. He convened a series of entomological meetings on all India basis to bring together all the entomologists of the country. |
| 1906 | 'Indian Insect Pests' by Professor Lefroy. |
| 1912 | Plant Quarantine Act was enforced. |
| 1914 | Destructive Insects and Pests Act was enforced. |
|  | T.B. Fletcher, the first Government Entomologist of Madras State, published his book "Some South Indian Insects". Another contribution of Fletcher is the publication of series under the "Catalogue of Indian Insects" which marked another era in the history of Indian Insect Taxonomy. |
|  | 'Indian Forest Insects of Economic Importance: Coleoptera' was published by the first Imperial Forest Entomologist E.P. Stebbing. |
| 1916 | Imperial Forest Research Institute was established at Dehradun, and E.P. Stebbing was appointed as Forest Zoologist. |
|  | The Natural History Section of the Indian Museum was formed as the Zoological Survey of India. |
| 1925 | Indian Lac Research Institute started at Namkum, Ranchi. |
| 1934 | Hem Singh Pruthi who succeeded Fletcher as Imperial Entomologist, made efforts for the foundation of 'Entomological Society of India' in 1938. Afzal Hussain was the first President of the Entomological society of India and the Vice-President were Hem Singh Pruthi and Ramakrishna Ayyar. The official publication of the Society is the 'Indian Journal of Entomology'. |

*Contd...*

| Year | Discovery |
|------|-----------|
| 1939 | Locust Warning Organisation was established |
| 1940 | Dr. T.V. Ramakrishna Ayyar published the book "Handbook of Economic Entomology for South India" and revised second edition was published in 1963. Other useful publications by Indian authors. |
| 1946 | Directorate of Plant Protection, Quarantine and Storage of GOI started. |
| 1960 | "The Desert Locust in India" monograph by Y.R. Rao. |
| 1963 | Text book of Agricultural Entomology by H.S. Pruthi. |
| 1968 | The Govt. of India enacted 'Central Insecticide Act' which came into force from 1st January, 1971. |
| 1968 | Dr. M.S. Mani's "General Entomology". |
| 1969 | Dr. Pradhan's "Insect Pests of Crops". |
| 1969 | "The monograph on Indian Thysanoptera" was published by Dr. T.N. Ananthakrishnan. |

## HISTORY OF INSECTICIDE DEVELOPMENT

| Year | Insecticide Development |
|------|-------------------------|
| 2500 BC | First records of insecticides, e.g., The Sumerians were using sulphur compounds to control insects and mites. |
| 900 | Arsenites in China (Inorganic compound). |
| 1690 | Tobacco used in Europe (Plant/natural product). |
| 1787 | Soaps used in Europe. |
| 1867 | Paris Green in US. |
| 1874 | DDT synthezized by Zeidler. |
| 1883 | Bordeau in France. |
| 1925 | Dinitro compounds (First synthetic organic insecticide). |
| 1939 | DDT insecticidal property discovered by Paul Muller of Switzerland. Paul Muller awarded Nobel Prize in 1948 for discovering insecticidal property of DDT. |
| 1941 | BHC in France and UK (in 1942) (BHC is presently called as HCH). |

*Contd...*

| Year | Insecticide Development |
|------|------------------------|
| 1944 | Parathion (Organo phosphate) discovered by Gerhard Schrader in Germany. |
| 1945 | Chlordane (Cyclodian compound) in Germany. |
| 1947 | Carbamate insecticides in Switzerland. |
| 1962 | Rachel Carson's "Silent Spring" appears (US) : "Silent Spring" helped to stimulate the implementation of the concept of Integrated Pest Management (IPM) in the late 1960's, and biological control was seen as a core component of IPM by some. |
| 1967 | First JH mimic (Juvenile Hormone mimic) used in US (Insect growth regulator). |
| 1970 | Development of synthetic pyrethroids (UK) (Fast degradation) (Effective at very low doses). |
| 1980 | Discovery of avermectins (derived from bacteria). Effective at low dose. Fast degradation. |
| 1990 | Discovery of newer groups like (1) Neonicotinoids (Imidacloprid), similar to natural nicotin, (2) Spinosyns, (e.g., Spinosad) derived from actinomycet. |

| | Generation | Year | Compounds |
|---|-----------|------|-----------|
| 1. | First generation insecticide | 1939-42 | BHC and DDT |
| 2. | Second generation insecticide | 1944-47 | Organophosphates and Carbamate |
| 3. | Third generation insecticide | 1967 | Hormonal insecticides, JH mimic insect growth regulators |
| 4. | Fourth generation insecticide | 1970s | Synthetic pyrethroids |

## History and Development of Biological Control

Early History: 200 A.D. to 1887 A.D.

| Year | Development of Biological Control |
|------|----------------------------------|
| Ancient times | **Chinese were the first to use natural enemies to control insect pests.**<br><br>The Chinese citrus growers placed nests of predaceous ants, Oncophylla smaradina, in trees where the ants fed on foliage-feeding insects. Bamboo bridges were constructed to assist the ants in their movements from tree to tree. |

*Contd...*

| Year | Development of Biological Control |
|------|----------------------------------|
| | Ants were used in 1200 A.D. for control of date palm pests in Yemen (south of Saudia Arabia). Nests were moved from surrounding hills and placed in trees. |
| | Usefulness of ladybird beetles recognized in control of aphids and scales in 1200 A.D. |
| | **1300 A.D. to 1799 A.D—BC was just beginning to be recognized** |
| 1602 | In 1602 Aldrovandi recorded observations of parasitic larvae of Apanteles (Cotesia) glomeratus exiting from cabbage butterfly (Pieris rapae) and spinning external cocoons. However, he misinterpreted these cocoons as butterfly eggs. It wasn't until 1670 that Martin Lister correctly interpreted insect parasitism in a letter published in the Philosophical Transactions of the Royal Society of London. |
| 1700 | Antoni van Leeuwenhoek in 1700 described the phenomenon of parasitoidism in insects. He drew a female parasitoid laying eggs in aphid hosts. |
| 1706 | Vallisnieri (1706) first correctly interpreted this host-parasitoid association and probably became the first to report the existence of parasitoids. |
| 1726 | The first insect pathogen was recognized by de Reaumur in 1726. It was a Cordyceps fungus on a noctuid. |
| 1762 | **First successful importation of an organism from one country to another-**<br>The mynah bird, Acridotheres tristis, was successfully introduced/imported from India to Mauritius (off coast of Madagascar) for control of the red locust, Nomadacris septemfasciata. |
| 1770 | Bamboo runways between citrus trees for ants to control caterpillars |
| 1888 | First well planned and successful biological control attempt made During 1888 citrus industry in California (USA) seriously threatened by cottony. |
| | *cushion scale, Icerya purchasi* |
| | Mr. C.V. Riley, a prominent entomologist suggested that the scale inset originated from Australia and natural enemy for the scale from Australia should be introduced into USA. |

*Contd...*

| Year | Development of Biological Control |
|------|-----------------------------------|
| | Mr. Albert Koebele was sent to Australia. He found a beetle called Vedalia (Rodolia cardinalis) attacking and feeding on seeds. He sent 12,000 individuals of Cryptochaetum iceryae and 129 individuals of Rodolia cardinalis (the vedalia beetle). |
| | Vedalia beetle (Rodolia cardinalis) was imported in November 1888 into USA and allowed on scale infested trees. Within a year spectacular control of scale insect achieved. Even till date this beetle controls the scale insect-After this successful attempt of biological control many such introduction of natural enemies were tried. |
| 1898 | First introduction of natural enemy into India. |
| | A coccinellid beetle, Cryptolaemus montrouzieri was imported into India from Australia and released against coffee green scale, Cocus viridis. Even today it is effective against mealybugs in South India. |
| 1902 | The Lantana Weed Project in Hawaii (1902). First published work on BC of weeds. Koebele went to Mexico and Central America looking for phytophagus insects which were sent to R.C.L. Perkins in Hawaii. |
| 1911 | Berliner described Bacillus thuringiensis in 1911 as causative agent of bacterial disease of the Mediterranean flourmoth. |
| 1920 | A parasitoid, Aphelinus mali introduced from England into India to control Woolly aphid on Apple, Eriosoma lanigerum. |
| 1929-31 | Fodolia cardinalis imported into India (from USA) to control cottony cushion scale Icerya purchasi on Wattle trees. |
| 1958-60 | Parasitoid Prospatella perniciosus imported from China. |
| 1960 | Parasitoid Aphytis diaspidis imported from USA. Both parasitoids used to control Apple Sanjose scale Quadraspidiotus perniciosus. |
| | "Silent Spring" helped stimulate the implementation of the concept of Integrated Pest Management (IPM) in the late 1960's, and biological control was seen as a core component of IPM by some. More emphasis was placed on conservation BC than classical BC. |
| 1964 | In 1964, Paul DeBach and Evert I. Schliner (Division of Biological Control, University of California, Riverside) publish an edited volume titled "Biological Control of Insect Pests and Weeds" which becomes a major reference source for the biological control community. This was basically a California based book with international application. |

*Contd...*

| Year | Development of Biological Control |
|------|----------------------------------|
| 1964 | Egg parasitoid Telenomus sp. imported from New Guinea to control Castor semilooper Achaea janata. |
| 1965 | Predator Platymeris laevicollis introduced from Zanzibar to control coconut Rhinoceros beetle, Oryctes rhinoceros. |
| 1983 | Frank Howarth published his landmark paper titled "Biological Control: Panacea or Pandora's Box" and significantly impacted classical BC efforts by concluding that classical BC of arthropods significantly contributed to extimction of desirable species (e.g., endemic). |
| 1990 | In the 1990's, two additional biological control journals appeared, "Biological Control - Theory and Application in Pest Management" (Academic Press) and "Biocontrol Science and Technology" (Carfax Publishing). Additionally, "Entomophaga" changed its name to "Biocontrol" in 1997. |
| Till 1988 | At global level 384 importations made against 416 species of insect pests. Out of them |
| | 164 species (39.4%) - Completely controlled |
| | 75 species - Substantially controlled |
| | 15 species - Partially controlled |
| | Regional Station of Commonwealth Institute of Biological Control (CIBC) established at Bangalore in 1957. Presently Project Directorate of Biological Control (PDBC) Bangalore looks after Biocontrol in India. Now it is renemed as National Bureau of Agriculture Important Insects. |
| 1947 | In 1947 the Commonwealth Bureau of Biological Control was established from the Imperial Parasite Service. In 1951 the name was changed to the Commonwealth Institute for Biological Control (CIBC). Headquarters are currently in Trinidad, West Indies. |
| 1955 | In 1955 the Commission Internationale de Lutte Biologique contre les Enemis des Cultures (CILB) was established. This is a worldwide organization with headquarters in Zurich, Switzerland. In 1962 the CILB changed its name to the Organisation Internationale de Lutte. |
| | Biologique contre les Animaux et les Plants Nuisibles. This organization is also known as the International Organization for Biological Control (IOBC). Initiated the publication of the journal "Entomophaga" in 1956, a journal devoted to biological control of arthropod pests and weed species. |

## History of Integrated Pest Management

| Year | Development of IPM |
|---|---|
| Early 1800 | First book on on pest control. |
| 1880 | First commercial spraying machine. |
| 1885 | Introduction of Bordaux Mixture by P.A. Millardate. |
| 1901 | Successful biological control of weed. |
| 1939 | Insctciadl properties of DDT were discovered by Paul Muller DDT was synthesized by Zeilder in 1874. |
| 1946 | Resistance to DDT was noticed in Housefly, Musa domestica. |
| 1952 | Michelbacher and Bacon coined the term "integrated control". |
| 1959 | Stern et al. (1959) defined integrated control as "applied pest control which combines and integrates biological and chemicalcontrol. |
| 1959 | Introduction of concepts of economic thresholds, economic levels and integrated control by V.M. Stern, R.F. Smith, R. van den Bosch and K.S. Hagen. |
| 1960 | First insect sex pheromone isolated, identified and synthesis in the gypsy moth. |
| 1962 | Rachel Carsons "Silent Spring". |
| 1966 | Geier (1966) coined the term "pest management". |
| 1967 | Introduction of the term Integrated Pest Management by R.F. Smith and R. van den Bosch. The relevance of ecology to IPM through the concept of "Life Systems" was introduced by L.R. Clark, P.W. Geier, R.D. Hughes and R.F. Morris. |
| 1969 | The term Integrated Pest Management was formalized by the US Academy of Sciences in 1969. |
| 1967 | Food and Agricultural Organization (FAO, 1967) defined IPM as "a pest management system, that, in the context of associated environment and population dynamics of the pest species, utilizes all suitable techniques and methods in as compatible a manner as possible and maintains pest populations at levels below those causing economic injury". |
| 1981 | IPM in India. |
| 1988 | Establishment of NCIPM by ICAR, New Delhi. |
| 1989 | In 1989, IPM Task Force was established and in 1990. IPM Working Group (IPMWG) was constituted to strengthen implementation of IPM at international level. |
| 1997 | Smith and Atkisson were awarded the World Food Prize for pioneering work on implementation of IPM. |

# 2

# ABBREVIATIONS

| Abbreviation | Full Form |
|---|---|
| Ach | Acetyl Choline |
| AchE | Acetyl Choline Esterase |
| ADI | Acceptable Daily Intake |
| AE | Aerosol Dispenser |
| AH | Activation Hormone |
| ASPW | American Spring and Pressing Works. |
| BHC | Benzene Hexa Chloride |
| BIPM | Bio-intensive Integrated Pest Management |
| BR | Briquette |
| BT | Bacillus thuriengensis |
| BV | Baculovirus |
| CA | Corpora Alata |
| CAC | Codex allmentarius commission |
| CC | Corpora Cardiaca |
| CCD | Colony Collapse Dosorder |
| CCP | Critical Control Point |
| CFU | Colony Forming Units |
| CG | Encapsulated granule |
| CPCB | Central pollution control board |
| CPV | Cytoplasmic virus |
| CS | Capsule suspension |
| Ct value | Critical value |
| D | Dusts |
| DBM | Diamond Back Moth |

| Abbreviation | Full Form |
|---|---|
| DC | Dispersible concentrate |
| DDD | Dichloro Diphenyl Dichloro ethane |
| DDT | Dichloro Diphenyl Trichloro ethane |
| DF | Dry Flowable |
| Dfls | Disease Free Larvae |
| DIP | Destructive insect pest |
| DP | Dispersible powder |
| DP | Dustable Powder |
| EBIPM | Ecologically based Bio-intensive Integrated Pest Management |
| EC | Emulsifiable concentrate |
| EC | Emulsifiable concntrate |
| EC50 | Effective Concentration 50 |
| ED50 | Effective Dose 50 |
| EDB | Ethylene Dibromide |
| EDCT | Ethylene Dibromide + Carbon Tetrachloride |
| EDI | Estimated Daily Intake |
| EF | Entomopathogenic Fungi |
| EG | Émulsifiable granule |
| EIL | Economic Injury Level |
| EMDI | Estimated Maximum Daily Intake |
| EPA | Enviornment Protection Agency |
| EPN | Entomopathogenic Nematode |
| ES | Emulsion for seed treatment |
| ET50 | Effective Time 50 |
| ETL | Economic Threshold Level |
| FAO | Food and Agriculture Organization |
| FG | Fine granule |
| G | Granule |
| GABA | Gamma Amino Butyric Acid |
| GAP | Good Agricultural Practices |

| Abbreviation | Full Form |
| --- | --- |
| GB | Granular Bait |
| GC | Gas Chromatography |
| GC-MS | Gas Chromatography-Mass Spectrometer |
| GEL | General Eqiulibrium Level |
| GLC | Gas Liquid Chromatography |
| GLP | Good Laboratory Practices |
| GMO | Genetically Modified Organism |
| GP | Flo-dust |
| GV | Granulosis virus |
| HACCP | Hazard Analysis Critical Control Point |
| HPLC | High Performance Liquid Chromatography |
| ICZ | International Congress of Zoology |
| ICZN | International Commission of Zoological Nomenclature/ International Code of Zoological Nomenclature |
| IGR | Insect Growth Regulator |
| IPCS | International Programme on Chemical Safety |
| IPM | Integrated Pest Management |
| IPPC | International Plant Protection Convention |
| ISPM | International Standard for Phytosanitary Measures |
| IT50 | Inhibition Time 50 |
| ITK | Indigenous Technical Knowledge |
| IRAC | Insecticide Resistance Action Committee |
| IU | International Unit |
| IUPAC | International Union of Pure and Applied Chemistry |
| JH | Juvenile Harmone |
| KT50 | Knockdown Time 50 |
| LC | Liquid Chromatography |
| LC50 | Lethal Concentration 50 |
| LD50 | Lethal Dose 50 |
| LE | Larval Equivalent |

| Abbreviation | Full Form |
|---|---|
| LOEL | Lowest-observed-effect level |
| LT50 | Lethal Time 50 |
| ME | Microencapsulator |
| MFO | Mixed Function Oxidase |
| MG | Microgranule |
| MH | Moulting Hormone |
| MPI | Maximum Permissible Intake |
| MRA | Multiple Residue Analysis |
| MRL | Maximum Residue Level |
| NABAII | National Bureau of Agriculturally Important Insects |
| NIV | Non- Inclusion virus |
| NOAEL | Non- Observed Adverse Effect Level |
| NOEC/NOEL | No Observed Effect Concentration/Level : the highest dose administered in a study, under defined conditions of exposure, that produces no detectable changes in the investigated species |
| NPV | Nucleo Polyhydrosis virus |
| NSC | Neuro Secretory Cells |
| OD | Oil Dispersion |
| OV | Oryctes virus |
| PC | Paper Chromatography |
| PEQ | Post entry quarantine |
| PHI | Post Harvest Interval |
| PIB | Poly Inclusion Bodies |
| PMA | Phenyl Mercury Acetate |
| POB | Poly Occlusion Bodies |
| PRA | Pest Risk Analysis |
| PTG | Pro Thoracic Gland |
| SAR | Systematic acquired resistance |
| SC | Suspension Concentrate |
| SP | Soluble Powder |
| SPM | Sanitary Phytosanitary Measures |

| Abbreviation | Full Form |
|---|---|
| TB | Tablet |
| TD | Threshold dose |
| TLC | Thin Layer Chromatography |
| TMRC | Theoretical Maximum Residue Contribution. |
| WDG | Water Dispersible Granule |
| WDP | Water Dispersible Powder |
| WHO | World Health Organization |

## INSTITUTES

| Abbreviation | Long Form | Head quarter |
|---|---|---|
| NCIPM | National Center of Integrated Pest Management | New Delhi |
| PDBC/NBAII | Project Directorate of Biological Control Now PDBC is renamed as National Bureau of Agriculturally Important Insects | Bangalore |
| DPPQS | Directorate of Plant Protection Quarantine and Storege | Faridabad |
| CPPTI | Central Plant Protection Training Institute | Hyderabad |
| CIB&RC | Central Insecticide Board and Registration Committee | Faridabad |
| NPPO | National Plant Protection Organization | New Delhi |
| CBRI | Central Bee Keeping Research Institute /Central Bee Research and Training Institute | Pune |
| ILRI | Indian Lac Research Institute | Namkum, Ranchi |
| IOBC | International Organization of Biological Control | France |
| CIBC | Commonwealth Institute of Biological Control | Trininad, West Indies |

# 3

# TERMINOLOGY

| Term | Terminology |
|------|-------------|
| Aestivation | Retardation of biological activity by an organism in response to periods of hot or dry weather. |
| Episternum | The anterior sclerite of the pleuron of the thorax. |
| Abbott's formula | Mathematical formula used to correct for mortality in the untreated check, e.g., Corrected % mortality = % Mortality in treatment − % Mortality in check × 100/(100 − % Mortality in check). |
| Abdomen | The third division of the insects body; it is usually composed of ten or eleven segments and beats functional legs in the adult stage. |
| Abduction | Drawing back or retraction. |
| Abductor muscle | Any muscle that opens out or extend in appendage or draws away from the body. |
| Abiotic | Nonliving, pertaining to physico-chemical factors such as light, hea etc. |
| Acaricide | Agent destructive to acardis (ticks and mites) and used in the treatment of mange. Also called MITICIDE. |
| Acarology | The study of ticks and mites. |
| Acceptable daily intake (ADI) | Acceptable daily intake of a chemical is the daily intake, which during an entire lifetime, appears to have no appreciable risk to the health of the consumer on the basis of all the known facts at the time of the evaluation of the chemical. Without appreciable risk is understood to imply 'as a matter of practical certainty'. It is expressed in milligrams of the chemical per kilogram of body weight. ADIs are derived from the results of long term feeding studies with laboratory animals. A safety factor of 100 is applied to express the non-observed adverse effect level (NOAEL) in the most sensitive animal studied. |

| Term | Terminology |
|------|-------------|
| Accessory glands | In the female, a pair of glands opening primarily on the venter of the ninth abdominal segment, secreting an adhesive substance or material forming a covering or a case (ootheca) for the eggs; in the male, mucous glands opening into the ejaculatory duct. |
| Accessory vein | A secondary longitudinal vein in an insect wing. |
| Acclimatization | Adjustment to environment change on the part of the individual, the physiological adjustment or increased tolerance shown by an organism to environmental change, nonstandard variant of acclimatization. |
| Acetyl choline (Ach) | A chemical conductor of a nervous impulse, formed at the ends of nerves, to conduct the impulse over the gap between nerves or between nerves and muscles or glands. After the response, Ach is destroyed by the enzyme cholinesterase. |
| Acron | In the arthropod embryo, the head region anterior to the tritocerebral somite; prostomium. |
| Acrosternite | The narrow marginal flange anterior to the antecosta of a definitive sternal plate that includes the preceding primary intersegmental sclerotization; characteristic of abdominal sterna of insects, but absent on thoracic sterna. |
| Acrostichal bristles | In Diptera, two rows of bristles on the middle of the dorsum of the mesothorax. |
| Acrotergite | The anterior precostal part of the tergal plate of a secondary segment; usually a narrow flange, but sometimes greatly enlarged, and frequently reduced or obliterated. |
| Additive | Non-pesticidal ingredient added to formulation for specific purposes to improve application, persistence, efficacy, etc. |
| Adjuvant | A additive, marketed especially for microbial products or commercially used with them. It is used to improve application, persistence, efficacy, etc. |
| Anatomy | It is the study of internal structure of organ, tissue and system of the body of living organism. |

| Term | *Terminology* |
|------|---------------|
| Anticoagulant | A chemical used in a bait to destroy rodents. It destroys the walls of the small blood vessels and keeps the blood from clotting. As a result, the animal bleeds to death. |
| Antidote | A practical immediate treatment including first aid in case of poisoning; a remedy used to counteract the effects of a poison. |
| Anus | Posterior part; end opening of the alimentary canal. |
| Aphid | A plant louse, also called greenfly. Aphids are small, soft-skinned, often green plant-bugs. They suck sap mainly from young leaves and shoots, causing reduced growth and vigour, and leave a sweet, sticky excretion (honey dew) which is eagerly eaten by ants and which occludes the leaf and shoot pores. Often found in large numbers on wild and cultivated plants. Many are pests, e.g., Bean Aphid. |
| Apiary | (1) A place where honeybees are kept. Specially, a group of hives. (2) Clonies of bees in hives and other bee-keeping equipment in a location and used for the purpose of producing honey worker bees or queen bees. |
| Apical | At the end, tip, or outermost part. |
| Apiculture | The science and art of raising honey bees; beekeeping; bee culture. |
| Apiology | The scientific study of bees, especially honey bees. |
| Apneustic | Without specific external breathing organs, either spiracles or gills, the tracheal system usually absent or rudimentary. |
| Apodeme | The hallow invagination of the body wall that strengthens the exoskeleton and provides areas for muscle attachments or any cuticular outgrowth of the body wall usually formed in a multicellular matrix but sometimes in a single cell is known as apodeme. |
| Apophysis | Any tubular or elongate process of body wall, external or internal is the apophysis; it is a solid structure. |

| Term | Terminology |
|---|---|
| Apodous | Without feet; footless. |
| Appendage | Any part, piece or organ attached by a joint to the body or to any other main structure. |
| Appendicualte | Bearing appendages. |
| Applied control/ | The measures adopted by human beings to control |
| artificial control | pests.Applied entomology/ The application of pure entomology to the control |
| economic entomology | of insect pests. |
| Apposition eye | In diurnal insects, an eye which absorbs oblique rays of light in the pigmented cells of the ommatidia. |
| Apposition image | In diurnal insects an image built up in eyes by apposed points of light falling side by side and not overlapping.Apterous Without wings; wingless.Apterygota A subclass of Insecta. It includes insects, such as springtails, bristle-tails and silverfish, which are primitively wingless and to not undergo metamorphosis. It does not include bugs, fleas and lice that are secondarily wingless owing to their parasitic habit. Because of many similarities, they are probably descendent of crustaceans. |
| Arolium | A cushion like pad located between the tarsal claws and comprising part of the pretarsus. |
| Aorta | The fore part of dorsal blood vessels of insects. In contrast to the hind part (heart), the arota is without valves. It discharges into the head region. |
| Arrhenotoky | A facultative type of parthenogenetic reproduction in which only male progeny are produced. |
| Articulation | (1) A movable point of contact between two sclerotic parts of the body is called as articulation. (2) A joint as between two segment or structures. (3) Movable point or place where two segments or parts are joined. |
| Artificial infestation | Infestation of plants with pests by placing them directly on plants by hand or by releasing them into a cage, in contrast to natural infestation where a natural field population is depended on for testing a pesticide. |

| Term | Terminology |
|------|-------------|
| Asexual | Without sex; having the reproductive organs incompletely developed and procuring eggs or young by cell-budding; parthogenetic. |
| Aspirator | A device with which insects may be picked up by suction. |
| Atomiser | Devise for breaking liquid stream into a very fine droplets by the high pressure of compressed air. |
| Atomization | Process of breaking a liquid into a fine spray. |
| Atrial orifice | The external opening of the spiracular atrium. (Porta atrii). |
| Atrium | The spiracular chamber formed by a secondary invagination of the body wall external to the primary tracheal orifice. |
| Attractants | (1) Materials that attract insects or other animals and cause them to eat or contact poison bait or sprays, and consequently cause their death. (2) Substance that attracts certain insects by some mechanism like, olfactory, gustatory, vision, tactile etc. (3) Substances which elicit a positive directive response; chemicals having positive attraction for animals such as insects, usually in low concentration and at considerable distances. (4) substances or devices capable of attracting insects or other pests to areas where they can be trapped or killed. |
| Autocidal control | The use of an insect species against itself, usually through some means of genetic modification, to suppress or eradicate its natural population (Knipling, 1960). |
| Autolysis | The self destruction of biological cells after death as a result of the action of their own enzymes. |
| Autoparasitism (adelphoparasitism) | A special type of hyperparasitism in which the female develops as a primary as a primary parasitoid, but the male is a secondary parasitoid through females of its own species (Flanders, 1937). |
| Axilla (pl. axillae) | A triangular or rounded sclerite laterad of the scutelium, and usually just caudad of the base of the front wing. |

| Term | Terminology |
|------|-------------|
| Axillary cord | The thickened, corrugated posterior edge of the articular membrane of the wing base, continuous with the posterior marginal fold of the alinotum. |
| Axillary plate | The posterior sclerite of the wing bgase in Odonata, supporting the subcostal, radial, cubital and vannal veins. |
| Axillary region | The region of the base containing the axillary sclerites. |
| Axillary sclerites | The sclerites of the axillary region in the wing flexing insects, partly differentiated in Ephemerida, represented by the axillary plate in Odonata. |
| Axon | The process of a nerve cell that conducis impulses away from the cell body. |
| Bait | (1) Components of food used for enticing the insects. (2) Foodstuff used for attracting pests; usually mixed with a poison to form a poison bait. (3) An edible material that is attractive to the pest, which normally contains a pesticide unless used as a prebait. |
| Bait shyness | (1) The tendency for rodents, birds, or other pests to avoid a poisoned bait. (2) The discarding of a poison bait by the animals after the initial mortality of their follows. |
| Barrier application | The use of pesticide or another agent to stop pests from entering a container, area, field or building. |
| Basal treatment | A treatment applied to the stems or trunks of plants at and just above the ground line. |
| Basement membrane | Thin noncellular membrane forming the inner lining of the hypodermis of the body wall. |
| Basicosta | The proximal sub-marginal ridge of the inner wall of a leg segment. |
| Basicostal suture (bcs.) | The external groove of a leg segment forming the basicosta. |
| Basicoxite | The usually narrow basal rim of the coax proximal to the basicostal suture and its internal ridge (Coxomarginate). |

| Term | Terminology |
|------|-------------|
| Basisternum | That part of the sternum anterior to the sternacostal suture. |
| Basitarsus | The proximal or basal segment of the tarsus. |
| Bee | A furry insect of the order Hymesoptera. There are solitary and social bees. The latter live in colonies. Each colony has one Queen bee together with large numbers of female 'workers' and a fewer male 'drones'. The most highly socialized is the Honey-bee (Apismellifera) kept in hives by beekeepers for commercial Honey production. Bumblebees (Bombus spp.) have much smaller colonies. A growing practice is the hiring and placing of Beehives in Orchards to promote pollination and fruit setting. |
| Bee farming | The keeping of bees for honey and beeswax they produce and for their pollinating of plants. |
| Beehive | A box in which bees are kept to produce Honey. Several types of single and double hives are in common use, the former favoured for commercial production. Hives contain two separate chambers, a lower brood chamber in which the Queen lays and rears her Brood, above which is the honey chamber or supers containing frames in which the bees store surplus honey which is periodically removed. |
| Beeswax | A mixture of organic compounds secreted by special glands on the last four segments on the ventral side of the worker bee's abdomen; used by bees for building comb. Byproduct of beekeeping industry; produced from honeycombs, mostly from wild bee combs; yellowish to grey brown solid, brittle when cold, turns plastic by heat of the hand; insoluble in water and soluble in ether, chloroform and fixed and volatile oils; used for cosmetics, candles and a number of industrial products. |
| Behavioural resistance | A mechanisms in insects of resistance to insecticides which involves a behavioural change whereby contact with the insecticide is avoided. |

| Term | Terminology |
|------|-------------|
| Bilateral symmetry | Symmetry in which the body can be divided by a median plane into mirror images of one another. |
| Binomial | Pertaining to two names; the zoological system of nomenclature consisting of a generic and a specific name. |
| Binomial nomenclature | The method of naming plants and animals introduced by Linnaeus, a Swedish naturalist. Every plant or animal has two Latin names, a generic name designating its genus and a specific name indicating the species. |
| Bioaccumulation of pesticides | Accumulation of the toxicant inside dynamic ecological system depending on (i) physicochemical characteristics of insecticide; (ii) concentration vs. dilution or degradation in the system; (iii) rate of food consumption and body size and (iv) dynamics of accumulation, i.e., balance of uptake and elimination. |
| Bioassay | (1) It is measurement of potency of any stimulus (physical, chemical, biological, physiological or psychological) by means of reactions it produces in living matter. (2) Estimation of amount of toxicant in a sample by measurement of its effect on test organisms; usually estimation of amount necessary in a standard type of application to produce 50% kill. |
| Biocide | (1) A chemical which has be wide range of toxic properties, usually to members of both the plant and animal kingdoms. (2) General toxicant used for killing. |
| Biological control | (1) Biological pest suppression in its narrow, classical since usually restricted to the introduction, by man of parasitoids, predators and or pathogenic microorganisms to suppress populations of plant or animal pests; (2) The control of pests by employing predators, or disease; the natural enemies are encouraged and disseminated by man. |
| Biological magnification | When a small organism feeds on several micro-organisms containing residues of a pesticide, the pesticide may get deposited in its body. A |

| Term | Terminology |
|------|-------------|
| | predatory animal which feeds on several such small organisms, in gests a larger quantity of the pesticide. In this way, the larger predatory animals and the human beings at the top of the food chain consume quantity of the pesticide much larger than that was present in the microorganisms. This increase in chemical concentration as it passes through the foodchain is called biological magnification or bioconcentration. |
| Biological oxygen demand (BOD) | The quantity of oxygen used by micro-organisms in the biochemical oxidation of organic matter and oxidisable in organic matter by aerobic biological action. |
| Biomagnification of pesticides | In certain aquatic animals and plants the residue levels of pesticides are much higher than in other plants or water itself, e.g., in snails and phytoplankton. It is because they accumulate the lipid soluble pesticides both from water and foodchain. This process is called biomagnification of pesticide/residues. |
| Bionomics | The study of the habits, breeding and adaptations of living forms. |
| Biotic potential | Maximum possible rate of increase (multiplication) of an organism in the absence of any limiting factor or stress. Also called Absolute Reproductive Potential. |
| Biotype | A biological strain of an organism, morphologically indistinguishable from other members of its species but exhibiting distinctive physiological characteristics particularly in regard to its ability to successfully utilize pest-resistant host organisms or to act as an effective beneficial species. |
| Blanket application | The application of spray or dust over an entire area rather than on rows of beds. |
| Blastoderm | The cells produced through cleaves of the zygote. |
| Blastogenesis | The origin of different caste traits from variation in either the ovarian environment of the egg or the nongenetic contents of the egg (opposed to genetic control of caste and trophogenesis). |

| Term | Terminology |
|------|-------------|
| Blastokinesis | Movement or migration of the embryo within the egg. |
| Blastomeres | The cleavage cells, or cells produced by the division of the egg or its nucleus that form the blastoderm. |
| Blastopore | The mouth of the gastrulation cavity. |
| Blastula | The early stage of the embryo in which the only cell layer is the blastoderm. |
| Brachypterous | Insects with short wings which do not cover their abdomen fully. |
| Broad spectrum insecticide | Non-selective, having about the same toxicity of most insects. |
| Brood cell | A special chamber or pocket built to house immature stages. |
| Brood comb | A wooden frame enclosing a wax sheet from which bees build the cells of the honey comb, placed in the brood chamber of a hive. |
| Buccal cacity | The first part of the stomodaeum, lying just within the mouth; its dilator muscles arising on the clypeus, and inserted before the frontal ganglion and its connectives. |
| Build up | Accumulation of a pesticide in soil, animals, or in the foodchain. |
| Caecum (pl., caeca) | (1) A tube like blind structure which arises from the anterior end of the mid-gut in certain insects. (2) A blind sac, one of a group of appendages opening into the anterior region of the ventriculus. |
| Cannibalistic | Feeding on other individuals of the same species. |
| Cardiac valve | A valve at the junction of the proventriculus and ventriculus. |
| Caste | (1) Any set of individuals of a particular morphological types, or age group or both, that performs specialized labour in the colony. Any set of individuals in a given colony that are both morphologically distinct and specialized in behaviour. (2) A form or type in a social insect such as .......... Termites and ants. |

| Term | Terminology |
|------|-------------|
| Caterpillar | (1) Worm-like larva of a butterfly or certain other insects with well developed head, 3 thoracic and 10 abdominal segments, a pair of 5 jointed legs on each thoracic segment and 5 pairs of legs (prolegs) on III, IV, V, VI and X abdominal segments. |
| Category | It designates rank or level in a hierarchic classification. It is a class, the members of which are all taxa assigned a given rank. |
| Caudal | Pertaining to the posterior region or to the anal end of the insect body. |
| Cephalic | Pertaining to the head or attached to the head. |
| Cephalic lobes | The head lobes of the embryo, comprising the region of the prostomium and usually that of the tritocerebarl somite. |
| Cephalothorax | A body region in which head and thorax are fused, e.g., Spider. |
| Cerci | An appendage, usually paired of the tenth abdominal segment; usually segmented and slender. |
| Cervical sclerites, cervicalia | The sclerites of the neck, particularly one or two pairs of lateral neck plates (Kehlplatten) joining the head to the prothoracic episterna. |
| Cervix | The neck; including probable the posterior non-sclerotized part of the labial somtie and the anterior part of the prothorax. |
| Chaetotaxy | The arrangement and nomenclature of the setae or bristles on any part of the exoskeleton, especially in the Diptera; the study of such an arrangement. |
| Chelicerae | The first pair of appendages of adult Chelicerata, innervated from the tritocerebral ganglia of the brain; equivalent to the second antennae of Crustacea. |
| Chemoreceptor | A sense organ sensitive to chemical properties of matter (a taste receptor or an odour receptor). |
| Chemosterilant | Chemical used for making an insect sterile without killing it. |

| Term | Terminology |
|------|-------------|
| Chemotherapy | The treatment of a desired plant or animal with chemicals to destroy or control a pathogen without seriously harming the plant. |
| Chironomid | An insect of the order Diptera whose young ones, blood red and worm like, are invariably found at the bottom of ponds living in tubes. |
| Chitin | (1) A nitrogenous polysaccharide occurring in the cuticle of arthropods and certain other invertebrates. Probably occurs naturally only in chemical combination with protein. (2) A nitrogenous polysaccharide found in the exoskeleton of insects and in the cell walls of many fungi. It provides mechanical strength and resistance to chemicals. |
| Chronic effect | A slow and long continued effect. |
| Chronic poisoning | Resulting from long periods of exposure to a chemical. |
| Chronic toxicity | Ability to cause injury or death from prolonged exposure. |
| Cibarium | The food pocket of the extra oral or pre-oral mouth cavity between the base of the hypopharynx and the under surface of the clypeus. |
| Clasper | One of a pair of appendages at the tail end of a caterpillar by which it clings to leaves, twigs and other surfaces. |
| Classification | (1) The process of assigning pesticides into groups according to common characteristics. (2) The systematic arrangement of insects (or other animals or plants) in series showing their relation or agreement in structure, life habits or other characters forming the basis of classification. |
| Cleavage | In embryological development, the division of the originally single-celled egg into a number of cells called blastomers. |
| Cleptobiosis | The relation in which one species robs the food stores or scavenges in the refuse piles of another species, but does not nest in close association with it. |

| Term | Terminology |
|------|-------------|
| Cleptoparasitism | The parasitic relation in which a female seeks out the prey or stored food of another female, usually belonging to a different species and appropriates it for the rearing of her own offspring. |
| Cline | It is a gradual and nearly continuous change of a character in a series of contiguous populations; a character gradient. |
| Clypeus | That part of the head in front of the frons and to which the labrum is attached |
| Cochineal | A red dye that is extracted from the bodies of scale insects (Dactylopius coccus). About 500 g of the dye is produced from 70,000 insects. |
| Cockle | A small gall on wheat formed due to the attack from a nematode, Auguina tritici. |
| Cocoon | (1) The silky protective case of many larval forms before they become pupae; silky or other covering formed by many amimals for their egg. (2) A protective silky sheath spun by many insect larvae. |
| Coelome | A body cavity formed of the coelomic sacs only. |
| Coevolution | The evolution of a specific relationship between two or more distinctly different species. |
| Collophore | A tube like structure located on the ventral side of the first abdominal segment of spring tails Collembola. |
| Commensalisms | A living together of two or more species, none of which is injured thereby, and at least one of which is benefited. |
| Compatibility | The ability of a pesticide to mix with other pesticides without any undersirable effects, such as reduction in the pesticidal activity, increase in the plant injury, promotion and excessive run-off from the treated surface. |
| Compound eye | An eye composed of many individual elements or ommatidia, each of which is represented externally by a facet. |
| Commissure | A transverse tract of nerve fibers connecting the two ganglia of a segment or the lateral centers within a median ganglion. |

| Term | Terminology |
|---|---|
| Copulation | The union of the sexes for reproduction. |
| Cornea | The outer surface of the compound eye as a whole and of each individual facet; cuticular lens. |
| Cornicle | A par of tubes present on 5th abdominal segment of aphids secretary channels, excreting waxy fluid (not honey-dew), for self protection against predators. |
| Corpora allata | Paired endocrine organs located just behind the brain; the source of juvenile hormone. |
| Cosmopolitan | An species occurring very widely throughout the major regions of the world. |
| Costa | A longitudinal vein extending along the anterior margin of the wing. |
| Cover crop | A subsidiary, fast growing and usually annual crop that is grown with a more slowly growing biennial or perennial crop. |
| Crochets | Hooked spines at tip of the prolegs of lepidopterous larvae. |
| Crop | A thin walled dilation of oesophagus in certain insects for storing food for sometime before passing it on. |
| Cross veins | Short veins between the lengthwise veins and their branches, numerous in net veined wings, in others generally few and located in definite positions. |
| Cryptic colouration | Colouration of organism that blends with background. |
| Damage threshold | Pest population above which crop losses are visible. |
| Deme | It is a local population of a species : the community of potentially interbreeding individuals at a given locality. |
| Degradation | Process by which a pesticide is broken down to simpler structures through biological, or abiotic mechanisms. Synonyms include breakdown and decomposition. |
| Dendrons or dendrites | Finely ramifying branches given off from a nerve cell. |

| Term | Terminology |
|------|-------------|
| Density dependent | Refers to mortality factors or processes in the environment which destroy an increasing percentage of the subject population as the numerical population density increase and Vice Versa (Smith, 1935). |
| Density dependent factor | In ecology a factor which changes in intensity with changes in population density (e.g., lethal effects of a factor intensify as population numbers increase). |
| Density independent | Refers to mortality factors or process in the environment which destroy a relatively constant percentage of the subject population regardless of changes in its density (Smith, 1935). |
| Density independent factor | A factor regulating activity or population of living organisms but whose influence is independent of the population density of that particular organism (concerned organism). |
| Density, population | Number of individuals of a species that habitate or live in per unit area (generally per square kilometer) of any geographical region. |
| Deposit | The amount of initially laid down chemical/active ingredient/pesticide on the surface is called deposit. |
| Dermal toxicity | (1) Toxicity of a material as tested on the skin. (2) Dermal toxicity is the passage of pesticides into the body through the unbroken skin. |
| Deterrent | A chemical substance that deters feeding or oviposition of an insect. |
| Detoxification | Ridding a toxicant of its toxicity by chemical or physical means; metabolism (mostly catabolism or degradation) or xenobiotics in an organism or plant into ultimate harmless products. |
| Deutocerebrum | The part of the arthropod brain containing the first antennal nerve centers. |
| Deutonymph | The third instar of a mite. |
| Diapause | (1) It delays growth and development which is not immediately referable to the adverse environmental conditions. It is, thus, not easily |

| Term | Terminology |
|------|-------------|
|  | altered by change to a more favourable environment. It is an adaptive phenomenon or inborn rhythin which usually facilitates survival during unfavorable periods. (2) A state of an animal, such as an insect, in which a reduction of growth proceses or maturation occurs which is not necessarily caused by immediate environmental influence, does not depend for its continuance on unsuitable conditions and is not easily or quickly altered by change to a more favourable environment, frequently related to change in photoperiod. Once the state of diapause comes to an end, normal growth and development are resumed. (3) A temporary stoppage of development of an insect embryo or larva. |
| Diconlylic joint | A joint with two points of articulation between the adjacent leg segments. |
| Digestion | Enzymatic hydrolysis of food into absorbable size molecules. |
| Dilator muscles | Muscles extending from the body wall to the alimentary canal; called also suspensory muscles. |
| Dimorphism | (1) In caste systems, the existence in the same colony of two different forms, including two size classes, not connected by intermediates. (2) A difference in size, form, or colour, between individuals of the same species, characterizing two distinct types. |
| Direct pest | An organism which causes immediate and direct damage to a marketable item, such as a fruit, even at low population density. |
| Dislodgeable residues | Portion of pesticide residue on treated vegetation, which is readily removable and may be used as an index for risk to farm workers is the dislodgeable residue. Generally measured by the residue removed when leaf discs are shaken briefly in water. |
| Dispersal | Movement of individuals out of a population (immigration). |
| Dissipate | To get rid of by spreading out. Gases and vapours dissipate through the air. |

| Term | Terminology |
|------|-------------|
| Dissipation and persistence | In nature disappearance of residues takes place in 2 steps. The first step is the initial phase in which the disappearance of the residue is fast. This phase is called "Dissipation". The second phase, in which there is a slow decrease in the amount of residue, is known as "Persistence". The main difference between these is that the dissipation follows the law of 'first-order kinetics', whereas the persistence does not follow this law because of the storage of translocated insecticides and degradation of various rates. |
| Diverticula (pl., diverticulae) | Caecal tubes or pouches; branches from alimentary canal; extensions or evaginations. |
| Dormancy | (1) A state of quiescence or inactivity. (2) Alive but not growing; quiescent; with suspended animation or development. |
| Dormant spray | A spray applied during the dormant season when the plant growth is inactive. |
| Dorsal | Top or uppermost; pertaining to the back or upper. |
| Dorsal diaphragm | The membranous and muscular sheets of tissue extending from the dorsal blood vessel to the laterodorsal parts of the body wall, separating the dorsal sinus from the pervisceral sinus. |
| Dorsal trachea | The dorsal segmental trachea origination at a spiracle. |
| Dorsal vessel | The dorsal blood vessel, consisting of the pulsatile heart and the nonpulsating aorta. |
| Dorso-pleural line | The line of separation between the dorsum and the pleural region of the body, often marked by a fold or groove. |
| Dorsum | The entire back of an animal above the pleural regions; or specicifically, when qualified by the designation of a segment, the back region of a segment. |
| Dosage | A dose is a measured quantity, as of medicine, taken at one time or in one period of time. Dosage, therefore, is the amount of medicine in a dose. Used in connection with pesticides, it refers in general to rate of application. |

| Term | Terminology |
| --- | --- |
| Drift | (1) Movement of spray or dust material by wind or air currents outside the intended area, usually as fine droplets, during or shortly after application. (2) Spray or dust carried by natural air currents beyond the target area. |
| Drone | (1) The male bee, hatched from an unfertilized egg; larger than the worker bee; does not gather honey; has no sting. (2) A male social bee, especially a male honey fee or bumblebee. |
| Duct | A tube for carrying secretion from a gland to the point of discharge. |
| Ductless glands (endocrine glands) | The glands, not joined by tabular organs, that secrete hormones in the body which are transmitted throughout the body via the blood stream. |
| Dust | (1) Fine, dry particles of earth or other matter which can be wafted by wind. (2) A pesticide applied in dry stare; application of such dust to plants, animals, fowls. (3) A pesticide formulation in dry, finely divided form, containing a low concentration of an active ingredient to be used without further preparation or dilution. |
| Dust formulation | Mixing of an insecticide when it is to be sold as a dust with a special quality talcum or some other ingredient in a known proportion to from a fine powder. Amount of insecticide in such a mixture generally five to ten percent. |
| Duster | A mechanical device, consisting of a blower, dust reservoir, and a nozzle, used in the application of insecticidal or fungicidal dust so that the dust is emitted in a blast of air. Mainly hand-driven; also power-driven. |
| Dusting | The application of insecticide or fungicide to crops in a dry, powdery state. It is less efficient than spraying, usually requiring about twice the amount of chemical per unit area. |
| Dyar's Rule | An empirical law based on the observation that the head capsule and certain other parts of caterpillars grow in geometric proression, |

| Term | Terminology |
| --- | --- |
| | increasing in size by a constant ratio at each moult. Thus if the lograthim of the size of the head is plotted against the number of instar, the resulting graph will be a straight line. Although not always accurate, this rule has sometimes been of use in deducting facts about the life history of an insect in cases where other is lacking, especially indicating when one instar has been overlooked. |
| Ecdysis (pl. ecdyses) | (1) Molting; the process of shedding the skin. (2) The molting (shedding of the skin), of larval arthropods from one stage of development to another the final moult leading to the formation of the uparium or chrysalis. |
| Ecdysone | Insect molting hormone; a chemical secreted by the prothoracic glands after each molt or at the termination of diapause, which stimulates growth and development of various somatic tissues and other physiological and morphological changes necessary to the continued repetition of the molting cycle. |
| Eclosion | Process of an individuals emerging from pupa hatching from an egg. |
| Ecological niche | (1) The place an organism occupies in its biotic relationships and physical environment as determined by its particular structural adaptations, physiological adjustment and developed behavioural patterns. (2) The way of life of an organism including food, habitat requirements, and so forth. |
| Ecology | The study of organisms in relation to their environment, including both the physical and biotic components. |
| Economic entomology | That branch of insect study directed towards preventing man's losses and increasing his gains through manipulation of insect populations. |
| Economic injury level | The lowest population density of a pest that causes economic damage, the pest density that causes damage equal to the cost of preventing the damage. |

| *Term* | *Terminology* |
| --- | --- |
| Economic threshold | (1) A population density concept which allows the determination of the point at which pest numbers are sufficient to cause economic injury unless suppressive action is taken (Headley, 1972). (2) That point of pest infection where application of a control measure would return more money than the cost of the control procedure. (3) Pest population level at which control measures should be started to prevent the pest population from reaching economic injury level. |
| Ecosystem | (1) Interacting system of living organisms in an area and their physical environment. (2) The functional ecological system of a given area comprised of the community of plant and animal populations and the nonliving environment. |
| Ectoderm | The embryological layer which gives rise to the nervous system, integument and several other parts of an insect. |
| Ectognathous | Insects with exerted mouthparts. |
| Ectoparasites | Plants and animals that live and feed on the out side of an animal or plant. Most are annoying cause injury; or carry disease organisms. Examples lice, fleas, some fungi. |
| Ectoparasitoid | An insect parasite which develops externally on its arthropod host. |
| Ejaculatory duct | Tube through which the spermatozoa pass from the vasa differentia. |
| Embolium | A narrow part of the corium, separated by a suture, along the costal margin in front wing of Hemiptera. |
| Empodium | A single pad - like or filiform median structure often present between the claws of insects. |
| Emulsifiable concentrate | (1) A liquid formulation, usually containing a pesticide, which will form an emulsion when mixed with water or certain other liquids. (2) Produced by dissolving the toxicant and an emulsifying agent in an organic solvent. A solvent substantially in soluble in water is usually selected since water miscible solvent have not in general |

| Term | Terminology |
|------|-------------|
| | proved satisfactory. Strength usually stated in pounds of toxicant per gallon of concentrate. (3) Liquid formulation that immediately disperses as fine droplets in water to form an emulsion. (4) A liquid formulation of insecticide which contains an emulsifier so that water may be added to form an emulsion. |
| Emulsifying agent | Substance when added to emulsion of insecticide helps resist tendency of oil droplets and other water unmiscible liquids to coalesce when mixed with water and thus help stabilize emulsion. Used for making stock emulsions. |
| Emulsion | (1) A finely divided, fatty or resinous substance held in suspension in an aqueous liquid. (2) A mixture of liquids that do not dissolve in each other. (3) Stable dispersion of oil droplets in as aqueous solution. |
| Encapsulation | (1) A method of formulating pesticides, in which the active ingredient is encased in a material (often poly vinyl) resulting in sustained pesticidal release and decreased hazard. |
| Endocrine | Secreting internally, applied to organs whose functions to secrete into blood or lymph a substance which has an important role inmetabolism. |
| Endocuticle | It is a thicker, softer and flexible innermost layer of the cuticle. |
| Endoderm | The innermost layer of cells produced in an early embryo that becomes part of the lining of the alimentary canal. |
| Endoparasite | A parasite living within the body of the host. |
| Endoparasitoid | An insect parasite which develops within the body of its arthropod host. |
| Endoskeleton | (1) It is the internal framework of rigid processes of the body wall. Such a framework in the head is called tentorium. (2) Chitirous processes extending inward into the body cavity from the body wall for attachment of muscles. |

| Term | Terminology |
|------|-------------|
| Enterio-caeca, gastric caeca (sing. Caecum) | Sac-like diverticula at the anterior end of the mid-gut. Entognathous (entognathy) Mouthparts hidden in the head. |
| Entomogenous | Growing in or on an insect, for example certain fungi. |
| Entomophagous | An animal (or plant) which feeds upon insects. |
| EPA | Environmental Protection Agency. Responsible for the protection of the environment in the united state. |
| Epicranial suture | The dorsal Y-shaped suture of the cranium, including the median coronal suture(cs) of the vertex and the divergent frontal sutures(fs) of the facial region. |
| Epicranium | A term variously applied to the entire cranium, to the cranium exclusive of the frons, or preferably to the upper part of the cranium. |
| Epicuticle | It is an outer film-like water proof layer of the cuticle. |
| Epidemic | A sudden widespread increase in the incidence of a disease or organism. |
| Epidermis | The cellular layer of the body wall that secretes a thick cuticle on its outer surface. It is made up of epithelial cells. |
| Epimorphic development | Immatures possess same number of body segments as adults. |
| Epiphallus | A sclerite in some Orthoptera in the floor of the genital chamber proximal to the base of the phallus (Pseudosternite). |
| Epipharynx | A mouth part structure attached to the inner surface of the upper lip. Sometimes it is elongated to form a part of the proboscis. |
| Epiphysis | A mobile pad or lappet like process on the inner aspect of the front tibia of some Lepidoptera. |
| Epipleuron (pl. epipleura). | The deflexed or inflexed portion of the elytron, immediately beneath the edge. |
| Epiproct | The dorsal part of the eleventh segment of the abdomen. |

| Term | Terminology |
|------|-------------|
| Epistomal suture | The frontoclypeal suture; a groove uniting the anterior ends of the subgenal sutures across the face forming internally a strong epistomal ridge (ER), typically straight, but often arched upward, sometimes absent. |
| Eradication | The complete elimination of either weed, insects, disease organisms or other pests from an area. |
| Eri silkworm | Attacus ricinis. Domesticate silk worm feeding on castor leaves. Reared in Assam and northern West Bengal. Produces eri silk. |
| Estimated daily intake (EDI) | Prediction of the daily intake of a pesticide residue based on the most realistic estimation in food items and the best available food consumption data for a specific population. Residue levels are estimated taking into account known uses of the pesticide, the portion of commodity treated and the quantity of contaminated commodities. The EDIs are expressed in milligrams of residue per person. |
| Estimated maximum daily intake (EMDI) | Prediction of the maximum daily intake of a pesticide residue, based on the assumption of average daily food consumption per person and maximum residues in the edible portion of a commodity, corrected for the reduction or increase in residues resulting from preparation, cooking or commercial processing. The EMDI expressed in milligrams of residues per person. |
| Eusternum | The intrasegmental ventral plate of a thoracic segment, exclusive of the spinasternum, but usually including the sternopleurites. |
| Exarate pupa | A pupa which has its appendages free and not glued to the body. |
| Excretion | The elimination of waste products of metabolism. |
| Exocuticle | The hard and usually darkened layer of the cuticle lying between endocuticle and epicuticle. |
| Exoskeleton | The segmented, external skeleton of an insect; the insects 'skin'. |

| Term | Terminology |
|------|-------------|
| Exotic | Not native to a place or country but introduced from foreign land. |
| Exotoxin | A soluble toxic substance produced by certain bacteria and found in their surrounding growth medium; Also called endo-toxin. |
| Exposure | When contact occurs with a pesticide through skin (dermal), mouth (oral), lungs (inhalation/respiratory), or eyes. |
| External genitalia | The reproductive structures concerned with mating and deposition of eggs. |
| Extraneous residue limit (ERL) | The ERL refers to a pesticide residue or a contaminant arising from environmental sources (including agricultural uses) other than the use of a pesticide or contaminant substance directly or indirectly on the commodity that is recommended to be permitted in or on a feed or food commodity. It is the maximum concentration of a pesticide residue or contaminant that is recommended by the Codes Alimentarius Commission to be legally permitted or recognized as acceptable in or on a food, agricultural commodity, or animal feed. The concentration is expressed in milligrams of pesticide residue or contaminant per kilogram of the commodity. |
| Exuviae | (1) The cuticular parts discarded at a moult are called as exuviae. (2) The cast skin of a larva or nymph. |
| Exuvial glands | Glands of the epidermis supposed to secrete the exuvial or molting liquid. |
| Facet | The lens which forms the base of the visual element, the ommatidium, visible on the surface of the compound eye. |
| Factitious host | An unnatural, but acceptable host used in laboratory propagation of beneficial organisms, e.g., Corepra. |
| Facultative | Incidental, not necessarily compelled to live under one type of environment. |

| Term | Terminology |
|------|-------------|
| Facultative parasitism | Here in reference to nematodes which may either parasitize healthy insects, or develop in some other way in the environment (e.g., mycetophagy) if no insect host is encountered. |
| Feces | Indigestible wastes discharged from the digestive tract. |
| Fecundity | (1) Capacity to produce off springs. (2) In insect ecology, the number of eggs per female that hatch or become larvae. |
| Femur (pl., femora) | The 3rd stout part of the leg which is articulated to the body through trochanter and coax. |
| Fiducial limits | The lower and the upper values for the estimate of LD50 or LD90 computed from the probit regression equation at a given reliability coefficient say 95% or 99%. |
| Filter chamber | A part of the alimentary canal in Homoptera in which the two ends of the ventriculus and the beginning of the intestine are bound together in a membranous and muscular sheath. |
| Flagellum | The part of the antenna distal to the pedicel, typically filamentous, but of various forms, usually subsegmented or multiarticulate. |
| Follicle cells | The inner epithelial cells of an ovarian egg tube. |
| Food chain | The direction of energy flow between species that feed upon one another. (2) All assemblage of organisms, including producers, consumers and decomposers, through which energy and materials may move in a community. |
| Fore wing | The anterior pair of wings in insects. |
| Foregut | The anterior portion of the alimentary tract, from the mouth to the midgut. |
| Formulation | It is righr combination of pesticide (active ingrediant) with various additives, adjuvants, accessories, according to a predetermined formula to produce a final consumer product. |

| Term | Terminology |
| --- | --- |
| Fossil | A relic or trace of a former living thing preserved in the rocks. |
| Frenulum | The spine or spines which arise from the base of the hind wing in many Lepidoptera and project beneath the fore wing. |
| Frons | That region of the head lying between the eyes and the clypeus. |
| Fumigation | The use of poisonous gases for the destruction of pests; mainly used against insects and rodent. |
| Furuca | The forked endosternal process of higher insects, formed of the sternal apophyses supported on a median inflection of the stermun. |
| Furcasternum | A distinct part of the sternum in some insects bearing the furca. (The term generally applied to the sternelum). |
| Furcula | A springing organ on the ventral surface of the abdomen of Collembola. |
| Galea | The outer enidite lobe of a maxilla, provided with a muscle arising in the stipes. |
| Gall | (1) Bile, secretion of liver. (2) A plant tumour found practically on any part of the plant; caused by irritation from fungus, bacterium or insect. |
| Gametocyte | A sex cell stage of plasmodium, the material parasite. |
| Ganglion (pl. ganglia) | Nerve center containing a cell mass and fibers; white disc-like bodies connected by a double cord. |
| Gastric caeca | The sac-like diverticula at the anterior end of the midgut. |
| Gastrulation | The formation of the endoderm, either by invagination of one wall of the blastula or by internal proliferation of cells from the blastoderm. |
| Genae | The lateral parts of the parietals, generally the areas behind and beneath the eyes. |
| Genital chamber | In the female a copulatory invagination cavity behind or above the eighth abdominal sternum containing the gonopore and the orifice of the spermatheca, ofter narrowed to form a pouchlike or coputubular vagina (Bursa copuatrix). |

| Term | Terminology |
|------|-------------|
| Genital claspers | Organs of the male genalia which serve to hold the female during copulation. |
| Genital ridge | One of the embryonic gonadial rudiments, a ridge like swelling of the splanchnopleure wall of the mesoderm containing the term cells. |
| Genital segment | Specfically the ninth segment of the abdomen in the male, though other segments are frequently associated with the ninth in the genital complex. (Gorosomite). |
| Genitalia | (1) The organs of reproduction collectively, but usually applied only to the external genitalia. (2) A term applied collectively to all the genital structures. (3) The reproductive organs; the external which enable the sexes to copulate and the females to deposit eggs, strictly these are the external genitala. |
| Germ band | The area of thickened cells on the ventral side of the blastoderm that becomes the embryo, (Embryonic rudiment, germ disc, primitive streak, Keimstreif, Keimscheibe, bandclctte, plaque ventrale, piastra germinativa). |
| Germ cells | Reproductive cells, or more specifically the early undifferentiated reproductive cells designed to become ova or spermatozoa. |
| Germarium | The end chamber of an ovarial or testicular tube, containing the primary oogonia or spermatogonia (Endkammer). |
| Germicide | A chemical or agent that kills microorganisms such as bacteria or prevent them from causing disease. |
| Gizzard or proventriculus | A part of the fore-gut immediately behind the crop. It is poorly developed in sucking insects. In chewing-insects, it is equipped with powerful muscles and teeth with which the harder particles of food can be crushed. |
| Glossa | The inner part of the extremity of the labium of insect mouthparts. |
| Gonad | The ovary or testis or their embryonic rudiments, formed by splanchni mesodermal cells covering the germ cells. |

| Term | Terminology |
| --- | --- |
| Gonoduct | Duct of gonad to the genital opening in Female, the oviduct and in male, the vasa defrentia. |
| Gonopore | In the male the external opening of the median ejaculatory duct. usually concealed in the endophallus, or one of the apertures of paired exit ducts. |
| Good laboratory practice | The formalized process and conditions, under which laboratory |
| (GLP) | studies on pesticides are planned, performed, monitored, recorded, reported and audited. Studies performed under GLP are based on the national regulations of a country and are designed to assure the reliability and integrity of the studies and associated data. |
| Granulosis | An insect viral disease characterized by the presence of minute granular inclusions (capsules) in the infected cells. |
| Gravid | An insect containing fertilized eggs. |
| Grub | A term applied in the general sense to insect larvae such as caterpillars and legless maggots, but sometimes used more specifically for the legged larvae of beetles. |
| Gula | A median ventral plate of the head of some insects, developed as a sclerotization of the gular region of the neck proximal to the posterior tentorial pits, continuous with the basal plate of the labium. |
| Gular sutures | The ventral ends of the postoccipital suture extended forward on the under side of the head in some prognathous insects. |
| Habitat | Physical place where an organism lives. |
| Haemocoele | The blood cavity or cavities of the embryo between the mesoderm and the other germ layers, probably a remainant of the blastocoele. |
| Haemocytes | The free floating cells of the blood or circulatory fluid. |
| Haemoglobin | The oxygen carrying red pigment of the blood. It is absent in most of the insects. |

| Term | Terminology |
|------|-------------|
| Haemolymph | The unpigmented blood of insect; the fluid in the body of some coelomates. |
| Half life | Time taken for half the depost of a pesticide chemical to disappeal or degrade. It is the time in which half the amount of initial deposit is eliminated by reacting or otherwise getting dissipated. Half-life values or 90-100% disappearance values are used to express the approximate rates of residue disappearance. These values are largely determined by the initial fast phase of dissipation through which the major portion of the initial deposit is often eliminated. Such values can serve as a general quick reference to the rate of insecticide disappearance. |
| Hamuli (sing, hamulus) | Minute hooks a series of minute hooks on the anterior margin of the hind wing, with which the front and hind wings are attached together in Hymenoptera. |
| Haustelum | The distal portion of the proboscis of some dipterous insects, adapted for sucking exposed liquid. |
| Head capsule | The welded together scleries of the head forming a hard compact mass. |
| Hemimetabola | Insects with simple metamorphosis, with immature stages aquatic and adults terrestrial; insects of the orders Odonata, Ephemeroptera and Pleccotera; young are called naiads. |
| Herbivorous | Feeding or living on grass or herbage. |
| Hermaphrodite | An individual, such as cottony cushion scale, Icerya purchasi, which contains both the male and the female reproductive organs. The individual is self fertilizing. |
| Heterometabola | Insects with simple metamorphosis including the paurometabola and hemimetabola. |
| Hexapoda | (1) The true insects. (2) Insects are also called hexapoda owing to the presence of six legs. |
| Hibernate | To enter a state of dormancy over the winter period. |

| Term | Terminology |
|------|-------------|
| Hibernation | A period of suspended animation in animals during seasonal low temperatures. |
| Hind-wings | Wings developed on the third thoracic segment. |
| Hind-gut | The posterior portion of the alimentary tract, between the midgut and anus. |
| Histopathology | The study of abnormal microscopic changes in the tissue structure of diseased organism. |
| Holometabola | Insects having compelx metamorphosis. The pupal stage is present. |
| Holometabolous | Undergoing a complete metamorphosis during development, with distinct larval pupal and adult stages, the Hymenoptera, for example are holometabolous (opposed to hemimetabolous). |
| Holopneustic | With all the spiracles open and functional. |
| Homeostasis | The maintenance of a steady state, especially a physiological or social steady state, by means of self regulation through internal feedback responses. |
| Homonym | The same name for two or more different things, e.g., Cotton bollworm. |
| Honeydew | (1) Certain insects, such as aphids and coccids, owing to almost continuous feeding on the juice of plants, take in more sugar and water than they need. The excess excreted as sweetish excretion is called honey dew; sweet thick fluid. (2) Sweet and sticky fluid discharged from the anus (not cornicles) by some Homoptera. |
| Host | (1) The term "host" means any plant or animal on or in which another lives for nrourishment, or protection. (2) Organism in on which an insect feeds; the organism may be plant (pest) or animal (parasite). |
| Host resistance | Here in the sense or Painter (1951) meaning the relative genetic ability of a certain variety of plant (or animal) to produce a larger crop of good quality that do ordinary varieties at the same level of pest infestation: practically speaking, based on three elements: non-preference, antibiosis and/or tolerance. |

| Term | Terminology |
|------|-------------|
| Humeral | Pertaining to the shoulder; located in the anterior basal portion of the wing. |
| Humeral angle | The basal anterior angle or portion of the wing. |
| Humeral bristles | Bristles on the humeral callus (Diptera). |
| Humeral cross vein | A cross vein in the humeral portion of the wing between the costa and the subcosta. |
| Hyper parasite | A parasite which attacks another insect which is itself a parasite. |
| Hypermetamorphosis | (1) A type of complete metamorphosis in which the different larval instars represent two of more different types of larvae. (2) A type of parasitic insect life cycle involving transformation through at least two distinctly different larval types, especially an active host-seeking first instar and several more passive parasitic instars. |
| Hyperparasite | A parasite whose host is also a parasite. |
| Hypodermis | The cellular layer of the body wall which secretes the cuticle. |
| Hypognathus | With the head vertical and the mouth parts located ventrally. |
| Hypopharynx | (1) Sensitive and sensory structure on the upper surface of labium, serving as organ of taste or true tongue. (2) A tongue-like lobe in the mouth. It is elongated and forms a part of proboscis in blood sucking insects, such as mosquitoes. |
| Ileum | The anterior part of the anterior intestine, between the ventriculus or pylorus and the colon. (Small intestine, Dunndarm.) |
| Imago | (1) The last or adult stage in insect metamorphosis; the perfect insect. (2) The adult insect. In termites, the term is usually applied only to adult primary reproductives. |
| Immature stage | Any stage of an insect prior to the imago or adult form. |
| Immigration | Movement of individuals into a population. |
| Immune | (1) Exempt from infection. (2) A state of not being affected by disease or poison; exempt from or protected against. |

| Term | Terminology |
|------|-------------|
| Incidence (of disease, fungal or insect attack) | The number of new cases of a particular malady within a given period of time (season, year) in a population under study. |
| Inclusion body | The proteinaceous or crystal-like structure produced in insect cells infected with certain viral pathogens; it occurs in various shapes and size and usually encloses a number of replicated virions. |
| Incompatibility | The inability of a pesticide to mix with other pesticides without producing undesirable effects. |
| Incompatible | Not capable of being mixed or used together, in the case of pesticides, the effectiveness of one or more is reduced; or they cause injury to plants or animals. |
| Incomplete metamorphosis | (1) Insect metamorphosis in which the young are hatched in general adult form and develop without the quiescent stage. (2) Hemimetabolic development; the immatures (nymphs) do not resemble the adults and a pupal stage is lacking. |
| Indirect pest | An organism which causes delayed or insidious degradation of marketable produce through generalized attacks on the host which may reduce growth, vigour or yield. |
| Induced resistance | Resistance of a plant to attack of pest or disease brought about by conditions under which it is growing and varying in degree according to these conditions; generally brought about by farm practices and nutritional means. |
| Infection | The development and establishment of a pathogen (e.g., a bacterium) in its host which will produce a disease. |
| Infestation | Pest that are found in an area or location where they are not wanted. |
| Ingested | Taken into the digestive system. |
| Inhalation | To make air into the lungs; to breathe in. |
| Inhalation toxicity | Poisoning through the respiratory system. |

| *Term* | *Terminology* |
|--------|---------------|
| Inhibitor | A chemical, usually of the regulator type, that prevents or suppresses growth or other physiological processes in plants. 2-amino-1, 2,4-triazole (ATA) is an example of a growth inhibitor. |
| Inoculation | The introduction of an infective agent (the inoculum) into or on living tissues producing a specific diseases. |
| Insect | Insects are tracheate arthropods, in which the body is divided into head, thorax and abdomen. They posses single pair of mandibles and two pairs of maxillae. The second pair fused medially to form the labium. The thorax carries pair of legs and usually one or two pairs of wings. The abdomen is divided in two ambulatory appendages and the genital opening is situated near the posterior end of the body. Post embryonic development is direct and metamorphosis usually occurs. |
| Insect growth regulator (IGR) | A general class of natural and synthetic chemical compounds associated with the control of growth and metamorphosis in insects and including the juvenoids plus several other chemicals that inhibit either JH or other physiological activities (Staal, 1975); Granett and Dunbar, 1975). |
| Insect parasites | Insects which lay eggs on the body of other insects, when the emerging larvae feed internally on them and kill them. |
| Insect pathology | The study of insect diseases, their causes and treatments. |
| Insect pest management | An ecologically based strategy of maintaining insect pest population below the economic injury level by the use of any or all control techniques that are economically, ecologically and socially acceptable. |
| Insect predators | Insects which catch other insects and feed on them. |
| Insect society | In the strict sense, a colony of eusocial insects (ants, termites, eusocial wasps; or bees). |
| Insect vector | An insect which carries virus, bacteriumor spores or mycelium of a pathogenic fungus and inoculates susceptible plants or animals. |

| Term | Terminology |
|------|-------------|
| Insect | (1) A class of arthropod with bodies divided into a head, thoraxd and abdomen, the head bearing a pair of feelers or antennae, the thorax three pairs of legs and wings. (2) An air-breathing animal of phylum Arthropoda, with a distinct head, thorax and abdomen; has one pair of antennae, three pairs of legs, usually two pairs of wings. Some infest plants and animals; some are insectivorous, some pollinate plants; some produce edible or otherwise useful products. (3) Any of the numerous small invertebrate animals generally having segmented bodies and for the most part belonging to the class insects, comprising six-legged, usually winged forms. |
| Insectary | A laboratory in which insects are bred in large numbers. Also called INSECTARIUM. |
| Insecticide | (1) A pesticide which kills insects. Most of those in farm use today are synthetic organic compounds and include chlorinated hydrocarbons, organo-phosphorous compound and carbamates. They are available in granular; liquid and powder forms and may be applied in various ways. (2) Chemical prepared with the express purpose of controlling insect pests; active ingredients can be gas, liquid or said, formulated into liquefied gases, smokes, dusts, water-dispersible powders, emulsion and solution. (3) A toxic chemical substance employed to kill and control insects. (4) A substance or mixture of substance intended to prevent or destroy any insects which may be present in any environment. |
| Insecticide resistance | The degree to which a species of insect tolerates a toxic substance. The degree of resistance is based on LD50 values of a population that has been previously exposed to insecticide, compared to those of an unexposed population. |
| Insectivorous | Feeding and living on insects. |

| Term | Terminology |
|------|-------------|
| Instar | (1) A stage of development of an insect between two moults of the hard exoskeleton, but excluding the egg. An adult butterfly is the sixth instar. The first four are caterpillars, of increasing size, followed by a pupa. (2) Any period between molts during the course of development. (3) A period or stage between molts in the larva, first instar is the stage between egg and the first molt. |
| Integrated pest management | A pest management system that, in the context of the associated environment and the population dynamics of the pest species, utilizes all suitable techniques and methods in as compatible a manner as possible to maintain the pest populations at levels below those causing economically unacceptable damage or loss. |
| Integument | The outer covering of the body of an insect. |
| Intercalary vein | A supplementary longitudinal wing vein lying between two preexisting veins. |
| Intermediate host | The host which harbors the immature stages or the asexual stages of a parasite. |
| Internal respiration | The process of oxidation accompanying metabolism in the cells of the body tissue. |
| International unit (IU) | An arbitrarily-set basis for comparing the efficacy of insect pathogenic Bacillus thuringiensis preparations. An international unit is one one-thousandth of the amount of insecticidal activity contained in one milligram of a preparation of the primary standard E-16 strain of B. thuringiensis, as measured by bioassay against certain lepidopterous larvae (i.e., comparative LD50). A B thuringiensis formulation with a potency of 1000 IU/mg is therefore, equal to the international standard. A secondary reference standard. A secondary reference standard B. thuringiensis strain (HD-1-S-1971) is in use in the USA, and has been assigned a potency of 18000 IU/mg against the cabbage looper. Trichophusia in (Hubner). |
| Intestine | The proctodaeum, or the part of the proctodaeum beyond the pylorus. (The term applied also to the entire alimentary canal.) |

| Term | Terminology |
| --- | --- |
| Intima | The extension of the exoskeleton into the internal cavities of the body for lining the foregut, hindgut and tracheae. |
| Intrinsic articulation | Articulation in which the points making contact are sclerotic prolongation in the articular membrane. |
| Invagination | Pouch or sac occurred due to the infolding or indrawing of the outer surface. |
| Inviscate | (1) To trap insects in sticky substance or on a sticky surface. (2) To make a substance more sticky by mixing it with viscous matter. |
| IOBC | International Organization for Biological Control of Noxious Animals and Plants (an affiliate of the International Union of Biological Sciences). A global organization of governmental units and individuals interested in biological pest suppression, headquartered in Zurich, Switzerland. Major objectives include dissemination of information coordination and promotion of research and application of biological pest suppression. Publisher of the journal, Entomophaga. |
| Iridescent virus | A non-occuded insect virus characterized particularly by its fascinating optical properties; the stem form Bragg reflection, and give purified pellets of virus particles an opalescent appearance. |
| Johnston's organ | A sense organ similar to a chordotonal organ, located in the second antennal segment of most insects. |
| Jugum | A finger like process extending from the base of the front wing underneath the hind wind (Lepidoptera). |
| Juvabione (paper factor) | A JH-active chemical found in paper products manufa tured from American balsam firtrees, active against the European red bug. Pyrrhocoris apterus L., and accidentally discovered by Slama and Williams (1966). |

| Term | Terminology |
|------|-------------|
| Juvenile hormone (JH) | (1) A chemical produced by the corpora allata, one of the three major insect developmental hormones and responsible for determining the type of molt which will occur when one is directed by ecdysore. A high JH blood titer, additional larval or nymphal molts occur, low titer or complete absence of JH causes the pupal or imaginal transformation to occur. (2) The hormone, secreted by the corpora allata, that maintains the immature form of an insect during early molts. |
| Juvenile stage | Any or every post embryonic stage in the development of an organisms that precedes the sexually mature adult stage, e.g., larval and pupal stages of insects. |
| Juvenoid juvenile hormone analog (JHA) | Any synthetic or natural chemical of plant or animal origin exhibiting biological activity similar to the true juvenile hormone. |
| Kairomone | A trans-specific chemical messenger, the adaptive benefit of which falls on the recipient rather than the emitter, see ALLOMONE (Erown, et al., 1970). |
| KD50 | Fifty percent knockdown effect or insects. |
| Key | It is a tabulation of diagnostic characters of species o genera etc. in dichotomous couplets facilitating rapid identification. |
| Key character | In taxonomy a character of special utility in key is known as key character. |
| Key pest | An insect pest or disease normally present at some time during the growing season that causes economic damage to a crop. |
| Knock down effect | The rapid falling down of the insects treated with an insecticides. |
| Labellum (Pl. labella) | The modified tip of the tabium of a number of Diptera. |
| Labial glands | The usual 'salivary glands' of insects, opening by median duct between the base of the hypopharynx and the labium, or on the hypopharynx. |
| Labial palpus (Pl. labial palpi) | One of the pair of sensory appendages (feeler like and two to five segmented) of the insect labium. |

| Term | Terminology |
|---|---|
| Labial suture | The suture of the labium between the prementum and the postmentum, always distal to the mentum when the latter is present. |
| Labrum | Sclerite beneath the clypeus; serves as upper lip in insects that have chewing mouthparts. |
| Lacinia | The inner endite lobe of a maxilla, provided with a muscle arising in the stipes and often with a second muscle arising on the cranial wall. |
| Lance | An accessory in sprayers; a metal tube of uniform cross section on one end of which the spray nozzle is mounted while the other end is connected to the liquid container with a hose pipe or pressure tubing. |
| Larva | (1) An immature stage which is radically different in form the adult; characteristic of the homometabolous insects, including the hymenoptera. In the termites, the term is used in a special sense to designate an in mature individual without any external trace of wing buds or solider characteristics. (2) Young insect that comes out of an egg in early stage of morphological development and differs fundamentally in form the adult. |
| Larvicide | An agent which destroys insect larvae. |
| Lateral ocellus | The simple eye in holometabolus larvae. Also called stemma (pl. stemmata). |
| Lateral oviduct | In insect, one of the paired lateral ducts of the female genital system connected with the ovary. |
| Lateral plates | The lateral areas of the germ band after differentiation of the middle plate. |
| Lateral tracheal trunk | The usual longitudinal tracheal trunk on each side of the body closely connected with the lateral spiracles. |
| Laterosternite | The lateral part of a definitive thoracic sternum apparently derived from ventral are (sternopleurite) of the subcoxa. |

| Term | Terminology |
|------|-------------|
| LC50 | A lethal dose for 50 per cent of the test organisms. A means of expressing the toxicity of a compound present in air as a dust, mist, gas or vapour. It is generally expressed as micrograms per liter as a dust or mist but in case of a gas or vapour as ppm. The LC50 is the statistical estimate of the dosage necessary to kill 50 percent of a very large population of the test species , through toxicity on inhalation under stated condition or by law, the concentration which is expected to cause in 50 per cent of the test animals so treated. |
| Leg | Organ of locomotion used in walking. |
| Legal control | Control of pests through the enactment of legislation that enforces control measures or imposes regulations such as quarantines, to prevent the introduction or spread of pests. |
| Legal residue | Residue that is within safe levels according to the regulations. |
| Life cycle | Definite pattern of birth, growth, reproduction, old age and death. |
| Life history | Habits and changes undergone by an organism from the egg stage to its death as an adult. |
| Life system | A population and the environment which influences it. |
| Life table | A device for expressing in an orderly fashion, observations on the changing density of an insect population in time and space and the processes which direct those changes Especially in relation to the age specific distribution of mortality and its causes. |
| Ligula | The terminal lobes of the labium collectively, or a terminal part of the labium formed by the union of the lobes. |
| Lipophobic | Used as opposite of lipophilic and therefore also of hydrophobic. |
| Longevity | Life span of an individual; long life. |
| LT50 | The time required for a toxic substance to kill 50 percent of a test population. |

| Term | Terminology |
| --- | --- |
| Maggot | The larva of a fly. It is without legs and without well developed head capsule. The head end is pointed and the posterior end appears as if cutoff at the tip. |
| Mandibles | Appendages of the fourth body segment that become the first pair of mouthparts in mandibulate species. |
| Mandibualr glands | A pair of glands often present in insects opening mesally at the bases of the mandibles. |
| Mandibulate | With jaws suited for chewing mouth parts derived from mandibles. |
| Mandibulate soldier | A soldier which has mandibles used in colony defense. |
| Marginal cell | A cell in the distal part fo the wing bordering the costal margin (Diptera, Hymenoptera). |
| Marginal vein | A vein on or just posterior to the wing margin, the vein forming the posterior side of the marginal cell (Hymenoptera). |
| Maxilla (Pl. maxillae) | (1) One of the paired mouth part structures immediately posterior to the mandibles. (2) Upper jaw; part of jaw behind premaxilla. In insects and arthropods, the two accessory jaws or appendages immediately behind the mandibles. |
| Maxillary glands | Glands present in some insects opening mesally at the bases of the maxillae. |
| Maxillary palpus (pl. maxillary palpi) | A small feelerlike structure arising from the maxilla. |
| Maximum permissible intake (MPI) | The maximum permissible intake (MPI) is derived by multiplying ADI with the average body weight. |
| Maximum residue limit (MRL) | Amount of pesticide residue permitted to remain in a food. Expressed as parts per million (ppm). |
| Mechanical control | Control of pests by mechanical means as window screens, earth barriers, and so on. |
| Median effective dose (ED50) | The concentration of a chemical in half the test subjects. It is an indirect measure of the mean tolerance of a batch of the test subjects. The median lethal dose (LD50) is a special case in which death is the response. |

| Term | Terminology |
|---|---|
| Median effective time (ET50) | The time at which a response occurs in half the text subjects after exposure to a pathogenic (including toxicological) stimulus. Its symbol is ET50. Median lethal time is a special case, in which death is the response. In case of median survival Time (ST50) response as surviving individuals is considered. |
| Median lethal dose (of radiation, (M) LD50 | The dose of ionizing radiation which would kill 50% of a large batch of organisms within a specified period of time. |
| Median lethal time (LT50) | In a time dependent biological assay procedure, this is the period of exposure to a pathogenic (including toxicological) stimulus which will produce death in half the test subject. The length of exposure is direct measure of dosage and an increase in the period of exposure results in an increasing uptake and true dose increase in the same ratio. |
| Median survival time or ST50 | A restricted concept of median effective time, the time at which death occurs in half the test subjects, after exposure to a pathogenic (including toxicological) stimulus ST50 is not a direct measure of dosage, unlike LT50. |
| Mesenteron | The stomach of embryo, in insects regenerated from scattered endodermal cells or from intact endodermal remnants; becomes the epithelium of the adult stomach or ventriculus. |
| Mesothorax | The second segment of the thorax; bearing the first pair of wings in winged insects. |
| Metabola | Insects having a definite metamorphosis. |
| Metamorphosis | (1) Changes in form during the development of an insect from the egg to the adult. (2) the series of changes through which an insect passes in its growth from the egg through the larva and pupa to the adult; complete when the pupa is inactive and does not feed; incomplete when there is no pupa or when the pupa is active and feeds. (3) the change in form of an organism through various growth stages from larva to adult,, e.g., a caterpillar into a pupa and subsequently into a fully developed insect. |

| Term | Terminology |
| --- | --- |
| Metanotum | The dorsal sclerite of the metathotax. |
| Metapleuron (pl metapleura) | The lateral sclerite(s) of the metathorax. |
| Metapneustic | An insect respiratory system (particularly in some dipterous larvae) in which only the last abdominal pair of spiracles is functional. |
| Metascutellum | The sternum, or ventral sclerite, of the metathorax. |
| Metasoma | The hindmost of the three principal divisions of the insect body. In most insect groups it is the strict equivalent of the abdomen. In the higher Hymenoptera it is composed only of some of the abdominal segments, since the first segment (the 'propodeum') is fused with the thorax and has therefore become part of the mesosoma. |
| Metasternum | Ventral sclerite of the metathorax. |
| Metatarsus (pl., metatarsi) | The basal segment of he tarsus. |
| Metathorax. | The third segment of the thorax. It bears tee hind legs and the second pair of wings in winged insects. |
| Metepimemron (pl. metepimera). | The epimeron of the metathorax. |
| Microbial insecticide | (1) A pathogenic microorganism or its products (e.g., toxins) when used by man to suppress an insect population; use of the word "insecticide" is properly reserved only, for chemical agents of insect mortality, and microbial pathogen is a preferable term for materials whose active agent is a microorganisms. (2) Pathogen used for killing insects. Free from harmful residues, non-phytotoxic to crops of higher doses, harmless to beneficial insects. |
| Microenvironment | A small or restrict set of distinctive environmental conditions, such as dead animal or fallen log (at habitat). |
| Microorganism | A microscopic animal or organism. |
| Microphagous | Feeding on organisms of minute size. |

| Term | Terminology |
|------|-------------|
| Micropyle | A minute opening or group of openings into the insect egg through which the spermatozoa enter in fertilization. |
| Midgut | The middle part of the alimentary canal and the main site of digestion and absorption. |
| Migration | (1) A regular movement from one region to another by animals especially birds. (2) Movement of ions (and nutrients) in soil or during electrolysis. (3) In chromatography, movement of loaded sample in the direction of mobile phase during elution or development. |
| Mimicry | Superficial resemblances between two organisms that result in protection for at least one organism, the mimic. |
| Mite | (1) Tiny eight-legged animal with body divided into two parts. It has no antennae (feelers). During the nymphal stage it has six legs. (2) A sub-class of the Arachnida allied to ticks, found in many habitats (*e.g.*, the soil, decaying organic matter) and parasitic on both plants and animals. |
| Molt (moult) | The casting off of the outgrown skin or exoskeleton in the process of growth. Also the cast-off skin itself. The word is further used as an intransitive verb to designate the performance of the behaviour. |
| Monocondylic joint | A joint with a single point of articulation between the adjacent leg segments. |
| Monophagous | Feeding upon only one kind, for example one species or one genus of plants. |
| Monophyly | The derivation of a taxon through one more lineages from one immediately ancestral taxon of the same or lower rank. |
| Moribund (insects) | Dying, near death. The insects with no sign of recovery. |
| Morphology | It is science of form. It deals with the study of form of structure of organism, specially outer form, inner structure and development of living organism. |
| Mortality | Death-rate. |

| *Term* | *Terminology* |
|---|---|
| Moulting | The process of casting-off the old tough outer layer (cuticle) of the body wall to accommodate the growth of the body. |
| Multi-dose poison | Chronic poisons, such as inticoagulants like WAR-FARIN. |
| Multiparasitism | A condition resulting from the simultaneous use of a single host individual by two or more species of primary parasitoids. |
| Multivoltine | Having two or more complete generations annually. |
| Mutagenic | Capable of producing genetic change. |
| Mutualism | Symbiosis that benefits the members of both of the participating species. |
| Natality | Birth-rate. |
| Natural control | (1) The reduction of pest populations by the forces of nature, such as climatic factors, parasitoids, predators and diseases. (2) The process of dynamic equilibrium which maintains the characteristics mean density of a wild population within particular upper and lower limits, over a period of time, by a complex combination of all the additive, conditioning and subtractive processes impinging upon that wild population. |
| Nematicide | A material, often a soil fumigant used to control nematodes infesting roots of crop plants. |
| Nerve | Any one of the fibrous tracts of the peripheral nervous system, whether a single fiber or a group of fibers; a nerve trunk. |
| Nerve cord | Cord composed of delicate filaments of tissue functioning to transmit stimulus to or from a ganglion or from of to any part of the body. |
| Nerve fiber | The axon or other branches of a neurocyte. |
| Nervous system | All nerve cells and tissues in animals, including the brain, spinal cord, ganglia, nerves, and nerve centers. |

| Term | Terminology |
|------|-------------|
| Neuroblasts | The primitive nerve cells differentiated from the ectoderm. |
| Neurone | The entire nerve cell including all its processes. |
| Neuropile | The medullary substance, or mass of fibrous tissue within a ganglion. |
| Neurosecretory | Pertaining to the secretion of hormones by nerve cells. |
| Niche | The precise constellation of environmental factors into which a species fits or which is required by a species. |
| Nocturnal | (1) Pertaining to the night hours. (2) Active by night. |
| Nomenclature | It is the application of distinctive names to each of the groups recognized in classification. |
| No-observed effect concentration or level (NOEC/NOEL) | Highest concentration or amount of pesticide in the test system that causes no biological effect to the target organism. |
| Non-observed adverse effect level (NOAEL) | It is the highest dose of substance that does not cause any detectable toxic effects in experimental animal studied. It is expressed in milligrams per kilogram of body weight per day. |
| Non-preference | (1) Here used in the sense of Painter (1951) as an insect pests' response to a potential host (variety) which displays various characteristics that discourage its use for oviposition, food or shelter. (2) In this case, the resistance of a plant to the attack of a pest is due to the lack of one or more of the preferred factors in the plant. |
| Nozzle | (1) Mouthpiece of a spout through which liquids are discharge in a stream or as a spray. (2) Devices which control drop size, rate, uniformity, thoroughness, and safety of a pesticide application. The nozzle type determines the ground pattern and safety of a pesticide application. Examples: Flat fan, even flat fan, cone, flooding, off-set, atomizing, broadcast, and solid stream nozzles. |
| Nuptial flight | The mating flight of the winged queens and males. |

| Term | Terminology |
|------|-------------|
| Nymph (pl. nymphae) | (1) Immature stage of an insect species, similar to adult but with wings and sex organs not fully developed. (2) In general entomology, the young stage of any insect species with hemimetabolous development. In termites, the term is used in a slightly more restricted sense to designate immature individuals which possess external wing buds and enlarged gonads and which are capable of developing into functional reproductive by further molting. (3) In insects, an immature stage (instar) which is similar to the adult except for absence of functional wing and sexual organs. Distinct form larva of other insects, which is entirely different from the adult. |
| Obligate parasitism | Here in reference to nematodes which must develop parasitically and cannot reproduce and complete growth away from a host. |
| Obligate pathogen | A disease causing microorganisms which requires a living host in which to grow and reproduce. |
| Occipital | The area of the cranium between the occipital and postoccipital sutures; its dorsal part is the occiput proper (Oc), its lateral parts the postgenae (Pge). |
| Occipital condyles | Processes on the margin of the postocciput to which the lateral neck plates are articulated. |
| Occipital ganglion | A single or paired postcerebral ganglion of the stomodaeal nervous system (Pharyngeal, oesophageal, or postcerebral, ganglion). |
| Occipital suture | A transverse suture in the posterior part of the head which separates the vertex from the occiput dorsally and the genae from the postgenae laterally. |
| Occiput | Portion of head behind the vertex, eyes and gena. |
| Occluded viruses | Viruses characterized by enclosure of the mature replicated virions in the matrix of proteinaceous or crystal-like inclusion bodies, such as capsules or polyhedra; also called INCLUSION VIRUSES. |
| Ocellus | One of the three simple eyes of adult insects, located on or near the centre line of the dorsal surface of the head. The ocelli should be distinguished from the laterally placed compound eyes. |

| Term | Terminology |
|------|-------------|
| Ocular sclerite | A narrow band of the cranial wall encircling the compound eye within the ocular suture. |
| Ocular suture | The line of inflection in cranial wall around the compound eye, forming internally a circumocular ridge. |
| Oesophagus | A tubular part of the fore-gut between the mouth and the crop. |
| Oligophagous insects | Insects which restrict their feeding to a small number of usually similar plants, e.g., the spotted boll-worm. |
| Oligophagus parasite | A parasite which can develop upon a few of the closely related host species. |
| Oligopod | Larva coming out from embryo hatched in advanced stage of development, bearing a general resemblance to adult, e.g., Thysanoptera. |
| Ommatidium. | One of the basic visual units of the insect compound eye. The ommatidia are bounded externally by the facets that together make up the glassy, rounded outer surface of the eye. |
| Omnivorous | Feeding generally on animal or vegetable food, or on both. |
| Oocytes | The egg cell differentiated from the oogonium, before maturation. |
| Oogonium | The first stage in the differentiation of an egg cell from a primary female germ cell. |
| Oral toxicity | (1) Toxicity of a compound when given by mouth. Usually expressed as the number of mg of chemical per kg of body weight of animal when given orally in single dose that kills 50 per cent of the animals. The smaller the number, the greater the toxicity. (2) Ability of a pesticide chemical to sicken or kill an animal or human if eaten or swallowed. |
| Organelles | Organized microstructures inside, cells to which certain biochemical processes are confined, e.g., mitochondria, chloroplasts. |
| Ovariole | One of the egg tubes which collectively form the ovary. |

| Term | Terminology |
| --- | --- |
| Ovary | A mass of tubes lying one on each side of the body cavity of the female insect in which the eggs are developed. |
| Ovicide | (1) A chemical which kills eggs before they hatch or one so used. Almost restricted to action on eggs of phytophaguous mites (red spider). The word is almost universally accepted in this sense, but the classical scholar insists that it should mean "sheep (Latin, ovis) killer and that on egg-killer should be ovacide". (2) Pesticides or agents used to destroy insect, mite or nematode eggs. |
| Oviduct | One of the paired tubes through which the egg passes from the ovary into the vagina; oviductus lateralis. |
| Oviparous reproduction | Reproduction by means of eggs, which hatch outside the body. |
| Ovipositor | The extension tube on the abdomen of a female insect which she uses to place her eggs in suitable location. Often capable of remarkable drilling power. Hence oviposition the act of egg-placing. |
| Oviparity | Reproduction by laying eggs. |
| Ovoviviparity | Reproduction in which eggs hatch soon after they are laid. |
| Palaeo-entomology | Study of extinct fossil insects. |
| Palpifer | Small lobe or sclerite articulated to the stipes which bears the maxillary palpus. |
| Palpiger | Small sclerite on the labium which bears the palpus. |
| Paraglossa (pl. Paraglossae) | One of a pair of lobes at the apex of the labium. Laterad to the glossae. |
| Paranotum (pl. Paranota) | Notal flanges in fossil insects. |
| Paraproct | One of the a pair of lobes bordering the anus lateroventally. |
| Parasite | (1) An organism which lives in or on another (the host) from which it derives food, for all or part of its life (2) An organism that lives in an organic relationship within or upon another living |

| Term | Terminology |
|------|-------------|
| | organism taxonomically unrelated to it from which it derives part or whole of its nourishment. Causes diseases in animals. Many live on plant. Also plant that climb about another plant. (3) A plant or animal that harms another living plant or animal (the host) by living and feeding on or in it. Some of our worst pests are parasites which cause disease or injury to animals and plants grown by man or to man himself. |
| Parasitism | (1) Symbiosis in which members of one species exist at the expenses of members of another species, usually without going so far as to cause their deaths. (2) A qualitative term referring to a kind of symbiosis in which one party (the parasite) lives at the expenses of the other (the host), contributing nothing to the relationship and frequently destroying the host in the process. |
| Parasitoid | (1) An insect parasite of an arthropod; parasitic only in its immature stage, destroying its host in the process of its development and free living as an adult: Also see PARASITE, predator. (2) A parasitic insect which lives in or on a larger host organism and eventually kills it. (3) A parasite that slowly kills the victim, this event occurring near the end of the parasites larval development. The term is also used as an adjective. (4) A parasite that usually kills its host. |
| Parthenogenesis | Reproduction by the development of are egg without its being fertilizer by a sperm. |
| Pedicel | The "waist" of the ant. It is made up of either one segment (the petiole) or two segments (the pedicel plus the postpetiole). |
| Penetrated residue | The surface residue becomes penetrated residue by migration into the sub-strata. |
| Penis | One of the paired intro-mittent organs of certain insects, or the usual median phallic organ. |
| Periodicity | A regular cyclic or repetitive behaviour of an organ cell or organism in a unit time. |

| Term | Terminology |
| --- | --- |
| Peripheral nervous system. | The outlying parts of the nervous system in distinction to the central ganglia and connectives, including the sensory neurocytes and their axons, and the axons of the motor neurons. |
| Peripneustic | With none or only a few spiracles closed in each lateral series. |
| Peripneustic respiration | The prevalent type in insects, consisting of a row of spiracles on each side of the body. |
| Periproct | The terminal piece of the body containing the anus, anterior to which segmentation takes place (Telson). |
| Peritrophic membrane | (1) The delicate membrane surrounding the food in the mid-gut; present in many insects. (2) A cylindrical membranous envelope surrounding the food in the ventriculus, and sometimes extending into the proctodaeum; geerated from the ventricular epithelium, either from all or the length of the latter or from a ring of specialized cells at its anterior end. |
| Permissible level | A concentration expressed in ppm of residue in or on arnod when first offered for consumption. It is calculated from (i) the Accepted Daily Intake (ADI), (ii) the food factor, and (iii) the average weight of the consumer. The ADI is the daily dose which during an entire life time will not cause any injury. It is set at a safe level as determined by data obtained in animal feeding studies. A safety of 100 is used in translating animal data to human beings. It is expressed in milligram of the pesticide per kilogram of body weight per day (mg/kg/day). The food factor refers to the total diet made up to a food or of various foods. |
| Persistence | (1) The term applied to chemicals that remain active for a long period of time after application. (2) The quality of an insecticide to persist as an effective residue due to its volatility and chemical stability. (3) The property of a chemical to remain effective as a residue. (4) Property of chemical to remain active for long period after application. |

| Term | Terminology |
| --- | --- |
| Pest | Any specie, strain or biotype of plant, animalor pathogenic agent injurius to plant or plant products. It includes insects, mites, nematodes, fungi, bacteria, viruses, phytoplasma, weeds, etc. |
| Pest control | The use of disinfectant, herbicide, pesticide, insecticide, management or cultural practice to control pests. |
| Pest control programme | An organized programme to keep down the pest population in an area below pest densit. |
| Pest density | Population level at which a pest species causes economic damage. |
| Pest spectrum | The complete range of pests attacking a particular crop. |
| Pesticide residue | Pesticide remaining on or in a plant, plant product air, water, or treated area following a time lapse after application. Pesticide residues means any specified substances in food, agricultural commodities, or animal feed resulting from the use of a pesticide. The term includes any derivatives of a pesticide, such as conversion products, metabolites reaction products, and impurities considered to be toxicological significance. |
| Pesticide tolerance | The amount of pesticide residue which may legally remain in or on a food crop. |
| Pharynx | (1) The back part of the mouth; a slight enlargement of the beginning of the oesophagus. (2) The part of the stomodaeum between the mouth or buccal cavity and the oesophagus, the dorsal dilator muscles of which arise on the frons and the dorsal part of the cranium and are inserted posterior to the frontal ganglion and its connectives; usually not extending beyond the nerve ring of the head, but in some insects there is a posterior pharynx behind the brain. |
| Phenon | It is a sample of phenotypically similar specimens; a phenotypically reasonably uniform sample. |
| Phenotype | The physical manifestation of a genetic trait. |

| Term | Terminology |
| --- | --- |
| Pheromone | (1) A chemical substance, usually a glandular secretion, which is used in communication within a species. One individual releases the material as a signal and another responds after tasting or smelling it. (2) A substance secreted by an insect to the exterior causing a specific reaction in the receiving insects. (3) A special chemical substance which, secreted by one individual, influences the behaviour of other individuals of the same species. |
| Phragma (pl. phragmata) | A plate-like apodeme of the dorsal wall of the thorax. |
| Phragmata (Sin. phragma) | Expanded cuticular plates of tergum at the base of antecostal region. It give attachment to dorsal longitudinal muscle helping in downstroke of wing. |
| Phylogeny | Evolutionary history of species. |
| Physiology | Study of various functions of living beings. |
| Phytophagous | Feeding on plants or on materials of plant origin. |
| Phytotoxic | Toxic to plant, usually restricted to higher (green) plants. Hence phytotoxicity, which is used generally to include the action of intentional herbicides but the some authors restricted to adverse side-effects on the crop of fungicides, insecticides or formulation agents. |
| Plant resistance (to pest) | The ability of plant to grow and produce economically despite the presence of a pest. Such plants possess some chemicals or juices which inhibit the parasite from feeding on it. |
| Pleura (Sing. Pleuron or pleurum) | The lateral areas of the thoracic segment. |
| Pleural | Pertaining to the pleural, or lateral sclerites of the body; lateral. |
| Pleural apophysis | The internal of the pleural ridge. |
| Pleural region | The podial region, or ventrolateral parts of the body on which the limbs are implanted, metamerically divided into segmental pleural areas. |

| Term | Terminology |
|------|-------------|
| Pleural ridge | The endopleural ridge formed by the pleural suture, bracing the pleuron above the leg, or between the coxal articulation and the wing support. |
| Pleura sulcus | Groove running dorsally from coax; often referred to as pleural suture. |
| Pleura suture | The external groove of the pleural ridge, separating the episternum from the epimeron. |
| Pleura wing process | The wing support of the pleuron at the upper end of the pleural ridge. |
| Pleurite | Any minor scelrite of the pleural area of a segment, or one of the component sclerites of a pleuron. |
| Pleuron ventral line | The line of separation between the pleural region and the venter, lying mesad of the limb bases, but obscured when the latter are fused with the sterna. |
| Poison bait | (1) A poison mixed with bran, molasses, grain, or other attractant so as to serve as a bait and kill insect pests and rodents. (2) It consists of a relatively small quantity of a poison combined with a food material of the pest. (3) An attractant foodstuff for insects, moluscs, or rodents, mixed with an appropriate toxicant. |
| Pollen brush. | A brush of short stiff hairs used in collecting pollen (bees). |
| Polyembryony | The development of several individual organisms from a single egg. |
| Polygenic | A characteristic, such as resistance, conferred by more than one gene. |
| Polyhedral viruses | Viruses which cause insect disease characterized by the presence of polyhedral (many sided) inclusions in the infected cells of the host. |
| Polyphagous insects | Insects, such as locusts and grasshoppers, which feed on a variety of plants. |
| Population dynamics | The study of numerical changes in populations of living organisms in time and space and of the processes which cause such variations. |
| Pore canal | The channel of the cuticula beneath the seta or other external part of many sense organs. |

| Term | Terminology |
|------|-------------|
| Posterior intestine or rectum | The terminal section of the rotodaeum commonly termed the "tectum", but usually divided into an anterior rectal sac (rsc), and a posterior rectum proper (rect). (Mastdarm) |
| Posterior notal wing process | A posterior lobe of the lateral margin of the alinotum supporting the third axillary sclerite of the wing base. (Hinterer Galenkfortsatz). |
| Postmentum | The basal portion of the labium proximad of the labial suture. |
| Postnotum | An intersegmental plate of the dorsum of the thorax associated with the tergum of the preceding segment, bearing the antecosta and usually a pair of phragmatal lobes (Phragmanotum). |
| Postoccipital suture | The posterior submarginal groove of the cranium having the posterior tentorial pits in its lower ends; internally it froms the postoccipital ridge (PoR) on which are attached the dorsal prothoracid and neck muscles of the head. |
| Postocciput | The extreme posterior rim of the cranium behind the postoccipital suture, probably a sclerotic remnant of the labial somite. |
| Poststernite | The postcostal lip of a definitive sternal plate that includes the intersegmental sclerotization. |
| Post tergite | The narrow postocostal lip of a postnatal thoracic plate. |
| Potency | The strength of something. Example : how deadly a poison is. |
| Potentiation | The joint action of two pesticides to bring about an effect greater than the sum of their individual effects. |
| Pre-oviposition period | The period of time between the emergence of an adult female insect and the start of its egg laying. |
| Predator | (1) An animal (or, rarely, a plant, free living organism) which feeds upon other animals, called its prey. The prey is killed and consumed by its predator which generally devours many individuals during its life (contrast parasite). There are borderline cases between predation and |

| Term | Terminology |
| --- | --- |
| | parasitism, e.g., species of wasp which inject a permanent paralyzing drug into a caterpillar before laying eggs init. (2) An insect or other animal that attacks, feeds on and destroys other insects or animals. Predators help to reduce the number of pests which cause disease, damage and destruction. Pregenital. Anterior to the genital segments of the abdomen. |
| Prescutum | The anterior area of the mesonotum or metanotum between the antecostal suture and the prescutal suture, when the latter is present. |
| Presternum | A narrow anterior area of sternum sometimes set off from the basisternum by a submarginal suture of the eusternum (Not the acrosternite). |
| Pretarsus | The terminal parts of the leg distal to the tarsus, including median remnants of the dectylopodite, and the lateral claws, or ungues; in most larvae a simple claw-like segment. |
| Prey | An animal when fed upon by a predator. A predator of one species can itself be prey to another. |
| Primary parasite | A parasite which established itself in or upon a free living host, that is the host is not a parasite. |
| Primary parasitoid | An insect parasite of any arthropod which is not itself parasitic. |
| Proboscis | (1) The lengthened mouthparts of an insect specially adapted for piercing, sucking or other specialized modes of feeding. (2) The extended beak like mouth parts of sucking insects. |
| Procephalon | The region of the definitive head anterior to that of the gnathal segments, representing the primitive protocephaion, formed in the embryo from the cephalic lobes and the tritocerebral segment; bearing the labrum, the mouth, the eyes, the first antennae and the second antennae when the last are present. |
| Proctodaeum | The hind intestine, including the Malpighian tubules. |
| Proctodaeum | The posterior ectodermal part of the alimentary canal (Hind gut, hind intestine, Hinterdarm, Enddarm). |

| Term | Terminology |
| --- | --- |
| Proctodeal valve | In insects, a valve in the anterior end of the hindgut that serves as an occlusor mechanism. |
| Proctodeum | (1) The hindgut and Malpighian tubules of an insect. (2) Posterior division of alimentary canal formed from invagination. |
| Proleg | A fleshy abdominal leg found in caterpillars (Lepidoptera, sawtlies), bearing a characteristic arrangement of crochets. |
| Pronotum | A large, rounded, shield like plate covering the front part of the body; the dorsal cuticular covering of the first thoracic segment. |
| Propneustic | That type of respiration in which only the first pair of spiracles is functional. |
| Propodeum | In higher Hymenoptera the first abdominal segment when it is fused with the alitrunk. Same as epinotum. |
| Prosternum | The sternum, or ventral sclerite, of the prothorax. |
| Prothoracic gland | One of a pair of endocrine glands located in the prothorax near the prothoracic spiracles. |
| Prothorax | The first segment of thorax. It bears the anterior legs, but no wings. |
| Protocerebrum | The first part of the arthropod brain, containing the ocular and other association centres lying anterior or dorsal to the antennae (deutocerebral) centres. |
| Protonymph | The second instar of a mite. |
| Protopod | Primary larva, little more than prematurely hatched embryo, devoid of abdominal segmentation and having rudimentary cephalic and thoracic appendages, e.g., parasitic Hymenoptera. |
| Protopodite | The basal stalk of some crus acean limbs composed of the united coxopodite and basipodite. |
| Proventriculus | A specialized part of the stomodaeum immediately anterior to the ventriculus (Gizzard, gesier, chewing stomach, kaunagen). |

| Term | Terminology |
|------|-------------|
| Pseudopods | False legs on the abdomen of the larva of butterflies, moths and sawflies. |
| Pseudopupa | A coarctate larva; a larva in a quiescent pupalike condition oe or two instars before the true pupal stage (Coleoptera, Meloidae). |
| Pseudotrachea (pl. pseudotrachae) | A structure having the appearance of a trachea; a false trachea. |
| Pterothorax | The wing bearing portion of the thorax of winged insects, i.e., the meso and metathorax. |
| Ptillium | (1) An eversible sac of the fronts in dipterous pupae used for rupturing the puparium. (2) Eversible sac in some dipterous adults that is used to rupture the puparium. (3) In Diptera Cyclorrhapha, an organ capable of being inflated and thrust out thourgh the frontal suture, just above the bases of the antennae, at emergence from the pupa. |
| Pulsatile organs | Accessory organs that are sometimes present at the bases of the antennae and the wings for pumping blood into them. |
| Pupa | (1) A resting stage in insect metamorphosis between larva and young adult, during which movement and feeding ceases and a great change in form occurs. Also called chrysalis. (2) The pupa is the resting stage of many insects. The stage between the larva (caterpillars or maggot, etc.). |
| Puparium | (1) The tough encasement in which a larva changes to a pupa before developing into an adult insect. (2) the thickened, hardened last larval skin within which some flies pass the pupal stage. |
| Pyloric valve | A valvular fold usually situated in the pyloric region of the proctodaeum, but sometimes formed by the posterior end of the ventriculus. |
| Quarantine | (1) Isolation of an animal because of contagious disease may be enforced. (2) The place of isolation. (3) Prohibition to prevent the introduction or spread of any dangerous insect pest or plant disease. (4) All operations associated with the prevention of importation of unwanted organisms |

| Term | Terminology |
|---|---|
| | into a territory, or their exportation from it. (5) Regulation forbidding sale or shipment of plants or plant parts, usually to prevent disease, insect, nematode, or weed invasion of an area. |
| Quarantine pest | A pest of potential economic importance to the area endangered thereby and not yet present there, or present but widely distributed and being officially controlled. |
| Quiescent | State of relative biological inactivity induced by physical factors such as low temperature; inactive, resting, non-feeding or dormant. |
| Oesophagus | The narrow part of the alimentary canal immediately posterior to the pharynx and mouth. |
| Radius | The third vein of the wing; its base flexibly attached to the second axillary. |
| Raptorial legs | Legs adapted for catching prey. |
| Rectal glands | Padlike or papilliform structures on the inner wall of the rectal sac (not demonstrated to be glands). |
| Rectal gills | The gills within the rectum of nymphs in anisopteran dragonflies. |
| Rectum | In insects the posterior expanded part of the hindgut, typically pear-shaped. |
| Recurrent nerve | The median stomodaeal nerve extending posteriorly from the frontal ganglion. |
| Recurrent vein | One of two transverse veins immediately posterior to the cubital vein (Hymenoptera). |
| Reflex | A functional unit of the nervous system involving the entire pathway from a receptor cell to an effectors. |
| Repellent | (1) A compound that is annoying to a certain animal or other organism, causing it to avoid the area in which it is placed. (2) Substance which repels insects and keeps the treated object free from their attack; coaltar, creosote, oil of citreonella, examples. |
| Reproductive potential | The maximum possible rate of increase (multiplication) of an organism. |

| Term | Terminology |
|------|-------------|
| Residual activity | The insecticidal activity resulting from the presence of an insecticide and its metabolites on or in a treated site after an insecticide is applied. |
| Residual deposit | Amount of insecticide remaining on a surface after application. |
| Residual pesticide | A pesticide chemical that can destroy pests or prevent them from causing disease, damage, of destruction for more than a few hours after it is applied. |
| Residue | (1) The amount (often a trace) of pesticide and its metabolites left in food crop after treatment. (2) Its that part of a pesticide which remains in or on a plant or animal after application. It is expressed as ppm and is based on fresh weight of the sample. |
| Resistance | (1) The ability of an organism to suppress or retard the injurious effects of a pesticide. (2) It is development of the ability in a strain of insects to tolerate a dosage of a toxicant which would kill majority of individuals in a normal population of the same species at recommended application rates. |
| Resurgence | (1) An increase of a (pest) population, after a period of decrease, to a level higher than its original one, especially when the decrease is caused by an intended control operation. |
| | (2) Sudden rebounding of pest population after application of insecticide. |
| Retina | (1) The receptive apparatus of an eye. (2) The part of an eye upon which the image is formed. |
| Retina cells | The cells composing the retina. |
| Retinal pigment cells | Pigment cells in the retinal region on the eye. |
| Retinula | A group of two, three, or more light sensitive cells, which surround and secrete a longitudinal light receptor unit (optic rod or rhabdom). |
| Rhabdom | The light receptor unit or rod lying in the axis of a group of light-sensitive cells (retinula) below the crystalline cone of an eye; retinal rod' optic rod. |

| Term | Terminology |
|------|-------------|
| Rostrum. | (1) The beak or proboscis of Hemiptera. (2) The extended fore part of the head of an insect consisting of or bearing the mouth parts. The term is applied to insects such as weevils in which biting mouth parts are at the tip of snout as well as to insects with piercing/sucking mouth parts. |
| Royal chamber | The space reserved for the accommodation of kind and queen termites, or the space reserved for the queen of an ant colony. |
| Royal jelly | (1) Also called bee-milk. A special food given by worker bees to the larvae which are to be developed as queen bees. (2) A material supplied by workers to female larvae in royal cells which is necessary for the transformation of larvae into queens. Royal jelly is secreted primarily by the hypropharyngeal glands and consists of a rich mixture of nutrient substances, many of them possessing a complex chemical structure. |
| Salivary glands | Glands that open into the mouth and secrete a fluid digestive, irritant, or anticoagulatory properties. |
| Scape | Basal segment of insect antenna; leafless penduncle arising from a plant and bearing flowers. |
| Scent gland | A gland producing an odourous substance. |
| Sclerite | A plate of the body wall of an insect, bordered by sutures or membranes. |
| Sclerotin | The substance or substances forming the hard parts of the insects skeleton and deposited in the chitin. |
| Sclerozation | The hardening and darkening processes in the cuticle (involves the epicuticle and exocuticle). |
| Scolytoid larva | A fleshy larva resembling the larva of a scolytid beetle. |
| Scutellum | The area of the alinotum posterior to the suture of the V-shaped notal ridge, or the corresponding area when the ridge is incomplete or absent. |

| Term | Terminology |
|------|-------------|
| Scutoscutellar suture | The external suture of the V-shaped notal ridge of the alinotum, the arms divergent posteriorly, dividing the notum into scutum and scutellum. |
| Scutum | In ticks, the sclerotized plate covering all or most of the dorsum in males and the anterior portion in females, nymphs, and larvae of the lxodidae. |
| Sebaceous gland | A gland producing a greasy lubrication substance. |
| Second axillary | The pivotal plate of the wing base resting on the pleural wing process, connected with the base of the radial vein. |
| Secondary parasite | A parasite which establishes itself in or upon a host that is a primary parasite. |
| Secondary parasitoid | An insect parasite of a primary parasitoid. |
| Secondary segmentation | Any form of body segmentation that does not strictly conform with the embryonic metamerism; the usual segmentation of arthropods having a well- developed exoskeleton, in which the membranous intersegmental rings are the posterior parts of the primary segments. |
| Sectorial cross vein | A cross vein connecting two branches of the radial sector. |
| Segment | (1) A body segment may be primary, in which case the term is synonymous with metamere, or it may be secondary having its definitive limits different from those of the embryonic segment. Used also for the divisions of an appendage such as the leg or antenna. (2) A body segment is any of the successive annular subdivisions of the arthropod truncks, whether corresponding to the embryonic metameres or produced secondarily. (3) A subdivision of the body or of an appendage, between joints or articulations. (4) A ring or subdivision of the body or of an appendage between areas of flexibility associated with muscle attachment. |
| Segementation | (1) The body of insects and other arthropods is divided into a number of segments. (2) The embroyological process by which the insect body becomes divided into a series of parts or segments. |

| Term | Terminology |
|------|-------------|
| Seminal vesicles | Pouch-like structures in which seminal fluid of the male is stored. |
| Semio-chemicals | Substances that modify behaviour of insects by inducing in them the response for sex, feeding, fear or alarm etc. These are further classified into pheromones, secreted by insects for stimulating the insects of the same species (i.e., intraspecific stimulators) and allelochemicals, the substances used for interspecific communications. |
| Sense organ | Any specialized innervated structure of the body wall receptive to external stimuli, most insect sense organs are innervated setae. |
| Sensillum | A simple sense organ, or one of the structural units of a compound sense organ. |
| Sensory nerves | Afferent nerves carrying sensory information to the central nervous system. Most of the nerve poisons (e.g., organophosphate and organocarbamate insecticides) cutoff this information system resulting in paralysis of the organism. |
| Sensory neuron | A neuron leading from a receptor cell to the central nervous system. |
| Septicemia | A morbid condition caused by invasion and multiplication of microorganisms in the blood. This is the condition when pathogen multiplies in the blood and tissues throughout the body cavity. |
| Sex attractant | A volatile chemical substance produced by one sex of an insect (mostly female) to attract the opposite sex. |
| Sex pheromone | A substance secreted by one sex of an insect for attracting the opposite sex of its own species to facilitate mating. |
| Sex ratio | The ratio of males to females in a population. |
| Shelf life | It is a period of time during which the pesticide may be stored without becoming unit for use; storage life. |
| Sibling | One of two or more individuals having common parents. |
| Sibling species | Pairs or groups of closely related species which are reproductively isolated but morphologically identical or nearly so. |

| Term | Terminology |
|------|-------------|
| Silk gland | Gland in the silkworm and other insects secreting the substance produced as silk. |
| Skeleton | The sclerotized (hard) parts which form the external protective covering of the body and the internal points of attachment for the muscles. |
| Soldier | (1) A member of a worker sub-caste specialized for colony defense. (2) In termites, sterile males or females with large heads and mandibles; they function to protect the colony. |
| Solitary parasitoid | An insect parasitic which normally develops at a rate of one individual per arthropod host. |
| Speciation | Separation of one population into two or more reproductively isolated independent evolutionary units. |
| Species | (1) A group of individuals or populations that are similar in structure and physiology and are capable of interbreeding and producing fertile offspring; the largest group of individual within which effective gene flow occurs or could occur. (2) Groups of actually (or potentially) inter-breeding natural populations with are reproductively isolated from other such groups. |
| Sperm tube | One of the secondary divisions of the testis. |
| Spermatheca | (1) The sperm storage receptacle of the female insect. (2) Receptacle in the female which receives and stores the sperm of the male. |
| Spermathical gland | A gland associated with the spermatheca in the reproductive system of female insects. |
| Spermatid | An immature spermatozoon. |
| Spermatocyte | The sperm cell differentiated from a spermatogonium before maturation. |
| Spermiogenesis | Process of transforming spermatids into functional sperm cells. |
| Spinasternum | One of the spina-bearing intersegmental sclerites of the thoracic venter, associated, or united, with the sternum preceding; a spinasternum may become a part of the definitive prosternum or mesosternum, but not of the metasternum. |
| Spinneret. | A structure from which silk is spun. |
| Spinning glands | The glands which produce viscous secretion forming silk. |

| Term | Terminology |
|------|-------------|
| Spiracle | (1) It is one of the several small opening in the segments of insects connected to a system of internal tubes through which oxygen is inhaled and carbon dioxide is exhaled. (2) Small openings in the abdominal segment of insects connected to a system of internal tubes (trachea) through which oxygen and carbon dioside are exchanged with the atmosphere. (3) An external opening of the tracheal system through which diffusion of gases takes place. |
| Spray nozzle | The attachment of the end of the sprayer extension that breaks up spray liquids into fine drops so as to apply the material evenly on the plants. |
| Sprayer | A machine used to apply a spray by forcing the liquid through a nozzle under pressure. |
| Stadium. (pl. stadia) | The time interval between molts in a developing insect. |
| Sterile male technique | A method of pest control which involves the release of large number of sterile male insects in the populations. |
| Sterile-insect technique | A genetic method of pest suppression involving the mass release of compatible but sterile insects into a wild fertile pest population to overwhelm and suppress its reproductive capacity, often eventually to the point of extinction. |
| Sternellum | The area of the eusternum posterior to the bases of the sternal apophyses of the sternacostal suture. |
| Sternite | A ventral sclerite, in other words, a portion of the body wall bounded by sutures and located in a ventral position. |
| Sternopleurite | The infracoxal sclerotization of a generalized thoracic pleuron, generally united with the primary sternum in the definitive eusternal plate. |
| Sternopleuron (pl. sternopleural) | A sclerite in the lateral wall of the thorax, just above the base of the middle leg (Diptera). |
| Sternum | The ventral part of a body segment. |
| Stomach | The ventriculus, or mesenteron. |
| Stomodeum | The for gut of an arthropod. |

| Term | Terminology |
|------|-------------|
| Striated muscle | Muscle that is composed of fibers with alternate light and dark bands. |
| Stylet | (1) A hollow tube like organ, especially that possessed by aphids with which they probe the tissues of plants and through which the sugary contents of the phioem vessels is transmitted to the insect. (2) A hollow tube like organ, especially that possessed by aphid through which the sugary contents of the phloem vessels are sucked by the insect. |
| Subcosta | The usual second vein of the wing, associated basally with the anterior end of the first axillary sclerite. |
| Subcoxa | A secondary proximal subdivision of the coxopodite present in some arthropods, forming a support in the pleural wall of the body segment for the rest of the appendage (pleuron). |
| Subesophageal commissure | The two nerves that pass around the esophagus and connect the brain to the subesophageal ganglion. |
| Subesophageal ganglion | The knot-like swelling at the anterior end of the ventral nerve cord, just below the esophagus. |
| Subgenal areas | The usually narrow lateral marginal areas of the cranium set off by the subgenal sutures above the gnathal appendages, including the pliurce tonata and hypostomata. |
| Subgenal sutures | The lateral submarginal grooves of the cranium just above the bases of the gnathal appendages, forming internally a subgenal ridge (SgR) on each side, continuous anteriorly with the epistomal suture when the latter is present. |
| Subgenital plate | In the female the eighth abdominal sternum or the seventh when the eighth is reduced or obliterated. |
| Submarginal vein | A vein immediately behind and paralleling the costal margin of the wing (Chalcidoidea). |
| Submentum | The proximal sclerite of the labium, by means of which it is attached to the head. |

| Term | Terminology |
| --- | --- |
| Subspecies | A subdivision of a species, usually a geographic race. The different subspecies of a species are oridinarily not sharply differentiated, intergrade with one another, and are capable of inter-breeding. |
| Suctorial | Adapted for or capable of sucking, having sucker for feeding or adhering. |
| Sulci | The grooves on head which indicate the presence of strengthening ridges on inside called sulci. The are purely functional origin. They resist various strains imposed on the head capsule. They are constant in their occurrence and poison, however, they are variable in position in different species or absent. |
| Superparasitism | A condition resulting from the use of a single host individual by more individual parasitoids of the same species than it can successfully sustain to maturity because of nutritional limitations; |
| Superposition eye | An eye which permits the passage of light through the non-pigmented wall of one unit (ommatidium) to the neighbouring visual units. |
| Superposition image | Image produced by superposition or overlapping of points of light from a number of facets. |
| Supervised control | The control of pests as soon as it becomes necessary so that the pest does not increase its density above the economic threshold. Such a control necessitates keeping a regular record of the pest populations. |
| Susceptibility | Means the degree to which an organism is affected by a pesticide at a particular level of exposure. |
| Susceptible | Capable of being injured, diseased, or poisoned by a pesticide, not immune. |
| Susceptible species | A plant or animal that is affected by moderate amounts of a pesticide. |
| Suture | (1) The grooves marking the line of fusion o two distinct scelerites (plates) or two segments (2) A line-like external groove in the body wall or a narrow membranous area between sclerites; a line where adjacent parts have united. |

| Term | Terminology |
|------|-------------|
| Swarm | (1) Large number of insects, birds, small animals, etc. moving about in a cluster or irregular body; a large number of small motile organisms viewed collectively. (2) Departure of a number of bees from one hive to form another. |
| Sympatric | Inhabiting the same geographic region. |
| Symptom | Warming that something is wrong. Any indication of disease or poisoning in a plant or animal. This information is used to figure out what insect, fungus, other pest, or pesticide is causing the disease, damage, or destruction. |
| Synapse | (1) The region of contact between processes of two adjacent neurons, forming the place where a nervous impulse is transmitted from one neuron to another. (2) The central mechanism of intercommunication between the terminal fibres of two or more neurons. |
| Synergism | (1) Increased activity resulting from the effect of one chemical on another. (2) Increased pesticidal activity of a mixture of pesticides above that of the sum of the values of the individual components. (3) An activity of two or more agents which is greater than would be expected from summation of their single actions. Hence, synergist, which has come to mean a compound itself of low toxicity, which increases the action of a toxicant with economic advantage. (4) When the effect of two or more pesticides applied together is greater than the sum of the individual pesticides applied separately. |
| Synergist | (1) A chemical that when mixed with a pesticide increases its toxicity. The synergist may or may not have pesticidal properties of its own. (2) A chemical substance that when used with an insecticide, drug etc. will result in greater total effect than the sum of their individual effects. (3) It is a material that has a little or not toxicity, but is capable of increasing the toxicity of a pesticide. (4) Non-insecticidal or slightly insecticidal chemical added to an insecticide to enhance its toxicity and residual action. |

| Term | Terminology |
|---|---|
| Synonyms | Two or more names for the same thing. |
| Synapse | The site at which the axonic end of a neuron comes in close approximation with the dendritic end of an adjacent or a following neuron is known as synapse. |
| Systematics | According to Simpson, "systematics is the scientific study of the kinds and diversity of organisms and of any and all relationships among them", or more simply, 'systematics is the science of the diversity of organisms". |
| Systemic | (1) A compound that is capable of absorption into plant sap or animal blood. The systemic pesticides in this way render the sap or blood toxic to those pests that feed in or on the treated host. (2) A term used to describe certain pesticides that function by entering and becoming distributed within the plant as opposed to the pesticides that function by contact of the plants surface. (3) A pesticide absorbed through the plant surfaces (usually roots) and translocated through the plant vascular system. |
| Tactile organs, tactile receptors | The organs of touch. |
| Taenidium (pl. taenidia) | The spiral thickening of chitin strengthening the walls of air tubes (tracheae). |
| Tagoreceptor | An organ of touch having a single sense cell. |
| Tarsus | The foot of an insect, the one-to-five-segmented appendage attached to the tibia, or lower leg segment. |
| Taxis (pl. taxes) | A directed response involving the movement of an animal toward or way from a stimulus |
| Taxon | (1) Group of real organisms recognized as a formal unit at any level of a hierarchic classification. (2) It is a taxonomic group that is sufficiently distinct to be worthy of being distinguished by name and to be ranked in a definite category. Its pleural is taxa. |
| Taxonomy | (1) According to Simpson, "taxonomy is the theoretical study of classification, including its bases, principles, procedures and rules" (2) The |

| Term | Terminology |
| --- | --- |
| | arranging of species and groups thereof into a system which shall exhibit their relationship to each other and their places in a natural classification. |
| Tegmen (pl. tegmina) | The leathery fore-wings of insects, such as cockroaches and grasshoppers. |
| Tegula (tegulae) | A small scalerite sclerite at the extreme base of the costa of the fore wing (particularly in the Lepidoptera). |
| Telson. | A twelfth abdominal segment present in the embryos of some insects and in the adults of the primitive order Protura. |
| Tendon. | A slender chitinous plate, hand or strap to which muscles are attached for moving appendages. |
| Tentorial pits | The external depressions in the crania wall at the roots of the tentorial arms; the anterior tentorial pits (at) located in the subgenal sutures, or usually in the epistomal suture, the posterior tentorial pits (pt) in the lower ends of the post-occipital suture. |
| Tentorium (pl. tentoria) | The endoskeleton of the head, consisting of two or three apodemes. |
| Tergite | (1) A dorsal sclerite in other words a portion of the body wall bounded by sutures and located in a dorsal position (opposed to sternite). (2) A tergal sclerite; the dorsal surface of an abdominal segment. |
| Tergum (pl. terga) | Dorsal part of a body segment; often referred to as the notum in thoracic segments. |
| Terminal filament | The cellular end thread of an ovariole (Endfaden). |
| Termitarium | A termite nest. Also, an artificial nest used in the laboratory to house termites. |
| Thelytoky | Parthenogenetic reproduction in which only female progeny are produced. |
| Third generation pesticides | A term proposed by Williams (1967) for juvenoids and other natural or synthetic chemicals potentially or synthetic chemicals potentially useful for pest suppression by virtue of their effective interference with hormonally directed natural processes in insects. |

| Term | Terminology |
|------|-------------|
| Thorasic tergits | Sclerites (hardened plates) on the dorsal part of the thorax. Also called NOTA. |
| Thorax | (1) The middle part of an animals or insects body, between the head and the abdomen. (2) The second or intermediate region of the insect body, bearing the true legs and wings. It consist of three segments, the pro, meso, and meta throax (named in order from the front). In insects, the legs and wings are borne on the thorax. |
| Threshold | The lowest value of any stimulus, signal or agency which will produce a specified effect. |
| Threshold limit value | The concentration of a chemical, the exposure to which for a healthy worker for eight hours a day, day after day for a working life time, produces no harmful effect. |
| Tibia (pl. tibiae) | The fourth segment of the leg, between the femur and the tarsus; often having strong spines. |
| Tolerance (in pests) | Resistance of a pest to a pesticide, ability to survive normal dosage of a toxicant. |
| Tolerance level, median | The amount of a chemical in parts per million in food (oral tolerance) or in the environment (contact tolerance) which a sensitive species of an animal can tolerate for a short duration (4-7 days), without causing any noticeable harmful effect or injury. For example, TLM value for DDT against rainbow trout (Salmo gardener) is 0.01 ppm in water. |
| Tolerance level, pesticide | Pesticide tolerance has been defined by FAO and WHO as the maximum concentration of pesticide residue that is permitted in or on food at a special stage in the harvesting, storage or transport, marketing or preparation of food, upto the final point of consumption expressed in parts by weight of the food (ppm). These residue levels must be a result of good agricultural practices, toxicologically acceptable and be shown to present no hazards throughout the life of man or animal. |
| Tolerance limit | (1) It is the maximum amount of a pesticide residue (ppm) that is permitted by regulation to remain in or on a commodity. (2) It is the maximum |

| Term | Terminology |
|------|-------------|
| | concentration (in ppm) of pesticide residue that is permitted in or on food at a specified stage of harvesting, transport, marketing or preparation of food upto the final point of consumption. The term now has been replaced by MRL. |
| Tormogen (Tmg) | The epidermal cell associated with a seta which secretes the setal membrane or socket. |
| Toxicity | (1) The natural capacity of a substance to produce injury. Toxicity is measured by oral, dermal and inhalation studies on test animals. (2) Ability to poison. The toxicity after one exposure to a pesticide is called acute toxicity, whereas after small and repeated dose over a period of time is called chronic toxicity. |
| Toxicology | The branch of science dealing with the study of poisons in relation to living beings. In the field of agriculture, it includes the study of the resultant levels of toxicants (after consumption of treated food) in the animals, their long term effect on human beings and the environment (including pets, wildlife and ecystem). |
| Toxemia | Dissemination of toxin in blood. |
| Trachea | The windpipe; extends from the throat to the bronchi in mammals; each of the passages by which air is conveyed from the exterior in insects, arachnis. Duct, vessel, in plants. |
| Tracheal commissures | Transverse tracheal trunks continuous from one side of the body to the other. |
| Tracheal gills | Present in most aquatic insects, filiform or lamillate structures well supplied with tracheae and tracheoles. |
| Tracheal system | The part of the espiratory system composed of the tracheae and tracheoles. |
| Tracheoles | The terminal finer branches of the racheae without spiral thickenings. |
| Trail pheromone | A substance laid down in the form of a trail by one animal and followed by another member of the same species. |

| Term | Terminology |
| --- | --- |
| Transverse costal vein | A cross vein in the costal cell (Hymenoptera). |
| Transverse cubial vein | A transverse vein connecting the marginal and cubital veins (Hymenoptera). |
| Transverse marginal vein | A cross vein in the marginal cell (Hymenoptera). |
| Transverse median vein | A cross vein between the median or discoidal vein and the anal vein (Hymenoptera). |
| Transverse suture | A suture across the mesonotum (Diptera). |
| Trap crop | (1) A small planting of a susceptible and highly attractive host, planted early in the season, or removed in space from the main crop, in order to divert attack and infestation by pests and allow for their easy destruction. (2) Crop of plants (sometimes wild plants) grown especially to attract insect pests, and when infested either sprayed or collected and destroyed. Trap plants usually grown between the rows of the crop plants or else peripherally. |
| Trichogen | an epidermal cell that generates a seta. |
| Trochanter | Each of the several bony processes on upper part of thighbone; second joint of an insects leg. |
| Trochantin | The basal part of the trochanter when the latter is two segmented in the insect leg. |
| Trophic level | Placement or grouping of all organisms which secure their food at a common step away from the first level, e.g., (i) plants, (ii) herbivores, (iii) carnivores. |
| Trophocytes | Cells that elaborate nutritive materials; the cells of the fat body having a trophic in distinction from the urate cell. |
| Tympanal organs | Auditory organs on the abdomen or legs of certain insects. |
| Unisexual | Of one sex only, e.g., aphids where asexual (parthenogenic) females are known. |
| Univoltine | Having only one complete generation annually. |
| Unsclerotized | Not hardened by the deposition of sclerotin. |

| Term | Terminology |
|------|-------------|
| Vagina | (1) Structure formed by the union of the oviducts of the female reproductive organs and opening to the outside. (2) A part of the definitive egg passage in many insects posterior to the true oviductus communis, from the genital chamber. |
| Valvifers | The basal plates of the ovipositor, probably derived from the coxopodites of the gonopods, carrying the valvulae, including first valvifers (1 Vlf) of the eight abdominal segment and second valvifers (2Vlf) of the ninth segment. |
| Valvulae | The three pairs of processes forming the blades and ensheathing lobes of the ovipositor. The first and second valvulae (1VI, 2VI) are gonapophyses of the gonopods; the third valvulae (3VI) are distal outgrowths of the coxopodites of the ninth abdominal segment. |
| Venation | The arrangement of veins in the wings of insects. |
| Ventral | Pertaining to the underside of the body. |
| Ventral diaphragm | A membranous and muscular sheet present in some insects stretched between the lateral edges of the abdominal sterna, sometimes extending into the thorax, separating the ventral sinus from the perivisceral sinus. |
| Ventral nerve cord | The chain of connected ventral ganglia, morphologically beginning with the tritocerebral ganglia of the brain, in entomology the term usually applied to the thoracic and abdominal ganglia only. |
| Ventral sinus | The space of the definitive body cavity below the ventral diaphragm, containing the nerve cord. |
| Ventral trachea | The ventral segmental trachea originating at a spiracle. |
| Ventriculus | Midgut; mesenteron or stomach of an insect. |
| Vermiculate | With irregular sinuous markings (worm-eaten). |
| Variety | It is an ambiguous term of classical taxonomy used for a heterogenous group of phenomena including non genetic variations of the phenotype, morphs, domestic breeds and geographic races. |

| Term | Terminology |
|------|-------------|
| Vermiform larva | A legless (apodous), headless (acephalic), worm-like larva typical of some diptera. |
| Vertebrates | Animals with a spinal column or backbone, such as fishes, birds, mammals, and so on. |
| Vertex | The top of the head, between the eyes and anterior to the occipital suture. |
| Vestigial | Having the nature of a degenerate or atrophied organ, more fully functional in an earlier stage of development of the individual or species. |
| Virulence | (1) Power of pathogen to invade and injure a host tissue. (2) The disease-producing power of a micro-organism, i.e., the relative capacity of a micro-organism to invade and injure the tissues of its host (Steinhaus and Martignoni, 1967). |
| Virus | A sub-microscopic pathogen that requires living cells for growth and is capable of causing disease in plants or animals. Plant viruses are often spread by insects. |
| Viscera | The internal organs in the cavity of the body particularly those in abdomen. |
| Vitelline membrane | The wall of the egg cell; undivided in meroblastic cleavage, surrounding the blastoderm. |
| Vitellophags | Endoderm cells proliferated into the yolk and accomplishing a partial digestion of the latter. |
| Viviparous reproduction | Type of reproduction in insects, where embryo develops within the body of the insect and young ones produced similar to adults; found in some flies. |
| Vulva | The external opening of the genital chamber or vagina serving in most cases for both copulation and the discharge of the eggs; sometimes on the eighth abdominal segment, sometimes on the ninth. |
| Waggle dance | The dance whereby workers of various species of honeybees (genus Apis) communicate the location of food finds and new nest sites. The dance is basically a run through a figure eight pattern; with the middle, transverse line of eight containing information about the direction and distance of the target. |

| Term | Terminology |
|------|-------------|
| Waiting period | It is a time limitation in respect of harvest of the treated crop or slaughter of the treated animal; safety period. |
| Wing base | The proximal part of the wing between the bases of the veins and the body, containing the humeral and axillary sclerites. |
| Wing pads | The undeveloped wings of nymphs and naiads. They appear as two flat structures on each side |
| Xenobiosis | The relation in which colonies of one species live in the nests of another species and move freshly among the hosts, obtaining food from them by regurgitation or other means but still keeping their brood separate. |
| Xenophagus | Feeding on non-specific host plant. |
| Zero tolerance | By law, no detectable amount of the pesticide may remain on the raw agricultural commodity when it is offered for shipment. Zero tolerance are no longer allowed. |

# 4

# DISTINGUISH

### Differentiate between Apterygota and Pterygota

| S. No. | Character | Apterygota | Pterygota |
|---|---|---|---|
| 1. | Wings | Primarily wingless | Winged or secondarily wingless evolved from winged ancestors, e.g., Flea, head louse, bed bug. |
| 2. | Metamorphosis | Totally absent or slight. | Present. |
| 3. | Mandibular articulation in head | Monocondylic, i.e., single | Dicondylic, i.e., double |
| 4. | Pleural sulcus in thorax | Absent | Present |
| 5. | Pregenital abdominal appendages | Present | Absent |

### Differentiate between Exopterygota and Endopterygota

| S. No. | Character | Exopterygota | Endopterygota |
|---|---|---|---|
| 1. | Wing development | External | Internal |
| 2. | Type of metamorphosis | Incomplete (Hemimetabola) or gradual (Paurametabola) | Complete (Holometabola) |
| 3. | Pupal stage | Absent | Present |
| 4. | Immature stage | Naiad or Nymph | Larva |
| 5. | No. of orders | 16 | 9 |

### Differentiate between Insecta and Arachinida/Mites

| S. No. | Insecta | Arachinida/Mites |
|---|---|---|
| 1. | In - into, Sect-cut | Arachna - spider |
| 2. | One pair of antenna | Absent |
| 3. | 3 body region | 2 region - cephalothorax, abdomen |
| 4. | Mouth parts- feeding | Chelicerae - feeding part |
| 5. | 3 pairs of legs | 4 pairs of legs |
| 6. | Wings are present | Absent |
| 7. | Insects | Spiders, mites |

## Differentiate between Spiders and Insect

| S. No. | Spiders | Insect |
|---|---|---|
| 1. | The body is divided into two parts. Cephalothorax and abdomen. | The body is divided is divided into head, thorax and abdomen. |
| 2. | The have four pairs of leg. | They have three pair of leg. |
| 3. | Respiration by trachea and book lungs. | Respiration by spiracles and tracheae. |
| 4. | Antennae absent. | Antennae are present. |

## Differentiate between Ticks and Mites

| S. No. | Ticks | Mites |
|---|---|---|
| 1. | Hallers organ present in the first tarsus. | Hallers organ absent. |
| 2. | Recurved teeth present on hypostome. | Recurved teeth on hypostome absent. |
| 3. | Leathery textured body. | Soft body. |
| 4. | Stigmata situated behind coxae IV or laterad between coxae II and IV, each surrounded by a stigmal plate. | Stigmal opening either lateral (Mesostigmata); between coxae III and IV (Prostigmata); located near gnathosoma or trachea opens through acetabular cavities of legs (Astigmata). |

## Differentiate between Crustacea and Hexapoda or Insecta

| S. No. | Crustacea | Hexapoda or Insecta |
|---|---|---|
| 1. | Cephalothorax and abdomen | 3 region - Head, thorax and abdomen |
| 2. | Antennules | Antennae |
| 3. | Antenna | Absent |
| 4. | Mandibles | Mandibles |
| 5. | 1st pair of maxillae | 1st pair of maxillae |
| 6. | 2nd pair of maxillae | Labium |
| 7. | 3 pairs of maxillipedes | 3 pairs of legs |
| 8. | Cray fish, lobsters, crabs, shrimp, copepods and their relatives | Insects |

## Differentiate between Larva and Nymph

| S. No. | Larva | Nymph |
|---|---|---|
| 1. | Complete. | Incomplete metamorphosis. |
| 2. | The food of larva and adult is different. | The food of nymph and adult is same. |
| 3. | Differ from adult in their habits. | Same as adult in their habits and habitats. |
| 4. | Compound eyes absent. | Compound eyes present. |
| 5. | Example, moth, wasp, house fly. | Example, bugs,grasshopper and termite. |

## Differentiate between Cockroach and Grass hopper

| S. No. | Cockroach | Grass hopper |
|---|---|---|
| 1. | Legs are walking type. | Legs are jumping type. |
| 2. | Antennae is setaceous. | Antennae is filliform. |
| 3. | Cerci long and segmented. | Cerci short and unsegmented. |
| 4. | No special sound producing organ. | Special sound producing organ is present. |
| 5. | Ovipositor is not well developed. | Ovipositor is well developed. |

## Differentiate between Moth and Butterfly

| S. No. | Moth | Butterfly |
|---|---|---|
| 1. | Nocturnal in habit. | Diurnal in habit. |
| 2. | Wing expand at rest. | Wing are straight, erect and upward at rest. |
| 3. | Antenna is pectinate or bipectinate. | Antennae is cleavate or capitate. |
| 4. | Abdomen is thick and pointed. | Abdomen is thin and blunt. |

## Differentiate between Cockroach and Mantid

| S. No. | Cockroach | Mantid |
|---|---|---|
| 1. | Legs are walking type. | Legs are raptorial type. |
| 2. | Antennae-settaceus. | Antennae-bipectinate. |
| 3. | Mostly seavenger. | Carnivorous. |
| 4. | Head is concealed by pronotum. | Head is not concealed by pronotum. |

## Differentiate between Sucking louse and Biting louse

| S. No. | Sucking louse | Biting louse |
|---|---|---|
| 1. | Ectroparasite of mammal. | Ectoparasite of bird and some mammal. |
| 2. | Mouth parts are piecing and sucking type. | Mouth parts biting type (chewing). |

## Differentiate between Fore gut and Mid gut

| S.No. | Fore gut | Mid gut |
|---|---|---|
| 1. | Ectodermal in origion. | Ecdodermal in origion. |
| 2. | Lined with intima. | Lined with peritrophic membrane. |
| 3. | Layer of longitudinal muscles out side the epithelium followed by circular muscles. | Circular muscles out side the epithelium followed by longitudinal muscles. |

<h2 style="text-align:center">Differentiate between Spines and Spurs</h2>

| S.No. | Spines | Spurs |
|---|---|---|
| 1. | They are large, thorn like in form, rigidly fixed to the cuticle. | They are articulated with the cuticle by a membranous ring. |
| 2. | No basal articulation. | Basal articulation present. |
| 3. | Not movable. | Movable. |
| 4. | Example, Spines on hind tibiae of grasshopper | Example, Spurs on the legs of many insects-at the ends of tibiae, lateral claws, etc.hind tibiae of grasshopper. |

<h2 style="text-align:center">Differentiate between Phasmida order and Orthoptera order</h2>

| S.No. | Phasmida order | Orthoptera order |
|---|---|---|
| 1. | Prothorax is short while metathorax, elongated. | Prothorax is long while measo and meta thorah is small. |
| 2. | All the legs are similar to each other. | Third pair of leg modified into jumping and sound producing. |
| 3. | They change colour according to condition. | They do not change the colour. |
| 4. | Example, stick insect, leaf insect. | Example, grasshopper, mole cricket, locust. |

<h2 style="text-align:center">Differentiate between Anisoptera and Zygoptera</h2>

| S.No. | Anisoptera | Zygoptera |
|---|---|---|
| 1. | Eyes are prominent and close to each other | Eyes are buttoned shaped and away from each other |
| 2. | Sclerites are different | Sclerites are less distinct |
| 3. | Wings are attached with a broad base | Wings are attached with a narrow base, hence known as petiolate wings |
| 4. | Hind wings are smaller than fore wings | Hind wings are similar in size shape and venation of fore wing |
| 5. | Dragon flies are strong | Damsel flies are comparatively smaller |
| 6. | Wings are kept horizontally at rest | Wings are kept vertically when at rest |
| 7. | Discoidal cell is triangular | Discoidal cell is quadrangular |
| 8. | Naids respire by rectal gills | Naids respire by caudal gills |
| 9. | Oviposition is exophytic | Oviposition is exophytic |
| 10. | Penis is jointed | Penis is not distinctly jointed |
| 11. | Example, Dragonfly | Example, Damselfly |

## Differentiate between Symphata and Apocrita

| S.No. | Symphata (mustard saw fly) | Apocrita (Honeybee) |
|-------|----------------------------|---------------------|
| 1. | Abdomen is broadly attached | Abdomen is petiolated - deeply costricted between 1st & 2nd abdominal segment. |
| 2. | Fore tibia with : 2 spurs | –1 spur |
| 3. | Larva: Thoracic+Abdominal legs | Larva - apodous (no legs) |
| 4. | Family-Tenthredinidae : Mustard sawfly | Family - Apidae : Honeybee |
| 5. | Larva is a caterpillar and it belongs belongs to eruciform type | Larva is a grub it belongs to apodous eucephalous type |
| 6. | Stemmata are present | Stemmata are absent. |
| 7. | Ovipositor is saw like | Ovipositor is not saw like |
| 8. | They are phytophagous | They are generally parasitic |

## Differentiate between Heteroptera and Apocrita Homoptera

| S.No. | Heteroptera (Hetero-different; ptera-wing) | Homoptera (Homo-uniform; ptera-wing) |
|-------|--------------------------------------------|--------------------------------------|
| 1. | Basal portion of wings is thicker and apical portion is hemielytra type | Wings are uniformly thickened |
| 2. | Head is prorect | Head is deflexed |
| 3. | Wings are held flat over the back at rest and left and right side overlap of the abdomen | Wings are held roof like over the back and wings do not overlap. |
| 4. | Antennae well developed | Antennae reduced and cancelled |
| 5. | Prothorax is large | Prothorax is small |
| 6. | Pronotum usually greatly enlarged | Pronotum is almost alway small and collar-like. |
| 7. | Scent glands are present | Scent glands are not developed but wax glands present |
| 8. | Filter chamber is absent | Filter chamber is present |
| 9. | Metamorphosis incomplete | Incomplete but in some species complete |
| 10. | Scutellum well developed (triangular plate found between the wing bases). | Scutellum not well developed. |
| 11. | Gular region of the head well defined (midventral sclerotised part between labium and foramen magnum) | Gular region not clearly defined |
| 12. | Honey dew secretion uncommon | Honey dew secretion common |
| 13. | Both terrestrial and aquatic | Terrestrial |
| 14. | Herbivorous, predaceous or blood sucking | Herbivorous |

## Differentiate between White Ant and Red Ant

| S.No. | Particulars | White Ant | Red Ant |
|-------|-------------|-----------|---------|
| 1. | Order | Isoptera | Hymenoptera |
| 2. | Antennae | Moniliform | Filiform/club shaped |
| 3. | Mouth parts | Chewing and biting | Lapping and sucking |
| 4. | Wing coupling | Absent | Hamulate |
| 5. | Wings | Membranous, elongate | Membranous |
| | Workers | Sterile male or female | Sterile female |

## Differentiate between Phasis-Solitaria and Phasis-Gregaria

| S.No. | Phasis-Solitaria (Grasshopper) | Phasis-Gregaria (Locusts) |
|-------|-------------------------------|---------------------------|
| 1. | Nymphs are yellowish brown or gray in colour | Nymphs are black yellow or orange in colour |
| 2. | Feed individually | Move and feed in groups |
| 3. | Prothorax convexd, comparatively bigger in adults | Prothorax generally concave and compared to solitary |
| 4. | Adults have similar feeding habits as nymphs | Adult also feed in groups |
| 5. | Nocturnal as well as diurnal organs | Mostly diurnal forms |
| 6. | Example Rice grasshopper Hierogly-phousbanican | Example Desert locust-Schistocerea gergaria Migratory locust, Locusta migerataria Bombay locust-Pantanga - succinada |

## Differentiate between Survey and Survillence

| S.No. | Survey | Survillence |
|-------|--------|-------------|
| 1. | It is non-systematic and unscientific method | It is systematic and planned activity |
| 2. | It is done at a time or over extended period of time | It is extensive survey. Periodic stock and keep close watch |

## Differentiate between Infection and Infestation

| S.No. | Infection | Infestation |
|-------|-----------|-------------|
| 1. | Pathogen multiplies with in the body of the host | Pathogen do not multiply or reproduce with in the body of the host but it may pass part of its life cycle with the host. |
| 2. | Bacterial, viral or nematode infection. | Borers infests the crops |

## Differentiate between Emulsion and suspension

| S.No. | Particulars | Emulsion | Suspension |
|-------|-------------|----------|------------|
| 1. | Continous phase | Liquid | Liquid |
| 2. | Stationary phase | Liquid | Solid |

## Differentiate between Male and Female Sex Pheromone

| S.No. | Properties | Female sex pheromone | Male sex pheromone |
|---|---|---|---|
| 1. | Range | Acts at a long range. Attracts males from long distance | Acts at a short distance |
| 2. | Role of other stimuli | Play less role | Visual and auduitory stimuli play major role |
| 3. | Action elicited in the other sex | Atrracts and excites males to copulate | Lowers females resistance to mating |
| 4. | Importance in IPM | More important | Less important |

## Compare between True Parasites, Parasites and Predators

| S.No. | Properties | True Parasites | Parasites (=Parasitoids) | Predators |
|---|---|---|---|---|
| 1. | Size | Smaller than host | Same size as host | Larger than host |
| 2. | Feeders on host | Both larvae and adults | Only larvae, adults free living and vegetarians | Both larvae and adults |
| 3. | Number of hosts needed for maturity | One | One | More than one |
| 4. | Injury to host | Feed without killing | Paralyse to oviposit | Kill to devour |
| 5. | Time of activity | Diurnal or Nocturnal | Diurnal | Crepuscular |
| 6. | Host speciificity | Great | Great | Not so great great |
| 7. | Activity | Function at low host density, so efficient | Function at low host density, so efficient | Function at higher host density, so not as efficient |
| 8. | Suitability for biological control | Not suited | Best suited | Suited |

## Compare between Nematodes and Insects

| S.No. | Nematodes | Insects |
|---|---|---|
| 1. | Phyllum - Nematoda | Phyllum - Arhtopoda |
| 2. | Microscopic | Macroscopic - visible to naked eye |
| 3. | Body is un segmented | Body is segmented |
| 4. | Body is bilaterally symmetrical | Body is not bilaterally symmetrical |
| 5. | Animals with false coelom | Animals with coelom |
| 6. | Body is soft | Body is covered with hard exoskeleton |

### Differentiate between Entomopathogenic Nematode and Plant parasitic nematodes

| S.No. | Entomopathogenic Nematode (EPN) | Plant Parasitic Nematodes (PPN) |
|---|---|---|
| 1. | Parasitize insect pests and pass one of the stage of their life cycle in insect pests. They never parasitise or damage plants. | Parasitise mainly root system of plants and pass one of the stages of their life cycle in/on the root tissues of plant as endoparasitic/ ectoparasitic nematode. |
| 2. | Beneficial to agricultural crops by attacking insect's pests by killing them quickly | Harmful to agricultural crops damage and alter physiology of crop plants, produce abnormalities, knots, lesion on root system, introduce fungal pathogen inside roots and aggravate wilt and root-rot diseases. |
| 3. | Increases crop yield by killing crop pests as a biological control agent. | Inflicting 15% yield loss on an average in agricultural crops globally as a parasite. |
| 4. | Size 3-1.5 mm (very small). | Size up to 4 mm (small to medium) |
| 5. | Absence of stylet. | Presence of stylet. |
| 6. | EPN feed on bacteria and decomposing host insect. | PPN feed on the plant parts mainly roots to get their nourishment. |
| 7. | Life cycle generally completes within a weeks. | Life cycle generally complete in 20-30 days. |

### The important Plant Parasitic Nematodes

| Nematode | Scientific name | Host | Symptoms |
|---|---|---|---|
| **Endoparasites** | | | |
| Root knot Nematode | (Meloidogyne spp.) | Vegetable crops like,tomato, brinjal, bhendi, cucurbits, cruciferous, potato, pulses like gram, mung, cowpea, pigeonpea, then sugarcane, banana, citrus, paddy, tea, grapes, groundnut, tobacco etc. | Infected plants develop knots or galls on roots. Shoot system remains stunted, leaves become yellow. |
| Cyst Nematodes | (Heterodera spp., Globodera spp.) | Wheat, barley, pigeonpea, maize, paddy, sorghum, potato, etc. | Root system becomes stunted with development of bussy lateral roots. Shoot system is stunted with chlorotic leaves. |
| Reniform Nematode | (*Rotylenchulus reniformis*) | Vegetable castor fruits grapes, pineapple, papaya, cowpea, pulses, fibre crops, oilseeds, cereals, mullets | Stunted growth, yellowing of leaves, wilting, and deterioration in quality of fruits. |
| | | and corner crops cotton. | Infected roots shows necrosis and feeder roots are destroyed. |

*Contd...*

*Contd...*

| Nematode | Scientific name | Host | Symptoms |
|---|---|---|---|
| Root lesion nematodes, Burrowing nematode | (*Pratylenchus* spp.) (*Radophilus* spp.) | Citrus, banana, black pepper, betelvine, coconut, ginger, paddy, wheat, gram, coffee, sugarcane etc. | The above ground symptoms are plant stunting, yellowing of leaves, defoliation, poor fruiting, dieback. Roots show discrete elliptical lesions. The root system is reduced. The necrotic lesions are often colorized by secondary pathogen which cause rooting. |
| Citrus Nematode | (*Tylenchulus semipenetrans*) | Citrus, maize, sorghum, tobacco, sugarcane. | Above ground symptom in infected. Plants resemble those of malnutrition like chlorotic leaves, defoliation of twigs. Infested roots show extensive necrosis and become brown. |
| Bulb and leaf nematode | (*Ditylenchus* spp.) | Potato, onion, tulip, lucerne etc. | Attack bulbs and stems of host plants. and cause rotting. |
| Bud and leaf nematodes | (*Aphelenchoides* spp.) | Chrysanthemum, dahlia, zinnia, aster and strawberry etc. | They infect only the buds and foliage to infected leaves are deformed and dry prematurely. |
| Seed gall nematode | (*Anguina* spp.) | Wheat | They are endoparasites. Ovaries in infected plants are converted into nematode galls leaves in infected plants are twisted and crinkled and wheat grains are replaced by dark brown nematode galls associated with a bacterium in causing the tundu (yellow ear rot) disease of wheat. |

## Ectoparasites

| | | |
|---|---|---|
| Spiral Nematodes | (*Helicotylenchus/Rotylenchus*). | These nematodes complete their life |
| Stunt Nematode | (*Tylenchorhynchus*) | cycle in soil, feedon roots and cause |
| Stubby root nematodes | (*Trichodorus*) | injury to root tips, retard root development and cause small lesions |
| Ring Nematode | (*Criconemoides, Criconema*) | on roots. |
| Needle Nematode | (*Longidorus*) | |
| Dagger Nematode | (*Xiphinema*) | |
| Sheath nematode | (*Hemicydicophora*) | |

## Different vector mosquitoes identified with their characters

| Characteristic | Anopheles | Aedes | Culex | Mansonia |
|---|---|---|---|---|
| Human disease | Malaria, Filaria | Dengue, Yellowfever Haemirragic fever | Filarial and Japanese Encephalitis | Filaria |
| Important breeding site | Clean water, channels, ponds, pools, road side, rain water, nice fields | Artificial containers, dump tyres, swamps and rocky pools | Polluted water of drains, stagnant drains, pons, waste, water etc. | - |
| Biting during | Night | Day | Night | Both |
| Eggs | Laid singly Has floats | Laid singly No floats | Laid in rafts No floats | - |
| Larnae | Rest parallel to water surface | Rest at an angle to the surface | Rest at an angle to the surface | - |
| Adult | Proboscis and body in the same straight line | Proboscis and body at an angle to one another | Proboscis and body at an angle to one another | - |
| | Wings spotted | Wings uniform tip of female abdomen usually pointed | Tip of female abdomen usually blunt. | |

# 5

# SHORT EXPLANATIONS

Position of the insects in animal kingdom

Living things (Biology) are broadly classified into two kingdoms, Animal kingdom (Zoology) and plant kingdom (Botany).

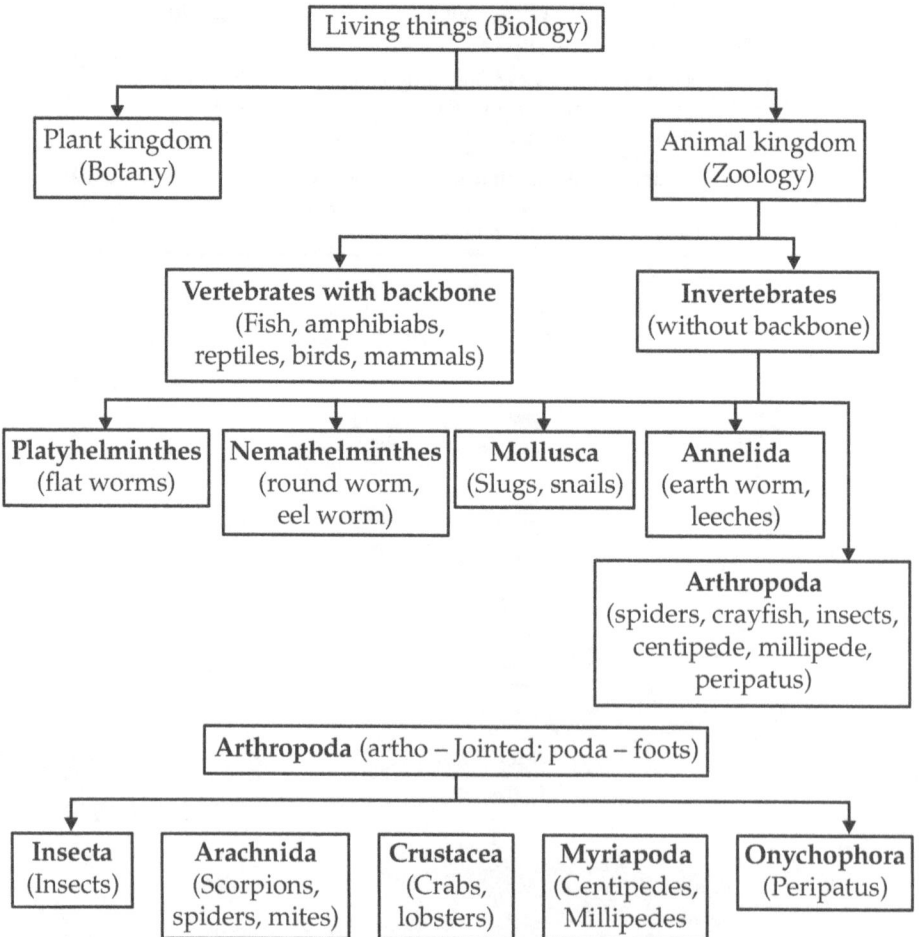

```
                    Living things (Biology)
            ┌───────────────┴───────────────┐
      Plant kingdom                    Animal kingdom
       (Botany)                          (Zoology)
                                 ┌──────────┴──────────┐
                    Vertebrates with backbone      Invertebrates
                      (Fish, amphibiabs,          (without backbone)
                     reptiles, birds, mammals)
  ┌──────────────┬───────────────┬───────────────┬──────────┴──────┐
Platyhelminthes  Nemathelminthes   Mollusca        Annelida
 (flat worms)    (round worm,   (Slugs, snails)   (earth worm,
                  eel worm)                          leeches)
                                            Arthropoda
                                   (spiders, crayfish, insects,
                                     centipede, millipede,
                                           peripatus)

            Arthropoda (artho – Jointed; poda – foots)
  ┌──────────┬───────────────┬───────────────┬───────────────┐
 Insecta    Arachnida        Crustacea       Myriapoda      Onychophora
(Insects)  (Scorpions,        (Crabs,       (Centipedes,    (Peripatus)
           spiders, mites)    lobsters)       Millipedes
```

Insects are members of the class Insecta in the invertebrate phylum Arthropoda, the largest in the Animal kingdom. Phylum Arthropoda is the large group of animals that contains the insects and their relatives. Out of 1.35 million living species of animals, 9,00,000 are insects. Arthropoda comprises some 85% of all living animals and of these the vast majority are insects. So far 1 million species were described and yet undescribed species are 3 to 4 millions.

**Characters of Arthropoda :** Arthropoda had descended from a polychaet annelid and are closely related to the phylum Mollusca.

1. Metameric segmentation, bilaterally symmetrical, triploblastic (division of insect body into primary body segment)

2. Bilateral symmetry and elongated body with moth and anus at opposite ends

3. External skeleton (made up of with Chitin)

4. Each body segment or most of the body segment has a pair of jointed legs

5. Double ventral nerve card

6. Open circulatory system

7. Sexes separate, dimorphism, sexual reproduction, fertilization internal

8. Body cavity is haemocoel

9. Muscular system (striate type)

10. Breathing by tracheal system or gills

11. Appendicular mouth parts

12. Excretion by green glands or malphigian tubules

## Phylum Arthropoda is Classified in to 7 classes

1. Onychophora (claw bearing), e.g., Peripatus

2. Crustacea (Crusta - shell), e.g., Prawn, crab, wood louse

3. Arachnida (Arachne - spider), e.g., Scorpion, spider, tick, mite

4. Chilopoda (Chilo - lip; poda - appendage), e.g., Centipedes

5. Diplopoda (Diplo - two; poda - appendage), e.g., Millipede

6. Trilobita (an extinct group)

7. Hexapoda (Hexa - six; poda-legs) or Insecta (In - internal; sect - cut), e.g., Insects.

## 1. Onychophora: Peripatus and its relatives

Small group of terrestrial animals and they show morphological similarities with both annelids and arthropods.

1. It is a elongate, worm like bodies with several bilateral pairs of lobe like legs

2. Paired nephridia (excretory organs) along the body that open near the leg bases.

1. Class: Trilobita, e.g., Trilobit These animals are now extinct.

   (i)   They are marine arthropods with body longitudinally moulded into three lobes, head trunk and tail.

   (ii)  Head is unsegmented and trunk is flexible.

   (iii) They possess a number of pairs of limbs, 4 pairs attached to the head and remaining the trunk region.

## Class Arachnida (arachnids) : Spiders, scorpions, ticks, mites, etc.

### *Arachnids possess*

- 2 body segments - cephalothorax and abdomen
- 8 legs
- 1 pair of chelicerae
- no antennae

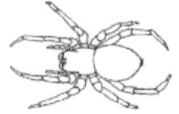

## Class Chilopoda (centipedes)

### *Chilipods possess*

- many body segments, body dorsoventrally flattened
- 1 pair of legs per body segment
- 1 pair of antennae
- 1st pair of legs modified into venomous "fangs"

## Class Diplopoda (millipedes)

### *Diplopods possess*

- Many body segments, cylindrical body
- 2 pair of legs per body segment
- 1 pair of antennae

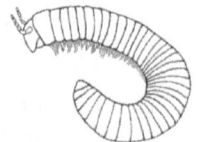

## Class Crustacea (crustaceans): crabs, shrimp, barnacles, sowbugs, etc.

### *Crustaceans possess*

- Several body segments - head, thorax and abdomen
- Segments may be fused
- Varied number of legs
- 2 pairs of antennae, respiration by gills.

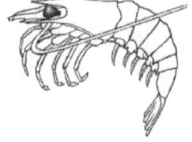

## Class Insecta (Insects); beetles, bugs, wasps, moths, flies, etc.

### *Insects possess*

- 3 body segments
- 6 legs
- 1 pair of antennae
- Diverse modifications to appendages

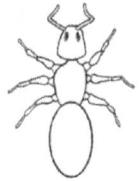

## THE CLASS INSECTA

### Characters of class Insecta

1. Body is divided into three regions
2. In head a pair of antenna and a pair of compound eyes are usually present.
3. Thorax is the centre of locomotion with, 3 pairs of five jointed legs and two pairs of wings.
4. Excretion is mainly through malpighian tubules.
5. Tracheal system of respiration well developed.
6. Brain is divided into protocerebrum, deutocerebrum and tritocerebrum.

The class Insecta has two subclasses *viz.*, Apterygota and Pterygota .

The class Insecta has 29 orders (4 in Apterygota and 25 in Pterygota)

The subclass Apterygota has 4 orders namely

1. Thysanura - Silverfish (Thysan-fringed, Ura-tail)
2. Collembola- Springtail or snowflea (coll-glue; embol-peg)
3. Protura - Proturans or Telsontail (Pro-first, Ura-tail)
4. Diplura - Diplurans or Japygids (Di-two; Ura-tail)

The sub-class Pterygota has two divisions, namely Exopterygota and Endopterygota based on the wing development.

## EXOPTERYGOTA GROUPS

1. Ephemeroptera - Mayflies Group I. Paleopteran orders (1, 2)
2. Odonata-Dragonfly, Damselfly
3. Plecoptera - Stonefly
4. Grlloblatodia - Rock crawlers
5. Orthoptera-Grasshopper, locust, cricket, mole cricket
6. Phasmida-stick insect, leaf insect Group II. Orthopteroid orders (3-11)
7. Dermaptera-Earwigs
8. Embioptera-Webspinners/Embids
9. Dictyoptera-cockroach, preying mantis
10. Isoptera - Termites
11. Zoraptera - Zorapterans
12. Psocoptera - Book lice
13. Mallophaga - Bird lice Group III. Hemipteroid orders (12-16)
14. Siphonculata - Head and body louse
15. Hemiptera - Bugs
16. Thysanoptera - Thrips

## ENDOPTERYGOTA

1. Neuroptera- Antilions, aphidlion, owl flies, mantispid flies.
2. Mecoptera - Scorpionflies.
3. Lepidoptera - Butterflies and moths.
4. Trichoptera - Caddisfly.
5. Diptera - True fly.
6. Siphonaptera - Fleas.
7. Hymenoptera - Bees, wasps, ants.
8. Coleoptera - Beetles and weevils.
9. Strepsiptera - Stylopids.

Economic classification of insects related to human being

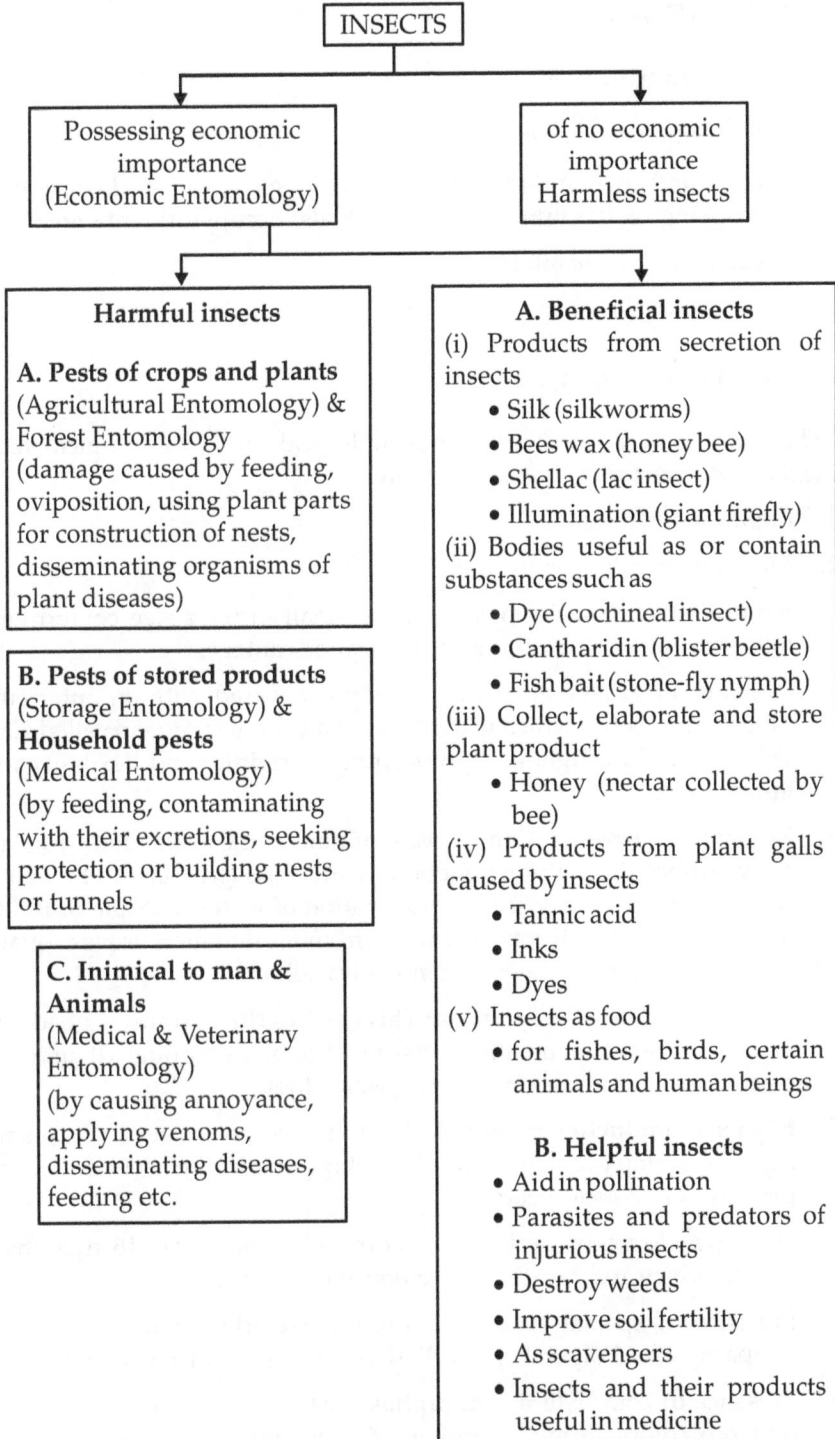

```
                    ┌──────────────┐
                    │   INSECTS    │
                    └──────────────┘
```

| Possessing economic importance (Economic Entomology) | of no economic importance Harmless insects |

---

**Harmful insects**

**A. Pests of crops and plants**
(Agricultural Entomology) &
Forest Entomology
(damage caused by feeding, oviposition, using plant parts for construction of nests, disseminating organisms of plant diseases)

**B. Pests of stored products**
(Storage Entomology) &
**Household pests**
(Medical Entomology)
(by feeding, contaminating with their excretions, seeking protection or building nests or tunnels

**C. Inimical to man &
Animals**
(Medical & Veterinary
Entomology)
(by causing annoyance, applying venoms, disseminating diseases, feeding etc.

---

**A. Beneficial insects**
(i) Products from secretion of insects
  • Silk (silkworms)
  • Bees wax (honey bee)
  • Shellac (lac insect)
  • Illumination (giant firefly)
(ii) Bodies useful as or contain substances such as
  • Dye (cochineal insect)
  • Cantharidin (blister beetle)
  • Fish bait (stone-fly nymph)
(iii) Collect, elaborate and store plant product
  • Honey (nectar collected by bee)
(iv) Products from plant galls caused by insects
  • Tannic acid
  • Inks
  • Dyes
(v) Insects as food
  • for fishes, birds, certain animals and human beings

**B. Helpful insects**
  • Aid in pollination
  • Parasites and predators of injurious insects
  • Destroy weeds
  • Improve soil fertility
  • As scavengers
  • Insects and their products useful in medicine

## INSECT DOMINANCE

### Measures of Dominance

1.  More number of species
2.  Large number of individuals in a single species:, e.g., Locust swarm comprising of 109 number of individuals, occupying large area.
3.  Great variety of habitats
4.  Long geological history

## REASONS FOR DOMINANCE

There are several structural, morphological and physiological factors responsible for insect dominance. They are:

1.  Capacity for flight
2.  More adaptability or universality
3.  **Smaller size:** Majority of insects are small in their size conferring the following physiological and ecological advantages.
4.  **Presence of exoskeleton:** Insect body is covered with an outer cuticle called exoskeleton which is made up of a cuticular protein called Chitin. This is light in weight and gives strength, rigidity and flexibility to the insect body.
5.  **Resistance to desiccation:** Insects minimise the water loss from their body surface through prevention of water loss (wax layer of epicuticle, closable spiracles, egg shell) conservation of water (capable of utilizing metabolic water, resorption of water from fecal matter, use less quantity of water to remove the nitrogenous waste).
6.  **Tracheal system of respiration:** This ensures direct transfer of adequate oxygen to actively breathing tissues. Spiracles through their closing mechanism admit air and restrict water loss.
7.  **Higher reproductive potential:** Reproductive potential of insect is high, e.g., Egg laying capacity (fecundity) of queen termite is 6000 - 7000 eggs per day for 15 long years.

    Short development period, e.g., corn aphid produces 16 nymphs per female which reaches the adulthood within 16 days.

    Presence of special types of reproduction other than oviparity and viviparity like Polyembryony, Parthenogenesis and Paedogenesis

8.  **Presence of complete metamorphosis:** More than 82 per cent of insects undergo complete metamorphosis (holometabolous insects) with four

stages. As the larval and adult food sources are different, competition for food is less.

9. **Presence of defense mechanisms:** By different defense mechanisms, insects escape from the enemies to increase their survival rate.

10. **Hexapod locomotion:** Insects uses 3 legs at a time during locomotion, while the remaining 3 legs are static, which gives greater stability.

## PEST-DEFINITION, CATEGORIES

**PEST -** Derived from French word 'Peste' and Latin term 'Pestis' meaning plague or contagious disease - Pest is any animal which is noxious, destructive or troublesome to man or his interests

- A pest is any organism which occurs in large numbers and conflict with man's welfare, convenience and profit.

- A pest is an organism which harms man or his property significantly or is likely to do so (Woods, 1976).

- Insects are pests when they are sufficiently numerous to cause economic damage (Debacli, 1964).

- Pests include insects, nematodes, mites, snails, slugs, etc. and vertebrates like rats, birds, etc. Depending upon the importance, pests may be agricultural forest, household, medical, aesthetic and veterinary pests.

## CATEGORIES OF PESTS

Based on occurrence following are pest categories

**Regular pest :** Frequently occurs on crop - Close association, e.g., Rice stem borer, Brinjal fruit borer.

**Occasional pest :** Infrequently occurs, no close association, e.g., Caseworm on rice, Mango stem borer.

**Seasonal pest :** Occurs during a particular season every year, e.g., Red hairy caterpillar on groundnut, Mango hoppers.

**Persistent pests :** Occurs on the crop throughout the year and is difficult to control, e.g., Chilli thrips, mealy bug on guava.

**Sporadic pests :** Pest occurs in isolated localities during some period., e.g., Coconut slug caterpillar.

## BASED ON LEVEL OF INFESTATION

**Epidemic pest :** Sudden outbreak of a pest in a severe form in a region at a particular time, e.g., BPH in Tanjore, RHC in Madurai, Pollachi.

**Endemic pest :** Occurrence of the pest in a low level in few pockets, regularly and confined to particular area, e.g., Rice gall midge in Madurai, Mango hoppers in Periyakulam.

## PARAMETERS OF INSECT POPULATION LEVELS

### General Equilibrium Position (GEP)

The average density of a population over a long period of time, around which the pest population over a long period of time, around which the pest population tends to fluctuate due to biotic and abiotic factors and in the absence of permanent environmental changes.

### Economic threshold level (ETL)

Population density at which control measure should be implemented to prevent an increasing pest population from reaching the ETL.

### Economic Injury Level (EIL)

The lowest population density that will cause economic damage.

### Damage Boundary (DB)

The lowest level of damage which can be measured. ETL is always less than EIL. Provides sufficient time for control measures.

## INSECT BODY WALL – ITS STRUCTURE AND FUNCTION

Insect body wall is called as Integument or Exoskeleton. It is the external covering of the body which is ectodermal in origin. It is rigid, flexible, lighter, stronger and variously modified in different body parts to suit different modes of life.

**Structure:** The body wall or integument consists of three distinct layers.

1. The cuticle. 2. The hypodermis or epidermis 3. The basement membrane

1. **CUTICLE :** It is an outer non cellular layer comprising of three sub-layers.

(A) **The Epicuticle :** Outer most layers which is very thin.

- It is outermost layer of 4 to few 100 micron.
- It is very thin and non chitinous in nature.
- it is secreted by epidermal or hypodermal cells.
- Much thicker

- devoid of chitin and of complex chemical composition and ultrastructure It is pierced by pore canals.
- It contains microfibers of chitin surrounded by a matrix of protein

  (a) **Inner epicuticle :** It contains wax filaments.

  (b) **Outer epicuticle :** It makes the contact with cuticulin.

  (c) **Cuticulin :** Non chitinous polymerised lipoprotein layer.

  (d) **Wax layer :** It contains closely packed wax molecules which prevents desiccation.

  (e) **Cement layer :** Outer most layer formed by lipid and tanned protein. It protects wax layer.

  (f) **Polyphenol layer :**

(B) **The procuticle :** It is relatively thicker layer and is multilaminar in structure. The procuticle after its formation undergo certain changes and may be differentiated into the following layers.

  (a) **The exocuticle :** is formed by the sclerotization of the outer layer and is also a seat of colour pigments, carotin, elanin and different types of setae.

  (b) **The endocuticle :** On the other hand, is formed by the remaining unchanged portion of the procuticle. It is thicker and flexible in nature.

**Exocuticle :** Outer layer

Darkly pigmented, hard and sclerotised layer of chitin and protein

**Endocuticle :** Inner layer

- It is the thickest layer of integument
- It is very soft and responsible for extensibilityof integument
- Not involved in tanning

## 2. EPIDERMIS OR HYPODERMIS

- Middle layer
- It cellular layer of body wall
- Secretes the parts of cuticle
- Produces moulting fluid which dissolves the old cuticle beforemoulting.
- It absorbs the digestion products of the old cuticle & repairs the wounds.
- It is an inner unicellular layer resting on basement membrane with the following function—

(i) Cuticle secretion    (ii) Digestion and absorption of old cuticle

(iii) Wound repairing (iv) Gives surface look

3.  **BASEMENT MEMBRANE**

- Very thin, amorphous glandular layer
- About 0.5 micron thick forms the innermost lining of the body wall and closely attached to epidermise.
- Provides base for attachment.
- Some dipterous larvae-tracheal cells lie in the membrane.
- The haemocytesthicken the membrane by addition of mucopolysaccharide.

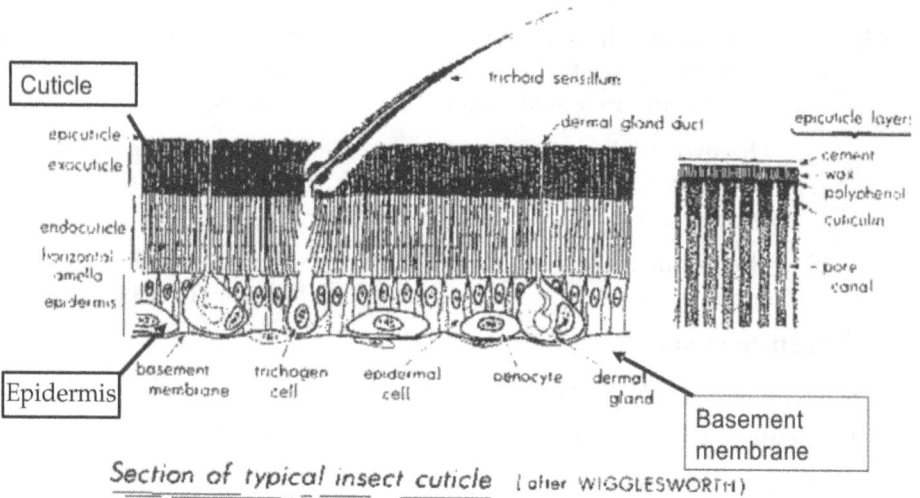

*Section of typical insect cuticle* ( after WIGGLESWORTH )

## Composition of Cuticle

(i)   **Chitin :** It is the main constituent of cuticle, which is nitrogenous polysaccharide and polymer of N-acetylglucosamine. It is water insoluble but soluble in dilute acids, alkalies and organic solvents.

(ii)  **Arthropodin :** An untanned protein, which is water soluble.

(iii) **Sclerotin :** Tanned protein, which is water insoluble.

(iv)  **Resilin :** An elastic protein responsible for the flexibility of wing sclerites.

   **Endoskeleton :** Cuticular in growth of body wall providing space for muscle attachment is known as endoskeleton. There are two types.

(i)   **Apodeme:** Hollow invagination of body wall (ridge like).

(ii)  **Apophysis:** Solid invagination of body wall (spine like).

## Cuticular Appendages

**Non-cellular:** Non-cellular appendages have no epidermal association , but rigidly attached. Eg. minute hairs and thorns.

**Cellular :** Cellular appendages have epidermal association and it may be unicellular, multicellular.

(i)    Unicellular structures:

    (a)    Clothing hairs, plumose hairs., e.g., Honey bee.

    (b)    Bristles, e.g., flies.

    (c)    Scales - flattened out growth of body wall, e.g., Moths and butterflies

    (d)    Glandular seta. et. caterpillar

    (e)    Sensory setae - associated with sensory neuron or neurons

    (f)    Seta-hair like out growth from epidermis. Epidermal cell generating seta is known as Trichogen, while the socket forming cell housing trichogen is known as Tormogen. Study of arrangement of seta is known as Chaetotaxy.

(ii)    Multicellular structures: Example, Spur - movable ; Spine- immovable.

## MOULTING (ECDYSIS)

**Ecdysis :** Periodical process of shedding the old cuticle accompanied by the formation of new cuticle is known as moulting or ecdysis. The cuticular parts discarded during moulting is known as exuvia. Moulting occurs many times in an insect during the immatured stages before attaining the adult hood. The time interval between the two subsequent moulting is called as stadium and the form assumed by the insect in any stadium is called as instar.

## STEPS IN MOULTING

1.    **Behavioural changes:** Larva stops feeding and become inactive.

2.    **Changes in epidermis:** In the epidermis cell size, its activity, protein content and enzyme level increases. Cells divide miotically and increases the tension, which results in loosening of cells of cuticle.

3.    **Apolysis:** Detachment of cuticle from epidermis.

4.    Formation of subcuticular space.

5.    Secretion of moulting gel in the sub-cuticular space which is rich with chitinase and protease.

6.    **New epicticle formation:** Cuticulin layer is laid over the epidermis.

7.    **Procuticle formation:** Procuticle is formed below the epicuticle.

8. **Activation of moulting gel:** Moulting gel is converted into moulting fluid rich in enzymes. This activates endocuticle digestion and absorption.

9. **Wax layer formation:** Wax threads of pore canals secrete wax layer.

10. **Cement layer formation:** Dermal glands secretes cement layer (Tectocuticle).

11. **Moulting:** This involves two steps.

    (i) *Rupturing of old cuticle:* Insect increases its body volume through intake of air or water which enhances the blood flow to head and thorax. There by the old cuticle ruptures along predetermined line of weakness known as ecdysial line.

    (ii) *Removal of old cuticle:* Peristaltic movement of body and lubricant action of moulting fluid helps in the removal of old cuticle. During each moulting the cuticular coverings of body, legs, internal linings of foregut, hindgut and trachea are discarded.

12. **Formation of exocuticle:** The upper layer of procuticle develops as exocuticle through addition of protein and tanning by phenolic substance.

13. **Formation of endocuticle:** The lower layer of procuticle develops as endocuticle through addition of chitin and protein. This layer increases in thickness.

**Control of Moulting:** It is controlled by endocrine glands like prothoracic gland which secrete moulting hormone. Endocrine glands are activated by prothoracico-tropic hormones produced by neurosecretory cells of brain.

## ORGANISATION OF INSECT BODY

### (STRUCTURE OF INSECT HEAD, THORAX AND ABDOMEN)

Insect body is differentiated into three distinct regions called head, thorax and abdomen. Grouping of body segments into distinct regions is known as tagmosis and the body regions are called as tagmata.

**Head :** First anterior tagma formed by the fusion of six segments namely preantennary, antennary, intercalary, mandibular, maxillary and labial segments. Head is attached or articulated to the thorax through neck or cervix. Head capsule is sclerotized and the head capsule excluding appendages formed by the fusion of several sclerites is known as cranium.

### ORIENTATION OR POSITION OF HEAD

The direction or position of head with rest of the body varies —

1. **Hypognathous {Hypo=below; gnathous=jaw}:** The mouthparts directed ventrally and the head right angle to the body.e.g. grasshopper, cockroach.

2. **Prognathous {Pro=in front; gnathous=jaw}:** The mouthparts directed anteriorly or forward and the head is straight, e.g., Soldier caste of Termites, Caterpillars, Grubs, etc.

3. **Opisthognathous {Opistho=behind; gnathous=jaw}:** The head is reflexed ventrally to such an extent that the mouthparts are directed posteriorly. The elongate proboscises lie between the front legs., e.g., Red cotton bug, aphids, etc.

## Sclerites of Head

(i) **Vertex :** Summit of the head between compound eyes.

(ii) **Frons :** Facial area below the vertex and above clypeus.

(iii) **Clypeus :** Cranial area below the frons to which labrum is attached.

(iv) **Gena :** Lateral cranial area behind the compound eyes.

(v) **Occiput :** Cranial area between occipital and post occipital suture.

**Sutures of Head :** The linear invaginations of the exoskeleton between two sclerites are called as suture (some times referred as sulcus).

(i) **Epicranial suture/ ecdysial line:** Inverted `Y' shaped suture found medially on the top of head, with a median suture (coronal suture) and lateral sutures (frontal suture).

(ii) **Epistomal suture/ Fronto clypeal suture:** Found between frons and clypeus. (epi -above; stoma- mouth parts).

(iii) **Clypeo-labral suture:** Found between clypeus and labrum (upper lip).

(iv) **Postoccipital suture:** Groove bordering occipital foramen. Line indicating the fusion of maxillary and labial segment.

## Functions of Head

(i) Food ingestion

(ii) Sensory perception

(iii) Coordination of bodily activities

**Thorax :** Second and middle tagma which is three segmented, namely prothorax, mesotthorax and metathorax. Meso and metathorax which bear wings are called as Pterothorax. Thoracic segments are made up of three sclerites namely, dorsal body plate tergum or nota, ventral body plate sternum and lateral plate pleuron.

**Thoracic nota :** Dorsal body plate of each thoracic segments are called as pronotum, mesonotum and metanotum respectively.

**Pronotum :** This sclerite is undivided and saddle shaped in grasshopper and shield like in cockroach.

**Pterothoracic notum :** Have 3 transverse sutures (antecostal, prescutal and scuto-scutellar) and 5 tergites (acrotergite, prescutum, scutum, scutellum and post-scutellum).

**Thoracic sterna :** Vental body plate of each thoracic segments are called as prosternum, mesosternum and metasternum. Thoracic sterna is made up of a segmental plate called eusternun and an intersternite called spinasternum. Eusternum is made up of three sternites viz., presternum, basisternum and sternellum.

**Thoracic pleura :** Lateral body wall of thoracic segment between notum and sternum. Selerites of pleuron is called as pleurite and they fuse to form pleural plate. Pleural plate is divided into anterior episternum and posterior epimeron by pleural suture. Pterothoracic pleuron provides space for articulation of wings and legs.

Thoracic appendages are three pairs of legs and two pairs of wings.Two pairs of spiracles are also present in the mesopleuron and metapleuron.

**Functions of Thorax :** Mainly concerned with locomotion.

**1. Abdomen :** Third and posterior tagma of insect body. This tagma is made up of 9-11 uromeres (segments) and is highly flexible. Abdominal segments are telescopic in nature and are interconnected by a membrane called conjunctiva. Each abdominal segment is made up of only two sclerites namely dorsal body plate (tergum) and ventral body plate (sternum). In grass hopper eight pairs of spiracles are present in the first eight segments, in addition to a pair of tympanum in the first segment. Eight and ninth abdominal segments bears the female genital structure and ninth segment bears male genital structure. Abdominal appendages in adult insects are genital organs and cerci.

**Function :** Concerned with reproduction and metabolism.

## CHEWING AND BITING TYPE OF MOUTH PARTS

Generalized/mandibulate/chewing and biting/chewing type of mouth part is the basic and most primitive type of mouth parts of insect present in Cockroach, Grasshopper, Beetles, etc. The mouth part consists of following five basic components.

1.  **Labrum :** The labrum forms an anterior or upper lip to the intergnathal cavity. Typically, it is a broad, transverse, flap like bilobed

plate attached to the clypeus by clypeolabral suture, allowing limited up and down movements. It is an unpaired appendage. The labrum is usually used to hold food. The notch is provided to hold the leaf blade while feeding.

2. **Epipharynx:** It is the inner lining of the labrum attached to the labrum and continues with the roof of the mouth and then in to the oesophagus. It acts as a gustatory organ or taste organ. It is the sensory organ of mouth.

3. **Mandibles:** It forms the first pair of jaws or the upper jaws. They lie behind labrum. These are solid sclerotized pieces articulated with head by ginglymus (groove) and condyle (round head). The ginglymus articulated with clypeus and condyle articulated with gena. Inner margins are toothed and used for cutting and crushing solid food. Each jaw moves horizontally by the powerful abductor (outer) muscles and adductor (inner) muscles.

4. **Maxillae:** It forms the second or posterior pair of jaw. They lie behind the mandibles. Each maxilla is composed of:

   (a) *Cardo :* It is the basal triangular segment attached to head at the lower side of post gena.

   (b) *Stipes :* It is the rectangular piece forming the body of maxilla. It articulates with cardo.

   (c) *Palpifer:* It is a small lateral sclerite on stipe to which maxillary palp is attached.

   (d) *Maxillary palp:* It is a sensory appendage articulated to stipes with palpfier. It is 1 to 7 segmented and has sensory hairs.

   (e) *Galea :* It is the outer lobe articulating with distal margin of stipes.

   (f) *Lacinia :* It is the inner lobe articulating the distal margin of stipes. It helps mandible to hold food while feeding, e.g., In coleopterous larvae, galea and lacinia are found together to form a single lobe-Mala.

5. **Hypopharynx :** The floor or oral cavity bears a tongue like prolongation called hypopharynx. It is attached to inner wall of labium, salivary duct opens behind it.

6. **Labium (second maxilla):** It consists of:

   **(a) Post mentum :** Proximal broad plate attached to head. It is divided into submentum and mentum.

   **(b) Prementum :** Distal plate attached to post mentum, labial suture exists between pre and post-mentum.

(c) **Palpiger** : Small sclerite on each lateral base of prementum to which labial palp articulates.

(d) **Labial palp** : It is 1 to 4 segmented, sensory appendage.

(e) **Paraglossae** : They are outer lobes arising from distal margin of prementum.

(f) **Glossae** : They are inner lobes arising from distal margin of prementum. Sometimes glossae and paraglossae are fused to form single median lobe called ligula,, e.g., grasshopper.

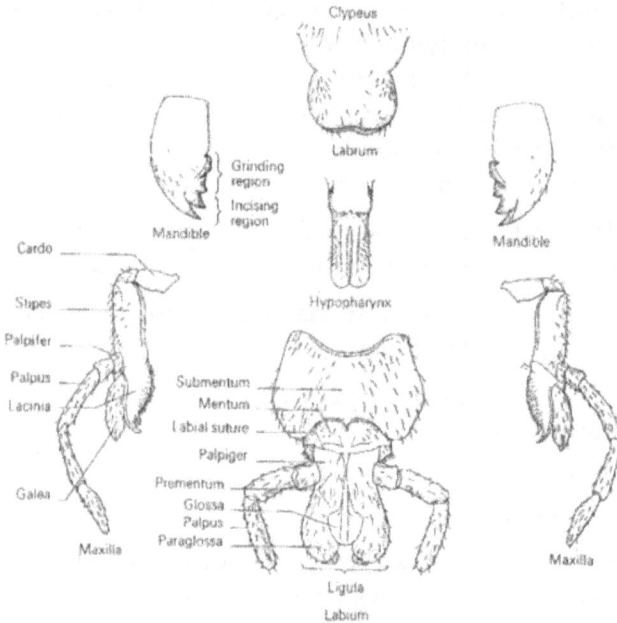

| Mouth parts | Piercing-sucking | Rasping-sucking | Chewing lapping | Sponging | Siphoning |
|---|---|---|---|---|---|
| Example | Plant bugs, Aphids, Jassids | Thrips | Honeybees | House fly | Moths and Butterflies, Cabbage butterfly |
| Labrum | Small, triangular, flap like structure. It covers labial beak. plates | The labrum and clypeus form mouth cone through which other parts are protruded. | Narrow plate attached withclypeus | Represnted by labrum epipharynx stylet on the anterior surphase of the haustellum (grooved on its anterior surphase) | Greatly reduced to a narrow plate. |

| | | | | | |
|---|---|---|---|---|---|
| **Mandibles** | Modified in to two very sharp chitinized stylets usually serrated at the apex. | Right mandible is absent. Left mandible is sclerotized in to a stylet for piercing the tissue. | Blunt, spatula like, useful for moulding wax in to cells for comb building. | Absent | Absent |
| **Maxillae** | Modified to form two stylets, tapers to a very fine points. Each of which is double grooved along its inner face to form microscopic closed tubes. | Modified in to a pair of stylets for piercing the tissue. | Long and greatly modified. Galea enlarged and concaved on the inner surphase. Maxillary palps minute. | Absent. Maxillary papls single segmented. | Galea are greatly elongated to form a proboscis. It form a slender hollow tube coiled up under the head when not in use. |
| **Hypopharynx** | — | Modified to form a stylet | Maxilolabial structures are modified to form the lapping tongue. Maxillae and labium are elongated and closely united to form a sort of LT. | Form a stylet on haustellum containing salivary glands. | — |
| **Labium** | Long, 4 segmented beak like hollow structure called proboscis, encloses 4 stylets in the dorsal groove. | — | The glossa are greatly elongated to form a hairy, flexible tongue. It terminates in to spoon shaped lobe called flabellum to lick the nectar. | Forms a fleshy, elbowed and rectratile proboscis which is differentiated into 2 regions- The basal rostrum and distal haustellum. | Represented by only the large, hairy or scaly, 3 segmented labial palps and a very small based plate. |

# TYPES OF ANTENNAE

The antennae are a pair of sense organs located near the front of an insect's head capsule. Although commonly called "feelers", the antennae are much more than just tactile receptors. They are usually covered with olfactory receptors that can detect odour molecules in the air (the sense of smell). Many insects also use their antennae as humidity sensors, to detect changes in the concentration of water vapor. Mosquitoes detect sounds with their antennae, and many flies use theirs to gauge air speed while they are in flight.

Each consists of a basal segment called the scape, followed by the pedicel and the remaining part is called the flagellum. The scape is inserted into a membranous region of the head wall and pivoted on a single marginal point, called the antennifer.

## Functions of Antennae

1. Olfactory reception

2. Johnston's organ helps in the regulation of oil speed in flying insects

3. Perception of near field sound-male mosquito, female Drosophila, worker honey bee

4. Hydrofuge hairs helps in the formulation of air funnel-Adult water beetle Hydrophilus

5. Assist the mandible in masticating the play - newly hatched grubs of Hydrophilus

6. To clasp the female during mating - Fleas and Collembola.

### Types of Antennae

| Type | Characteristic(s) | Example |
|------|-------------------|---------|
| Aristate | A single enlarged segment with lateral bristle-like outgrowth- "arista". | House fly |
| Bipectinate | Annuli with two rows of long processes, i.e., processes occur on both sides. | Silkworm- male (Tassar and Eri) |
| Capitate | Distal annuli enlarged abruptly (suddenly) so that the antenna has a pronounced knob at the end. | Khapra beetle |
| Clavate | Distal annuli enlarged gradually so that the antenna resembles a club. | Cabbage butterfly, Blister beetle. |
| Filiform | Annuli almost uniform in width. | Grasshopper |
| Flagellate | Basal annulus with fan-like process (es) | Male strepsipteran |
| Geniculate | Filiform and with elbowed pedicel or antenna bent suddenly at an angle resembling a bent knee. | Honey-bee, Wasp |
| Lamellate | Distal annuli drawn out into leaf-like plates (compactly arranged when not operating) | Dung roller |
| Moniliform | Annuli globose or ovoid so that antenna has the appearance of string of beads. | White-ant, Ground beetle. |
| Plumose | A whorl of hairs at the joints (hairs longer and numerous in the male and thus resembles a feather) | Mosquito (Male) |
| Serrate | Annuli with a row of small projections (triangular) at one side. | Pulse beetle, Mango stem borer |
| Setaceous | Annuli successively reducing in width towards the apex. | Cockroach, Crickets |
| Pectinate | Segments with longer processes like the teeth of a comb. | Mulberry silk moth, Mantid. |
| Pilose | Bases of each annuli gives whorl of hairs. | Female mosquito, Male of mango mealy bug. |
| Stylate | Antenna three segmented and the terminal segment ends with a style like process. | Robber fly, Horse fly |
| Hooked | Knobbed end of the antenna is hooked. | Skippers and sphingids |

## STRUCTURE AND MODIFICATIONS OF LEGS

**Structure :** In almost all insects all the three thoracic segments *viz.,* pro-, meso- and metathorax bear a pair of segmented legs. Each leg consists of five segments *viz.,* coxa, trochanter, femur, tibia and tarsus.

**Coxa (Pl. coxae) :** It is the first or proximal leg segment. It articulates with the cup like depression on the thoracic pleuron. It is generally freely movable.

**Trochanter :** It is the second leg segment. It is usually small and single segmented. Trochanter seems to be two segmented in dragonfly, dameselfy and ichneumonid wasp. The apparent second trochanter is in fact a part of femur, which is called trochantellus.

**Femur (Pl. femora) :** It is the largest and stoutest part of the leg and is closely attached to the trochanter.

**Tibia (Pl. tibiae) :** It is usually long and provided with downward projecting spines which aid in climbing and footing. Tibia of many insects is armed with large movable spur near the apex.

**Tarsus (Pl. tarsi) :** It is further sub-divided. The sub segment of the tarsus is called tarsomere. The number of tarsomeres vary from one to five. The basal tarsal segment is often larger than others and is named as basitarsus.

**Pretarsus :** Beyound the tarsus there are several structure collectively known as pretarsus. Tarsus terminates in a pair of strongly curved claws with one or two pads of cushions at their base between them. A median pad between the claws is usually known as arolium and a pair of pads, at their base are called pulvilli (Pulvillus-singular). Leg pads are useful while walking on smooth surface and claws give needed grip while walking on rough surface. When one structure is used, the other is bent upwards.

## Types or Modifications

Legs are modified in to several types based on the habitat and food habit of insect and used for a wide variety of functions.

1. **Ambulatorial:** (Ambulate - to walk; Walking leg), e.g., Fore leg and middle leg of grasshopper. Femur and tibia are long. Legs are suited for walking.

2. **Cursorial:** (Cursorial = adapted for running : Running leg), e.g., All the three pairs of legs of cockroach. Legs are suited for running. Femur is not swollen.

3. **Saltatorial:** (Salatorial = Leaping : Jumping Leg), e.g., hind leg of grasshopper.

4. **Scansorial:** (Scansorial = Climbing; climbing or clinging leg), e.g., all the three pairs of legs of head louse.

5. **Fossorial:** (Forrorial = Digging; Burrowing leg), e.g., Fore legs of mole cricket.

6. **Raptorial:** (Raptorial = predatory; Grasping leg), e.g., Forelegs of preying mantis.

7. **Natatorial:** (Natatorial = pertaining to swimming; Swimming leg), e.g., hing legs of water bug and water beetle.

8. **Sticking leg:** e.g., all the three pairs of legs of house fly.

9. **Basket like leg:** e.g., Legs of dragonfly and damselfly.

10. **Clasping leg:** e.g., Forelegs of male water beetle.

11. **Foragial leg:** (Forage = to collect food material), e.g., Legs of honey bee.

1. **Forelegs**

   (a) **Eye brush :** Hairs on tibia useful to clean the compound eyes.

   (b) **Antenna Cleaner :** Velum is a movable. Clasp present at distal end of tibia. Antenna comb semicircular notch lined with small spines.

   (c) **Pollen brush :** Bristles on basitarsus form the pollen brush which is useful to collect pollen from the head and mouth parts.

2. **Middle legs**

   (a) **Pollen brush :** Stiff hairs on basitarsus useful to collect pollen from middle part of their body.

   (b) **Tibial spur :** At the distal end of the tibia a movable spur is present which is useful to loosen the pellets of pollen from the pollen basket of hind legs and to clean wings and spiracles.

3. **Hind legs**

   (a) **Pollen basket :** The outer surface of the hind tibia is concave. The edges of the depressed area are fringed with long hairs. Pollen basket used to carry larger load of pollen and propolis.

   (b) **Pollen packer (Pollen press) :** Pecten is a row of stout bristles at the distal end of tibia. Auricle is a small poate fringed with hairs at the basal end of basistarsus. Pollen packer is useful to load pollen in pollen basket.

   (c) **Pollen comb :** Stiff spines present on the inner side of hind basitarsus. It is used to collect pollen from middle legs and from posterior part of body.

12. **Prolegs or False legs or Pseudolegs:** Example : abdominal legs of caterpillar.

There are two to five pairs of abdominal legs termed prolegs in caterpillar. Prolegs are thick, fleshy and not segmented. They are shed with last larval moult. One pair of prolegs on the last abdominal segment are called anal prolegs or claspers. The tip of proleg is called planta upon which are borne hooks or claws known as crochets which are useful in crawling or clinging to surface.

## Modifications of Legs

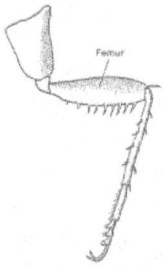

**Walking/Ambulateral**
(Cockroach)

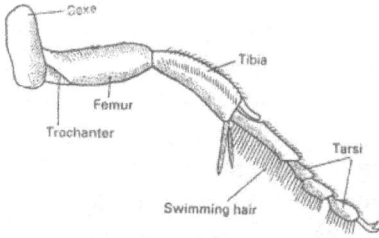

**Swimming/Natatorial**
(Water beetle)

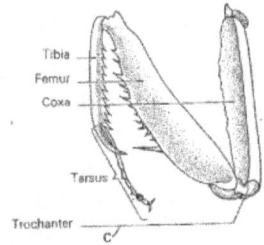

**Grasping/Raptorial**
(Praying mantid)

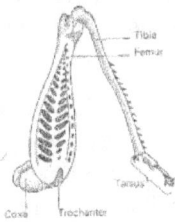

**Jumping/Saltatorial**
(Grasshopper)

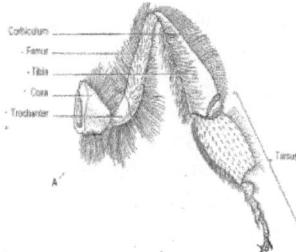

**Pollen Collecting type or Foragial**
(Honey bee)

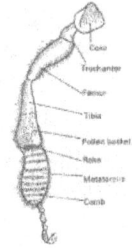

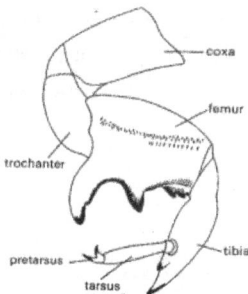

**Digging or Fossorial**
(Mole cricket)

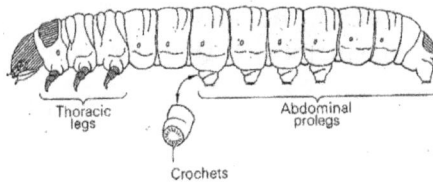

**Prolegs or False legs or Pseudo legs**
(Catterpillers)

## INSECT WING TYPES

1. **Tegmina (Singular : Tegmen) :** Wings are leathery or parchment like. They are protective in function. They are not used for flight., e.g., Forewings of cockroach and grasshopper.

2.  **Elytra (Sigular : Elytron) :** The wing is heavily sclerotised. Wing venation is lost. Wing is tough and it is protective in function. It protects hind wings and abdomen. It is not used during flight. But during flight they are kept at an angle allowing free movement of hind wings. Example, Fore wings of beetles and weevils.

3.  **Hemelytra (Singular : Hemelytron) :** The basal half of the wing is thick and leathery and distal half is membranous. They are not involved in flight and are protective in function. Example, Fore wing of heteropteran bugs.

4.  **Halteres (Singular : Haltere) :** In true flies the hind wings are modified into small knobbed vibrating organs called haltere. Each haltere is a slender rod clubbed at the free end (capitellum) and enlarged at the base (scabellum). On the basal part two large group of sensory bodies forming the smaller hick's papillae and the large set of scapel plate. They act as balancing organs and provide the needed stability during flight. Example, true flies, mosquito, male scale insect.

5.  **Fringed wings:** Wings are usually reduced in size. Wing margins are fringed with long setae. These insects literally swim through the air. Example, Thrips.

6.  **Scaly wings:** Wings of butterfly and moths are covered with small coloured scales. Scales are unicellular flattened outgrowth of body wall. Scales are inclined to the wing surface and overlap each other to form a complete covering. Scales are responsible for colour. They are important in smoothing the air flow over wings and body.

7.  **Membranous wings:** They are thin, transparent wings and supported by a system of tubular veins. In many insects either forewings (true flies) or hind wings (grass hopper, cockroach, beetles and earwig) or both fore wings and hind wings (wasp, bees, dragonfly and damselfly) are membranous. They are useful in flight.

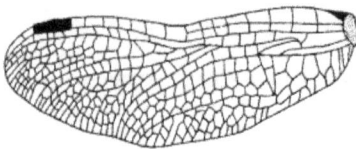

**Membranous wing**
(Dragonfly)

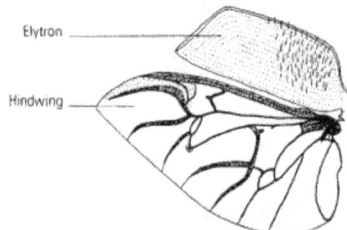

**Elytra**
(Beetles)

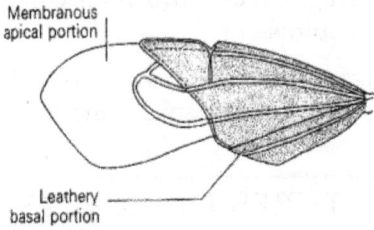

**Hemielytra**
(Bugs)

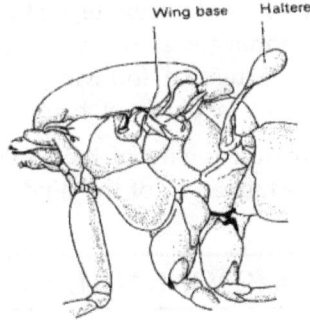

**Haltare**
(Housefly)

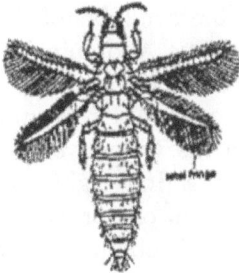

**Fringed wings**
(Thrips)

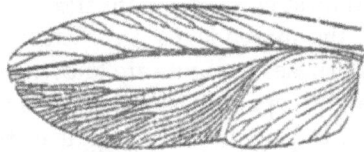

**Tegmina**
(Cockroach)

**Wing coupling:** Among the insects with two pairs of wings, the wings may work separately as in the dragonflies and damselflies. But in higher pterygote insects, fore and hind wings are coupled together as a unit, so that both pairs move synchronously. By coupling the wings the insects become functionally two winged.

## TYPES OF WING COUPLING

1. **Hamulate :** A row of small hooks is present on the coastal margin of the hind wing which is known as hamuli. These engage the folded posterior edge of fore wing. Example, bees.

2. **Amplexiform :** It is the simplest form of wing coupling. A linking structure is absent. Coupling is achieved by broad overlapping of adjacent margins. Example, butterflies.

3. **Frenate :** There are two sub types. Example, Fruit sucking moth.

    (i) **Male frenate :** Hindwing bears near the base of the coastal margin a stout bristle called frenulum which is normally held by a curved process, retinaculum arising from the subcostal vein found on the surface of the forewing.

(ii)  **Female frenate :** Hindwing bears near the base of the costal margin a group of stout bristle (frenulum) which lies beneath extended forewing and engages there in a retinaculum formed by a patch of hairs near cubitus.

4.  **Jugate :** Jugam of the forewings are lobe like and it is locked to the coastalmargin of the hindwings. Example, Hepialid moths.

# ABDOMINAL STRUCTURES IN INSECTS

## BASIC STRUCTURES

Segmentation is more evident in abdomen. The basic number of abdominal segments in insect is eleven plus a telson which bears anus. Abdominal segments are called uromeres. On eighth and nineth segment of female and nineth segment of male, the appendages are modified as external organs of reproduction or genitalia. These segments are known as genital segments. Usually eight pairs of small lateral openings (spiracles) are present on the first eight abdominal segments. In grasshoppers, a pair of tympanum is found one on either side of the first abdominal segment. It is an auditory organ. It is obliquely placed and connected to the metathoracic ganglia through auditory nerve.

## MODIFICATIONS

Reduction in number of abdominal segments has taken place in many insects. In spring tail only six segments are present. In house fly only segments 2 to 5 are visible and segments 6 to 9 are telescoped within others. In ants, bees and wasps, the first abdominal segment is fused with the metathorax and is called propodeum. Often the second segment forms a narrow petiole. The rest of the abdomen is called gaster. In queen termite after mating the abdomen becomes gradually swollen due to the enlargement of ovaries. The abdomen becomes bloated and as a result sclerites are eventually isolated as small islands. Obesity of abdomen of queen termite is called physogastry.

## ABDOMINAL APPENDAGES

(i)  Pregenital abdominal appendages in wingless insects:

(1)  **Styli (Stylus : Singular) :** Varying number of paired tube like outgrowths are found on the ventral side of the abdomen of silverfish. These are reduced abdominal legs which help in locomotion.

(2)  **Collophore or ventral tube or glue peg :** It is located on the ventral side of the first abdominal segment of spring tail. It is cylindrical.

It is protruded out by the hydrostatic pressure of haemolymph. It might serve as an organ of adhesion. It aids in water absorption from the substratum and also in respiration.

(3) **Retinaculum or tenaculum or catch :** It is present on the ventral side of the third abdominal segment. It is useful to hold the springing organ when not in use.

(4) **Furcula or Furca :** This is a 'Y' shaped organ. It is present on the venter of fourth abdominal segment. When it is released from the catch, it exerts a force against the substratum and the insect is propelled in the air.

(ii) Abdominal appendages in immature insects:

(1) **Tracheal gills:** Gills are lateral outgrowths of body wall which are richly supplied with tracheae to obtain oxygen from water in naiads (aquatic immature stages of hemimetabolous insects). Seven pairs of filamentous gills are present in the first seven abdominal segments of naiads of may flyand are called as lateral gills. Three or two leaf like gills (lamellate) are found at the end of adbomen of naiad of damselfly and are called as caudal gills. In dragonfly the gills are retained within the abdomen in a pouch like rectum and are called as rectal gills.

(2) **Anal papillae:** A group of four papillae surrounds the anus in mosquito larvae. These papillae are concerned with salt regulation.

(3) **Dolichasters:** These structures are found on the abdomen of antlion grub. Each dolichaster is a segmental protuberance fringed with setae.

(4) **Proloegs:** These are present in the larvae of moth, butterfly and sawfly. Two to five pairs are normally present. They are unsegmented, thick and fleshy. The tip of the proleg is called planta upon which are borne heavily sclerotised hooks called crochets. They aid in crawling and clinging to surface.

(iii) Abdominal appendages in winged adults :

(1) **Cornicles:** Aphids have a pair of short tubes known as cornicles or siphonculi projecting from dorsum of fifth or sixth abdominal segment. They permit the escape of waxy fluid which perhaps serves for protection against predators.

(2) **Caudal breathing tube:** It consists of two grooved filaments closely applied to each other forming a hollow tube at the apex of abdomen. Example, water scorpion.

(3) **Cerci (Cercus-Singular) :** They are the most conspicuous appendages associated normally with the eleventh abdominal segment. They

are sensory in function. They exhibit wide diversity and form.

Long and many segmented :- Example, Mayfly

Long and unsegmented :- Example, Cricket

Short and many segmented :- Example, Cockroach

Short and unsegmented :- Example, Grasshopper

Sclerotised and forceps like : Example, Earwig. Cerci are useful in defense, prey capture, unfolding wings and courtship.

**Asymmetrical cerci :** Male embiid. Left cercus is longer than right and functions as clasping organ during copulation.

(4) **Median caudal filament:** In mayfly (and also in a wingless insect silverfish) the epiproct is elongated into cercus like median caudal filament.

(5) **Pygostyles:** A pair of unsegmented cerci like structures are found in the last abdominal segment of scoliid wasp.

(6) **Anal styli:** A pair of short unsegmented structure found at the end of the abdomen of male cockroach. They are used to hold the female during copulation.

(7) **Ovipositor:** The egg laying organ found in female insect is called ovipositor. It is suited to lay eggs in precise microhabitats. It exhibits wide diversity and form. Short and horny : Example, Short horned grasshopper

**Long and sword like :** Example, Katydid, long horned grasshopper

**Needle like :** Example, Cricket

**Ovipositor modified into sting :** Example, Worker honeybee.

**Pseudoovipositor :** An appendicular ovipositor is lacking in fruit flies and house flies. In fruit flies, the elongated abdomen terminates into a sharp point with which the fly pierces the rind of the fruit before depositing the eggs. In the house fly the terminal abdominal segments are telescopic and these telescopic segments aid in oviposition. The ovipositor of house fly is called pseudoovipositor or ovitubus or oviscapt.

**Male genitalia :** External sexual organs of male insects are confined to ninth abdominal segment. In damselfly, the functional copulatory organ is present on the venter of second abdominal

## METAMORPHOSIS IN INSECTS

Metamorphosis is the change in growth and development an insect undergoes during its life cycle from birth to maturity. There are four basic types of metamorphosis in insects.

1. **Complete metamorphosis or Complex or Indirect Metamorphosis**

**(Holometabola):** Example, Butterfly, moth, fly and bees. These insects have four life stages namely egg, larva, pupa and adult. Majority of insects undergo complete metamorphosis. Larvae of butterflies are called caterpillar. Larva differs greatly in form from adult. Compound eyes are absent in larva. Lateral ocelli or stemmata are the visual organs. Their mouth parts and food habit differ from adults. Wing development is internal. When the larval growth is completed, it transforms into pupa. During the non-feeding pupal stage, the larval tissues disintegrate and adult organs are built up.

2. **Incomplete Metamorphosis or Direct or Simple metamorphosis (Hemimetabola) :** Example, Dragonfly, damselfly and may fly. These insects also have three stages in their life namely egg, young one and adult. The young ones are aquatic and are called as naiads. They are different from adults in habit and habitat. They breathe by means of tracheal gills. In dragonfly naiad the lower lip (labium) is called mask which is hinged and provided with hooks for capturing prey. After final moult, the insects have fully developed wings suited for aerial life.

3. **Ametabola (No metamorphosis) :** Example, Silver fish. These insects have only three stages in their life namely egg, young ones and adult. It is most primitive type of metamorphosis. The hatching insect resembles the adult in all respects except for the size and called as juveniles. Moulting continues throughout the life.

4. **Hypermetamorphosis :** This is specialized type of metamorphosis found in higher orders of endopterygote insects. It is a type of complete metamorphosis in which different larval instars represent two or more markedly different types of larvae. The first instar larva is active and usually compodeiform and the subsequent larval instars are vermiform or scarabaeiform, e.g., blister beetles.

5. **Paurometabola (Gradual metamorphosis) :** Example, Cockroach, grasshopper, bugs.

   The young ones are called nymphs. They are terrestrial and resemble the adults in general body form except the wings and external genitalia. Their compound eyes and mouth parts are similar to that of adults. Both nymphs and adults share the same habitat. Wing buds externally appear in later instars. The genitalia development is gradual. Later instar nymphs closely resemble the adult with successive moults.

# DIGESTIVE SYSTEM

The alimentary canal of insects is a long, muscular and tubular structure extending from mouth to anus. It is differentiated into three regions viz., Foregut, midgut and hindgut.

1. **Foregut:** It is ectodermal in origin. Anterior invagination of ectoderm

forms foregut (Stomodeum). Internal cuticular lining is present. Terminal mouthparts leads into a preoralcavity. Preoralcavity between epipharynx and hypopharynx is called as Cibarium. Preoralcavity between hypopharynx and salivary duct is Salivarium. Behind the mouth a well musculated organ called Pharynx is present which pushes the food into oesophagous. Pharynx acts as a sucking pump in sap feeders. Oesophagous is a narrow tube which conduct food into crop. Crop is the dilated distal part of oesophagus acting as food reservoir. In bees crop is called as honey stomach where nectar conversion occurs. Proventriculus or Gizzard is the posterior part of foregut and is musculated. It is found in solid feeders and absent in fluid feeders or sap feeders. Food flow from foregut to midgut is regulated through cardial or oesophageal valve. The internal cuticle of gizzard is variously modified as follows.

(i)     Teeth like in cockroach to grind and strain food.

(ii)    Plate like in honey bee to separate pollen grains from nectar.

(iii)   Spine like in flea to break the blood corpuscles.

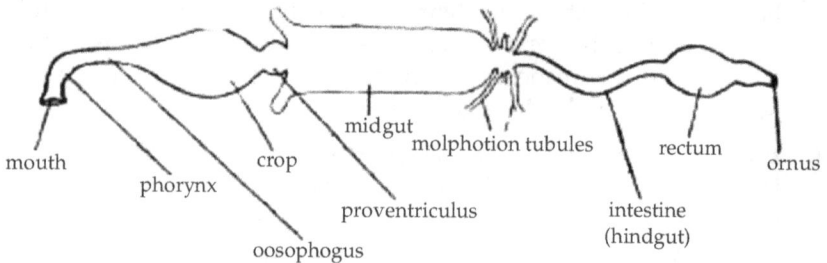

2. **Midgut:** It is endodermal in origin and also called as mesentron. This part contains no cuticular lining. Midgut is made up of three types of epithelial cells. (i) Secretory cells (Columnar cells) (ii) Goblet cells (aged secretory cells), (iii) Regenerative cells which replaces secretory cells. Important structures present in midgut are as follows:

(i)     *Peritrophic membrane:* It is the internal lining of midgut, secreted by anterior or entire layer of midgut epithelial cells. Present in solid feeders and absent in sap feeders. This layer is semipermeable in nature to digestive juices and digestion products. It lubricate and facilitate food movement. Envelops the food and protects the midgut epithelial cells against harder food particles.

(ii)    *Gastric caecae* **(Enteric caecae or Hepatic caecae)** : Finger like outgrowths found in anterior or posterior ends of midgut. This structure increases the functional area of midgut and shelter symbiotic bacteria in some insects.

(iii)   *Pyloric valve* **(Proctodeal valve)** : Midgut opens into hindgut through pyloric valve, which regulate food flow. In certain immature stages of insects midgut is not connected to hindgut till pupation., e.g., Honey bee grub.

(iv) ***Filter chamber :*** It is a complex organ in which two ends of ventriculus and the begining of hind gut are enclosed in a sac. This is useful to short circuit excess water found in liquid food in homopteran insects. This process avoids dilution of digestive enzymes and concentrates food for efficient digestion. Also helps in osmoregulation by preventing dilution of haemolymph.

3. **Hindgut:** It is ectodermal in origin and produced by the posterior invagination of ectoderm. Internal cuticular lining is present, which is permeable to salts, ions, aminoacids and water. The main functions of hindgut are the absorption of water, salt and other useful substances from the faeces and urine. Hindgut is differentiated into three regions viz., ileum, colon and rectum. In the larva of scarabids and termites, illeum is pouch like for housing symbionts and acts as fermentation chamber. Rectum contains rectal pads helping in dehydration of faeces and it opens out through anus.

**Gut physiology:** Primary functions of the gut is to digest the ingested food and to absorb the metabolites. Digestion process is enhanced with the help of enzymes produced by digestive glands and microbes housed in special cells.

## DIGESTIVE GLANDS

(a) **Salivary glands:** In Cockroach a pair of labial glands acts as salivary gland where the salivary ducts open into salivarium. In caterpillars mandibular glands are modified to secrete saliva, where the salivary glands are modified for silk production. Functions of saliva:

1. To moisten and to dissolve food.

2. To lubricate mouthparts.

3. To add flavour to gustatory receptors.

4. In cockroach the saliva contains amylase for the digestion of starch.

5. In honey bee saliva contains invertase for sucrose digestion.

6. In Jassid saliva contains lipase and protease for lipids and protein digestion. Jassid saliva also contains toxins which produces tissue necrosis and phytotoxemia on the plant parts.

7. In plant bug saliva contains pectinase which helps in stylet penetration and extra intestinal digestion.

8. In mosquito, saliva contains anticoagulin which prevents blood clotting.

9. In gall producing midges saliva contains Indole Acetic Acid (IAA).

10. In disease transmitting ectors the saliva paves way for the entry of pathogens.

(b) **Hepatic caecae and midgut epithelial cells:** It secretes most of the digestive juices. Two types of cells were involved in the enzyme secretion.

**Holocrine :** Epithelial cells disintegrate in the process of enzyme secretion.

**Merocrine :** Enzyme secretion occurs without cell break down.

(c) **Microbes in digestion:** In the insect body few cells were housing symbiotic microorganisms called as mycetocyte. These mycetocytes aggregate to form an organ called mycetome.

    (i) **Flagellate protozoa :** It produces cellulase for cellulose digestion in termites and wood cockroach.

    (ii) **Bacteria :** It helps in wax digestion in wax moth.

    (iii) Bed bug and cockroach obtain vitamin and aminoacids from microbes.

These microbes were transmitted between individuals through food exchange (mouth to mouth feeding) called trophallaxis and through egg called as transovarial transmission. In plant bug and ant lion grub partial digestion occurs in the host body prior to food ingestion called as extra intestinal digestion. In most of the insects digestion occurs in mid gut.

**Absorption:** In many insects absorption of nutrients occurs through microvilli of midgut epithelial cells by diffusion. Absorption of water and ions occur through rectum. In cockroach lipid absorption occurs through crop. In termites and scarabaeids (White grubs) absorption occurs through ileum. In solid feeders, resorption of water from the faeces occurs in the rectum and the faeces is expelled as pellets. In sap feeders (liquid feeders) the faeces is liquid like. The liquid faeces of homopteran bugs (aphids, mealy bugs, Scales and psyllids) with soluble sugars and amino acids is known as honey dew, which attracts ants for feeding.

## EXCRETORY SYSTEM

Removal of waste products of metabolism, especially nitrogenous compounds from the body of insects is known as excretion. The excretion process helps the insect to maintain salt water balance and thereby physiological homeostasis. Following are the excretory organs.

1. **Malpighian tubules:** Thin, blind-ending tubules, originating near the junction of mid and hindgut, predominantly involved in regulation of salt, water and nitrogenous waste excretion. This structure was discovered by Marcello Malpighi.

2. **Nephrocytes:** Cells that sieve the haemolmph for products that they metabolize (pericardial cells).

3. **Fat bodies :** A loose or compact aggregation of cells, mostly trophocytes, suspended in the haemocoel, responsible for storage and excretion.

4. **Oenocytes:** These are specialised cells of haemocoel, epidermis or fat body with many functions. One of the function is excretion.

5. **Integument:** The outer covering of the living tissues of an insect.

6. **Tracheal system:** The insect gas exchange system, comprising tracheae and tracheoles.

7. **Rectum:** The posterior part of hind gut.

Among the above organs, malpighian tubules are the major organ of excretion.

**Excretion and Osmoregulation:** Insect faeces, either in liquid form or solid pellets, contains both undigested food and metabolic excretions. Aquatic insects excrete dilute wastes from their anus directly into water by flushing with water. But, Terrestrial insects must conserve water. This requires efficient waste disposal in a concentrated or even dry form, simultaneously avoiding the toxic effects of nitrogen. Both terrestrial and aquatic insects must conserve ions, such as sodium ($Na^+$), potassium ($K^+$) and chloride ($Cl^-$), that may be limiting in their food or lost into the water by diffusion. Therefore the production of insect excreta (urine or pellets) is a result of two related processes: excretion and osmoregulation (maintenance of favourable osmotic pressure and ionic concentration of body fluid). The system responsible for excretion and osmoregulation is referred to as excretory system and its activities are performed largely by the Malpighian tubules and hindgut. However in fresh water insects, haemolymph composition is regulated in response to loss of ions to the surrounding water, with the help of excretory system and special cells. Special cells are called Chloride cells which are present in the hindgut, capable of absorbing inorganic ions from the dilute solutions. (e.g., Naids of dragonflies and damselflies).

**Malpighian Tubules:** The main organ of excretion and osmoregulation in insects are the malpighian tubules, acting in association with rectum or ileum. Malpighian tubules are outgrowths of the alimentary canal and consist of long thin tubules formed of a single layer of cells surrounding a blind-ending lumen, they are absent in spring tail and aphids, 2 numbers in scale insects, 4 in bugs, 5 in mosquitoes, 6 in moths and butterflies, 60 in cockroach and more than 200 in locusts. Generally they are free, waving around in the haemolymph where they filter out solutes. Each tubule is externally covered by peritonial coat and supplied with muscle fibres (aiding in peristalsis) and tracheloes. Functional differentiation of the tubules was seen, with the distal secretory region and proximal absorptive region.

**Physiology:** The malpighian tubules produce a filtrate (the primary urine) which is isosmotic but ionically dissimilar to the haemolymph and selectively reabsorbs water and certain solutes, but eliminates others. The malpighian tubules produces an iosmotic filtrate which is high in K$^+$ and low in Na$^+$ with Cl$^-$ as major anion. The active transport of ions especially K$^+$ into the tubule lumen generates an osmotic pressure gradiant for the passive flow of water.

Sugars and most amino acids are also passively filtered from the haemolymph via junctions between the tubule cells, where as amino acids and non-metabolizables and toxic organic compounds are actively transported into the tubule lumen. Sugar is resorbed from the lumen and returned to the haemolymph. The continuous secretory activity of each malpighian tubule leads to a flow of primary urine from its lumen towards and into the gut. In the rectum, the urine is modified by removal of solutes and water to maintain fluid and ionic homeostasis of the body.

**Nitrogenous excretion:** Terrestrial insects excrete waste products as uric acid or certain of its salts called urates, which were water insoluble and requires less amount of water for waste product removal. This type of excretion is known as Uricotelism. In aquatic insects ammonia is the excretory product, which is freely soluble in water and requires more amount of water for waste product removal. This type of excretion is known as Ammonotelism.

**Cryptonephry:** The distal ends of the Malpighian tubules are held in contact with the rectal wall by the perinephric membrane, which is concerned either with efficient dehydration of faeces before their elimination or ionic regulation. (e.g., Adult Coleptera, larval Lepidoptera and larval symphyta).

**Functions of malphighian tubule:** Excretory in function, mainly concerned with removal of nitrogenous wastes. The other accessory functions are as follows:

1. Spittle secretion in spittle bug

2. Light production in Bolitophila

3. Silk production in larval neuroptera

**Storage Excretion:** The excretory waste materials are retained within the body in different sites.

(i) Uric acid is stored as urates in the cells of fat body, e.g., American cockroach.

(ii) Uric acid is stored in the body wall, giving white colour, e.g., Red cotton bug.

(iii) Uric acid is stored in the male accessory glands to produce the outer coat of spermatophore, which is excreted during copulation.

(iv) Uric acid is stored in the wing scales giving white colour, e.g., Pierid butterflies.

(v) Waste products of pupal metabolism (Meconium) is stored and released during adult emergence.

# RESPIRATORY SYSTEM

Similar to aerobic animals, insects must obtain oxygen from their environment and eliminate carbon dioxide respired by their cells. This is gas exchange through series of gas filled tubes providing surface area for gaseous exchange (Respiration strictly refers to oxygen-consuming, cellular metabolic processes). Air is supplied directly to the tissue and haemolymph (blood) is not involved in the respiratory role. Gas exchange occurs by means of internal air-filled tracheae. These tubes branch and ramify through the body. The finest branches called tracheole contact all internal organs and tissues and are numerous in tissues with high oxygen requirements. Air usually enters the tracheae via spiracular openings positioned laterally on the body. No insect has more than ten pairs (two thoracic and eight abdominal).

Based on the number and location of functional spiracles respiratory system is classified as follows :

| | | |
|---|---|---|
| 1. | Holopneustic | 10 pairs, 2 in thorax and 8 in abdomen, e.g., grasshopper |
| 2. | Hemipneustic | Out of 10 pairs, one or two non-functional |
| 3. | Peripneustic | 9 pairs - 1 in thorax 8 in abdomen, e.g., Caterpillar |
| 4. | Amphipneustic | 2 pairs - One anterior, one posterior, e.g., maggot. |
| 5. | Propneustic | 1 pair -anterior pair, e.g., Puparium |
| 6. | Metapneustic | 1 pair - posterior pair, e.g., Wriggler |
| 7. | Hypopneustic | 10 pairs - 7 functional (1 thorax + 6 abdominal), 3 non functional, e.g., head louse |
| 8. | Apneustic | All spiracles closed, closed tracheal system, e.g., naiad of may fly |

## ORGANS OF RESPIRATION

Spiracles: Spiracles have a chamber or atrium with a opening and closing mechanism called atrial valve. This regulate air passage and minimise water loss. Each spiracle is set in a sclerotized cuticular plate called a peritreme. Tracheae are invaginations of the epidermis and thus their lining is continuous with the body cuticle. The ringed appearance of the tracheae is due to the spiral ridges called taenidia. This allow the tracheae to be flexible but resist compression. The cuticular linings of the tracheae are shed during moulting.

Tracheoles are less than 1 m in diameter and they end blindly and closely contact the respiring tissues. Taenidia and waxlayer is absent. Cuticulin layer is permeable to gases. It is intracellular in nature, but enclosed only in the cytoplasm of tracheal and cell called tracheoblast. Gaseous exchange occurs across tracheoles. There are four tracheal trunks viz., lateral, dorsal, ventral and visceral, helping in the passage of air. In the trachea, thin walled-collapsable sac like dilations are present, called as airsacs where taenidia is absent. Airsacs acts as oxygen reservoir. Provide buoyancy to flying and aquatic insects. Provide space for growing organs. Acts as sound resonator and heat insulators.

## MECHANISM OF RESPIRATION

Oxygen enters the spiracle and passes through the length of the tracheae to the tracheoles and into the target cells by a combination of ventilation and diffusion along a concentration gradient, from high in the external air to low in the tissue. Whereas the net movement of oxygen molecules in the tracheal system is inward (Inspiration), the net movement of $CO_2$ and water vapour molecules is outward (Expiration).

## RESPIRATION IN AQUATIC INSECTS

**1. Closed tracheal system :** In some aquatic and many endoparasitic larvae spiracles are absent and the tracheae divide peripherally to form a network. This covers the body surface, allowing cutaneous gas exchange, e.g., Gills : Tracheated thin outgrowth of body wall.

Lamellate gills - mayfly naiad

Filamentous gills - damselfly naiad

Rectal gills - dragonfly naiad 2. Open tracheal system:

(i)   **Air store :** Air bubble stored beneath wings acts as physical gill, *e.g.,* water bug.

(ii)  **Respiratory siphon :** Example, Wriggler

(iii) **Caudal breathing tube :** Example, Water scorpion

(iv)  **Plastron :** Closely set hydrofuge hairs of epicuticle hold a thin film of air indefinitely.

## CIRCULATORY SYSTEM IN INSECTS

Circulation in insects is maintained by a system of muscular pumps moving haemolymph through compartments separated by fibromuscular septa or membranes. The main pump is the pulsatile dorsal vessel. The anterior part may be called aorta and the posterior part the heart. The dorsal vessel is a simple tube, generally composed of one layer of myocardial cells and with segmentally arranged openings called ostia. The ostia permit the one-way flow of haemolymph into the dorsal vessel due to valves that prevent backflow. There may be up to three pairs of thoracic ostia and nine pairs of abdominal ostia. The dorsal vessel lies in the pericardial sinus, a compartment above a dorsal diaphragm (a fibromuscular septum - a separating membrane) formed of connective tissue and segmental pairs of alary muscles. The alary muscles support the dorsal vessel but their contractions do not affect heartbeat.

Haemolymph enters the periocardial sinus via segmental openings in the diaphragm and then moves into the dorsal vessel via the ostia during a muscular relaxation phase. Waves of contraction start at the posterior end of the body, pump the haemolymph forward in the dorsal vessel and out via the aorta into the head. Next the appendages of the head and thorax are supplied with haemolymph as it circulates posteroventrally and finally returns to the pericardial sinus and dorsal vessel.

Another important component of the insect circulatory system is the ventral diaphragm, a fibromuscular septum that lies in the floor of the body cavity associated with the ventral nerve cord. Circulation of the haemolymph is aided by active peristaltic contractions of the ventral diaphragm which direct the haemolymph backwards and laterally in the perineural sinus below the diaphragm. These movements are important in insects that use the circulation in thermoregulation. Ventral diaphragm also facilitates rapid exchange of chemicals between the ventral nerve cord and the haemolymph.

Haemolymph is generally circulated to appendages unidirectionaly by various tubes, septa, valves and pumps. The muscular pumps are termed accessory pulsatile organs and occur at the base of the antennae and legs. Antennal pulsatile organs releases neurohormones that are carried to the antennal lumen to influence the sensory neurones. Circulation occurs in the wings of young adult. In wing circulation is sustained by influxes of air into the wing veins, rather than any pulsatile organs. Pulses of air in the fine tracheal tubes of the veins push the haemolymph through the enclosed space of the veins.

The insect circulatory system shows high degree of co-ordination between dorsal vessel, fibro-muscular diaphragms and accessory pumps.

## HAEMOLYMPH AND ITS FUNCTIONS

Haemolymph is a watery fluid containing ions, molecules and cells. It is often clear and colourless but may be variously pigmented or rarely red due to haemoglobin in the immature stages of few aquatic and endoparasitic flies (e.g., Chironomid larva). Haemolymph performs the function of both blood and lymph. It is not involved in gas transporting function (respiration). Haemolymph contains a fluid portion called plasma and cellular fractions called haemocytes.

1.  **Plasma:** Plasma is an aqueous solution of inorganic ions, lipids, sugars (mainly trehalose), amino acids, proteins, organic acids and other compounds. pH is usually acidic (6.7). Density is 1.01 to 1.06. Water content is 84-92 per cent. Inorganic ions present are `Na' in predators and parasites, 'Mg' and 'K' in phytophagous insects. Carbohydrate is in the form of trehalose sugar. Major proteins are lipoproteins, glycoproteins and enzymes. Lipids in form of fat

particles or lipoproteins. Higher concentration of amino acids leads to a condition called aminoacidemia which effects the osmosis process. In high altitude insects glycerol is present which acts as a anti freezing compound. Nitrogenous waste is present in the form of uric acid.

2. **Haemocytes :** The blood cells or haemocytes are of several types and all are nucleate. Different types of haemocytes are as follows:

   (a) **Prohaemocyte :** Smallest of all cells with largest nucleus.

   (b) **Plasmatocyte (Phagocyte) :** Aids in phagocytocis

   (c) **Granular heamocyte:** Contains large number of cytoplasmic inclusions

   (d) **Spherule cell :** Cytoplasmic inclusions obscure the nucleus

   (e) **Cystocyte (Coagulocyte) :** Role in blood coagulation and plasma precipitation.

   (f) **Oenocytoids :** Large cells with ecentric nucleus

   (g) **Adipo haemocytes :** Round or avoid with distinct fat droplets

   (h) **Podocyte :** Large flattened cells with number of protoplasmic projections.

   (i) **Vermiform cells :** Rare type, long thread like.

## FUNCTIONS OF HAEMOLYMPH

1. **Lubricant :** Haemolymph keeps the internal cells moist and the movement of internal organs is also made easy.

2. **Hydraulic medium :** Hydrostatic pressure developed due to blood pumping is useful in the following processes.

   (a) Ecdysis (moulting)

   (b) Wing expansion in adults

   (c) Ecolosion in diptera (adult emergence from the puparium using ptilinum)

   (d) Eversion of penis in male insects

   (e) Eversion of osmeteria in papilionid larvae

   (f) Eversion of mask in naiad of dragonfly

   (g) Maintenance of body shape in soft bodied caterpillars.

3. **Transport and storage :** Digested nutrients, hormones and gases (chironomid larva) were transported with the help of haemolymph. It also removes the waste materials to the excretory organs. Water and raw materials required for histogenesis is stored in haemolymph.

4. **Protection:** It helps in phagocytocis, encapsulation, detoxification, coagulation, and wound healing. Non-celluar component like lysozymes also kill the invading bacteria.

5. **Heat transfer:** Haemolymph through its movement in the circulatory system regulate the body heat (Thermoregulation).

6. **Maintenance of osmotic pressure:** Ions, amino acids and organic acids present in the haemolymph helps in maintaining osmotic pressure required for normal physiological functions.

7. **Reflex bleeding:** Exudation of heamolymph through slit, pore etc. repels natural enemies, e.g., Aphids.

8. **Metabolic medium:** Haemolymph serves as a medium for on going metabolic reactions (trahalose is converted into glucose).

## NERVOUS SYSTEM

The basic component in the nervous system is the nerve cell or neuron, composed of a cell body with two projections (fibers) the dendrites that receive stimuli and the axon that transmits information, either to another neuron or to an effector organ such as a muscle. Axon may have lateral branches called Collateral and terminal arborization and synapse. Insect neurons release a variety of chemicals at synapses either to stimulate or to inhibit effector neurons or muscles. Acetylcholine and catecholamines such as dopamine are the important neurotransmitters involved in the impulse conduction. Neurons are of following types based on structure and function.

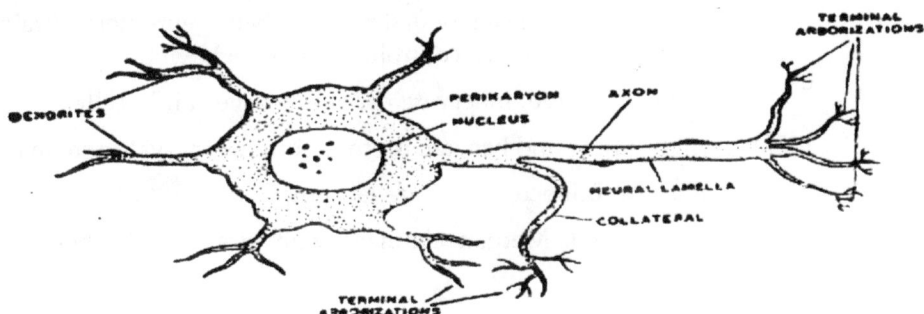

A. **On structural basis**

(i) **Monopolar :** Neuron with a single axon

(ii) **Bipolar :** neuron with a proximal axon and a long distal dendrite.

(iii) **Multipolar :** neuron with a proximal axon and many distal dendrites.

B. **Functional basis**

(i) **Sensory neuron :** It conducts impulse from sense organs to central nervous system (CNS).

(ii) **Motor neuron :** It conducts impulse from CNS to effector organs.

(iii) **Inter neuron (association neuron) :** It inter-links sensory and motor neurons.

The cell bodies of inter neurons and motor neurons are aggregated with the fibers inter connecting all types of nerve cells to form nerve centers called ganglia.

**Mechanism of impulse conduction:** Impulses are conducted by the neurons by two means.

**Axonic conduction:** Ionic composition varies between inside and outside of axon resulting in excitable conditions, which leads to impulse conduction as electrical response.

**Synaptic conduction:** Neurochemical transmitters are involved in the impulse conduction through the synaptic gap. Neurotransmitters and the type of reactions helping in the impulse conduction are as follows.

Nervous system can be divided in to three major sub-systems as

(i) Central nervous system (CNS)

(ii) Visceral nervous system (VNS)

(iii) Peripheral nervous system (PNS)

(i) *Central nervous system:* It contains double series of nerve centers (ganglia). These ganglia are connected by longitudinal tracts of nerve fibers called connectives and transverse tracts of nerve fibers called commissures. Central nervous system includes the following.

(a) *Brain:* Formed by the fusion of first three cephalic neuromeres. Brain is the main sensory centre controlling insect behaviour.

Protocerebrum: Large, innervate compound eyes and ocelli.

Deutocerebrum: Found beneath protocerebrum, innervate antennae.

Tritocerebrum: Bilobed, innervate labrum.

(b) *Ventral nerve cord:* Median chain of segmental ganglia beneath oesophagus.

(c) *Sub esophageal ganglia:* Formed by the last three cephalic neuromeres which innervate mandible, maxillae and labium.

(d) *Thoracic ganglia:* Three pairs found in the respective thoracic segments, largest ganglia, innervate legs and muscles.

(e) *Abdominal ganglia:* Maximum eight pairs will present and number varies due to fusion of ganglia. Innervate spiracles.

(f) *Thoraco abdominal ganglia :* Thoracic and abdominal ganglia are fused to form a single compound ganglia. Innervate genital organs and cerci.

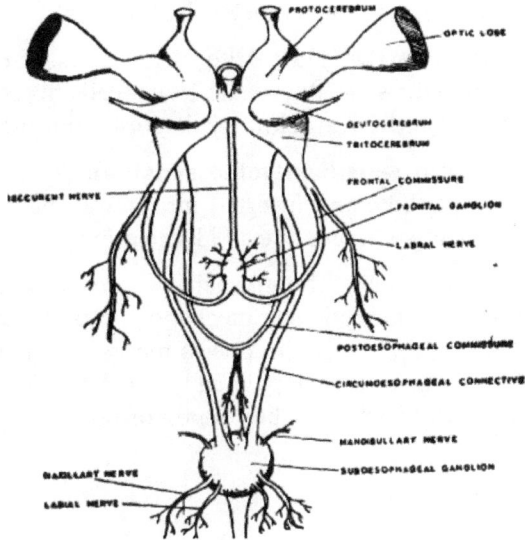

(ii) *Visceral nervous system:* The visceral (sympathetic) nervous system consists of three separate systems as follows: (1) the stomodeal/ stomatogastric which includes the frontal ganglion and associated with the brain, aorta and foregut; (2) Ventral visceral, associated with the ventral nerve cord; and (3) Caudal visceral, associated with the posterior segments of abdomen. Together the nerves and ganglia of these subsystems innervate the anterior and posterior gut, several endocrine organs (Corpora cardiaca and Corpora allata), the reproductive organs, and the tracheal system including the spiracles.

(iii) *Peripheral nervous system:* The peripheral nervous system consists of all the motor neuron axons that radiate to the muscles from the ganglia of the CNS and visceral nervous system plus the sensory neurons of the cuticular sensory structures (the sense organs) that receive mechanical, chemical, thermal or visual stimuli from an environment.

## 17. SENSE ORGANS

Sensilla are the organs associated with sensory perception and develop from epidermal cells. The different types of sense organs are:

1. Mechanoreceptors
2. Auditory receptors
3. Chemoreceptors
4. Thermo receptors and
5. Photo receptors.

1. **Mechano receptors (detect mechanical forces)**

    (i)   **Trichoid sensilla:** Hair like little sense organ. Sense cell associated with spur and seta. These cells are sensitive to touch and are located in antenna and trophi (mouth parts).

    (ii)  **Campaniform sensilla (Dome sensilla):** Terminal end of these sensilla is rod like and inserted into dome shaped cuticula. These cells are sensitive to pressure and located in leg joints and wing bases.

    (iii) **Chordotonal organ:** The specialized sensory organs that receive vibrations are subcuticular mechano receptors called chordotonal organ. An organ consists of one to many scolopidia, each of which consists of cap cell, scolopale cell and dendrite. These organs are interoceptors attached to both ends of body wall.

## Functions

    (i)   Proprioception (positioning of their body parts in relation to the gravity).

    (ii)  Sensitive to sound waves, vibration of substratum and pressure changes.

    (iii) **Johnston's organ:** All adults insects and many larvae have a complex chordotonal organ called Johnston's organ lying within the second antennal segment (Pedicel). These organs sense movements of antennal flagellum. It also functions in hearing in some insects like male mosquitoes and midges.

    (iv)  **Subgenual organ:** Chordotonal organ located in the proximal tibia of each leg, used to detect substrate vibration. Subgenual organs are found in most insects, except the Coleoptera and Diptera.

2. **Auditory receptors (detect sound waves)**

    (i)   *Delicate tactile hairs:* Present in plumose antenna of male mosquito.

    (ii)  *Tympanum:* This is a membrane stretched across tympanic cavity responds to sounds produced at some distance, transmitted by airborne vibration. Tympanal membranes are linked to chordotonal organs that enhance sound reception. Tympanal organs are located.

    - Between the metathoracic legs of mantids.
    - The metathorax of many nectuid moths.
    - The prothoracic legs of many orthopterans.
    - The abdomen of short horned grasshopper, cicada.
    - The wings of certain moths and lacewings.

3. **Chemoreceptors (detect smell and taste) :** Detect chemical energy. Insect chemoreceptors are sensilla with one pore (uniporous) or more pores

(multiporous). Uniporous chemorceptors mostly detect chemicals of solid and liquid form by contact and are called as gustatory receptor. Many sensor neurons located in antenna are of this type. Multiporous chemoreceptors detect chemicals in vapour form, at distant by smell and are acalled as olfactory receptor. Few sensory neurons located in trophi and tarsi are of this type. Each pore forms a chamber known as pore kettle with more number of pore tubules that run inwards to meet multibranched dendrites.

4. **Thermoreceptors (detect heat)** : Present in poikilothermic insects and sensitive to temperature changes. In bed bug it is useful to locate the host utilizing the temperature gradient of the host.

5. **Photoreceptors (detect light energy)**

    (a) **Compound eyes:** The compound eye is based on many individual units called ommatidia. Each ommatidium is marked externally by a hexagonal area called facet. Compound eye is made up of two parts called optic part and sensory part. Optic part contains a cuticular lens called corneal lens secreted by corneagenous cells and crystalline cone covered by primary pigment cells. Function of the optic part is to gather light. Sensory part contains six to ten visual cells called retinular cells covered by secondary pigment cells which collectively secrete a light sensitive rod at the centre called rhabdom. Rhabdom contains light sensitive pigments called rhodopsin. Each ommatidium is covered by a ring of light absorbing pigmented cells, which isolates an ommatidium from other. Nerve cells are clustered around the longitudinal axis of each ommatidium.

## Types of Ommatidia

(i) *Apposition type (light tight):* Due to the presence of primary pigment cells light cannot enter the adjacent cells. The mosaic image formed is very distinct. The image formed by the compound eye is of a series of opposed points of light of different intensities. This functions well in diurnal insects.

(ii) *Super position type:* Primary pigment cells are absent allowing light to pass between adjacent ommatidia. Image formed in this way are indistinct, bright and blurred. This type is seen in nocturnal and crepuscular insects.

    (b) *Lateral ocelli (Stemmata):* Visual organs of holometabolous larva. Structure is similar to ommatidium. It helps to detect form, colour and movement, and also to scan the environment.

    (c) *Dorsal ocelli:* Visual organs of nymph and it vary from 0-3 in numbers. It contains a single corneal lens with many visual cells individually secreting the rhabdomere. Dorsal ocelli perceive light to maintain diurnal rhythm and is not involved in image perception.

# REPRODUCTIVE SYSTEM

In insects male and female sexes are mostly separate. Sexual dimorphism is common where the male differ from the female morphologically as in bees, mosquito and cockroach. The other types are:

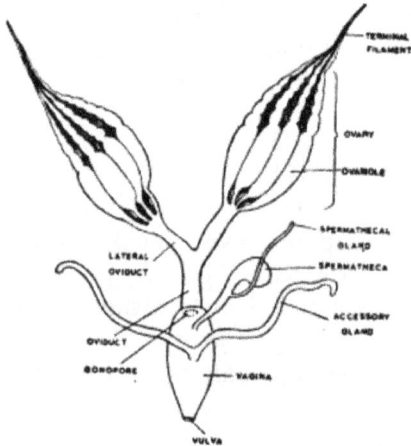

Female Reproductive system         Male Reproductive system

**Gynandromorph (Sexual mosaic) :** Abnormal individual with secondary sexual characters of both male and female, e.g., mutant Drosophila.

**Hermaphrodite :** Male and female gonads are present in one organism, e.g., Cottony cushion scale.

# FEMALE REPRODUCTIVE SYSTEM

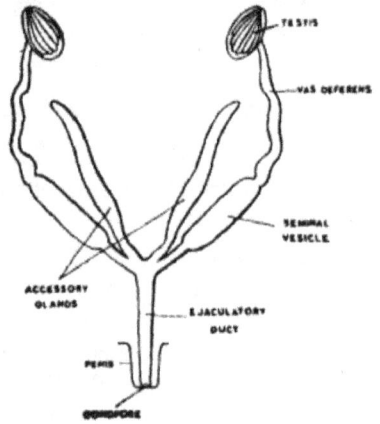

The main functions of the female reproductive system are egg production and storage of male's spermatozoa until the eggs are ready to be fertilized. The basic components of the female system are paired ovaries, which empty their mature oocytes (eggs) via the calyces (Calyx) into the lateral oviduct which unite to form the common (median) oviduct. The gonopore (opening) of the common oviduct is usually concealed in an inflection of the body wall that typically forms a cavity, the genital chamber. This chamber serves as a copulatory pouch during mating and thus is often known as the bursa copulatrix. Its external opening is the vulva. In many insects the vulva is narrow and the genital chamber becomes an enclosed pouch or tube referred to as the vagina.

Two types of ectodermal glands open into the genital chamber. The first is the spermatheca which stores spermatoza until they are needed for egg fertilization. The spermatheca is single and sac-like with a slender duct, and often has a diverticulum that forms a tubular spermathecal gland. The gland or glandular cells within the storage part of the spermatheca provide nourishment to the contained spermatozoa.

The second type of ectodermal gland, known collectively as accessory glands, opens more posteriorly in the genital chamber. Each ovary is composed of a cluster of egg or ovarian tubes, the ovarioles, each consisting of a terminal filament, a germarium (in which mitosis gives rise to primary oocytes), a vitellarium (in which oocytes grow by deposition of yolk in a process known as vitellogenesis) and a pedicel. An ovariole contain a series of developing oocytes each surrounded by a layer of follicle cells forming an epithelium (the oocyte with its epithelium is termed a follicle), the youngest oocyte occur near the apical germarium and the most mature near the pedicel. There are different types of ovarioless based on the presence or absence of specialized nutritive cells called trophocytes/nurse cells for nourishment of oocytes.

**Paniostic ovariole:** Lacks specialized nutritive cells so that it contains only a string of follicles, with the oocytes obtaining nutrients from the haemolymph via the follicular epithelium, e.g., Cockroach.

**Telotrophic ovariole (Acrotrophic) :** The trophocyte is present and its location is confined to the germarium and remain connected to the oocytes by cytoplasmic strands as the oocytes move down the ovariole, e.g., bugs.

**Polytrophic ovariole:** A number of trophocytes are connected to each oocyte and trophocytes moves down along with the ovariole, providing nutrients until depleted. Thus individual oocytes are alternated with groups of smaller trophocytes in the ovarioles, e.g., moths and flies.

Accessory glands of the female reproductive tract are often called as colleterial or cement glands, because their secretions surround and protect the eggs or cement them to the substrate, e.g., egg case production in mantis, ootheca formation in cockroach, venom production in bees.

## MALE REPRODUCTIVE SYSTEM

The main functions of the male reproductive system are the production and storage of spermatozoa and their transport in a viable state to the reproductive tract of the female. Morphologically, the male tract consists of paired testes, each containing a series of testicular tubes or follicles (in which spermatozoa are produced) which open separately into the sperm duct or vas deferens. This vas deferens expands posteriorly to form a sperm storage organ or seminal vesicle. Tubular paired accessory glands are formed as diverticula of the vasa deferentia. Some times the vasa deferentia themselves are glandular and fulfil the functions of accessory glands. The paired vasa deferentia unite where they lead into ejaculatory duct (the tube that transports the semen or the sperm to the gonopore). Accessory glands are 1-3 pairs associated with vasa deferentia or ejaculatory duct. Its function is to produce seminal fluid and spermatophores (sperm containing capsule).

## TYPES OF REPRODUCTION

1.  **Oviparity:** Majority of female insects, are oviparous, lay eggs. Embryonic development occurs after oviposition by utilizing the yolk, e.g., Head louse, moths.

2.  **Viviparity:** Unlike oviparous, here initiation of egg development takes place within the mother. The life cycle is shortened by retention of eggs and even developing young within the mother. Four main types of viviparity are observed in different insect groups.

    (i)   *Ovoviviparity:* Fertilized eggs containing yolk are incubated inside the reproductive tract of the female and hatching of egg occur just prior to or soon after oviposition, e.g., Thrips, some cockroaches, few beetles, and flesh fly. Fecundity of this group is low.

    (ii)  *Pseudoplacental viviparity:* This occurs when a yolk deficient egg develops in the genital tract of the female. The mother provides a special placenta-like tissue, through which nutrients are transferred to developing embryos. There is no oral feeding and larvae are laid upon hatching, e.g., aphids, some earwigs, psocids and polytenid bugs.

    (iii) *Haemocoelous viviparity:* This involves embryos developing free in the female's haemolymph with nutrients taken up by osmosis. This form of internal parasitism occurs only in sterpsiptera and some gall midges.

    (iv)  *Adenotrophic viviparity:* This occurs when a poorly developed larva hatches and feeds orally from accessory gland (milk gland) secretion within the uterus of the mother. The full grown larva is deposited and pupates immediately, e.g., Tsetse flies, louse, ked, bat flies.

3.  **Parthenogenesis :** Reproduction without fertilization is parthenogenesis. Different types of parthenogenesis are as follows:

    (a)  *Based on occurrence*

        (i)   Facultative (not compulsory), e.g., bee.

        (ii)  Obligatory or constant (compulsory), e.g., stick insect

        (iii) Cyclic/sporadic: alternation of gamic and agamic population, e.g., aphid.

    (b)  *Based on sex produced*

        (i)   **Arrhenotoky:** Produce male, e.g., bee

      (ii) **Thelytoky:** produce female, e.g., aphids

      (iii) **Amphitoky / deuterotoky:** Produce both male and female, e.g., Cynipid wasp.

  (c) *Based on meiosis*

      (i) **Apomictic :** no meiosis occurs

      (ii) **Automictic :** meiosis occurs, but diploidy is maintained

4. **Polyembryony:** This form of asexual reproduction involves the production of two or more embryos from one egg by subdivision. Mostly observed in parasitic insects (e.g., Platygaster). Nutrition for a large number of developing embryo cannot be supplied by the original egg and is acquired from the host's haemolymph through a specialized enveloping membrane called trophamnion.

5. **Paedogenesis**: Some insects cut short their life cycles by loss of adult and pupal stages. In this precocious stage gonads develop and give birth to young one by parthenogenesis, i.e., reproduction by immature insects.

(i) Larval paedogenesis] e.g., Gall midges (ii) Pupal paedogenesis, e.g., Miaster sp.

# GLANDULAR SYSTEM

Glandular system is otherwise called as secretary system and is divided in to two major groups based on the presence or absence of ducts.

A. **Exocrine glands (glands with duct)**

1. *Salivary glands:* Salivary glands are modified labial glands which secrete saliva and open beneath hypopharynx.

2. *Mandibular glands:* Secrete saliva in caterpillars when salivary glands are modified into silk glands. In queen bee it secretes queen substance.

3. *Maxillary glands:* Secretions are useful to lubricate mouthparts.

4. *Pharyngeal glands:* Secrete bee milk or royal jelly in nurse bee.

5. *Frontal glands:* Secrete sticky defensive fluid in nasute termites.

6. *Setal glands:* Glandular seta (Scoli) secrete irritant fluid in hairy/ slug caterpillar.

7. *Tenant hairs:* Secrete sticky fluid found in pulvilli of legs and helps in ceiling walking in house flies.

8. *Moulting glands:* Modified glandular epidermal cells, secrete moulting fluid necessary for moulting.

9.    *Stink glands (Repugnatorial glands):* Secrete bad smelling substance, e.g., Stink bugs, bed bugs.

10.    *Osmeteria (Forked gland):* Eversible gland in the thorax of papilionid larva with defense function, e.g., Citrus butterfly larva.

11.    *Androconia (Scented scales):* Secretions of glandular scales of male pierid butterflies to attract the opposite sex.

12.    *Pheromone glands:* Found in abdominal terminalia of one sex and its secretions are released outside to attract opposite sex of the same species.

13.    *Wax glands:* Dermal glands producing wax in bees and mealy bugs.

14.    *Sting glands:* Modified accessory glands secreting venom in worker bees and wasps.

15.    *Lac glands:* Dermal glands secreting resinous substances in lac insect.

16.    *Milk glands:* Modified accessory gland nourishing larva developing in uterus, e.g., Sheep ked.

## B. Endocrine glands (glands without duct)

1.    *Neurosecretory cells:* A pair of median neuro-secretory cells and lateral neurosecretory cells are present. The axons of these neurosecretory cells form two pairs of nervi corpora cardiaci ending in carpora cardiaca. This structure influence the functioning of other endocrine glands.

2.    *Corpora cardiaca:* It consist of paired bodies fused in middle and have both nervous tissues and glandular tissues. It acts as a conventional storage and release organ for neurosecretory cells. It controls heart beat and regulate trehalose level in haemolymph.

3.    *Corpora allata:* It is a paired gland attached to corpora cardiaca and secretes juvenile hormone (JH) there by inhibit metamorphosis. It is needed for egg maturation and functioning of male accessory glands. Practically JH analogues interfere with insect development. Precocene is an anti-JH which induce precocious metamorphosis and death in insects.

4.    *Prothoracic glands:* Paired gland present in ventrolateral part of prothorax of larva and is degenerated in adults. It secretes the moulting hormone ecdysone. Neurosecretory cells activate prothoracic glands to secrete ecdysone.

5.    *Weismann's ring:* Formed by the fusion of carpora cardiaca, carpora alleta, prothoracic glands and hypocerebral ganglion to secrete puparium hardening hormone. Present in maggots of Dipteran flies.

## INSECT EGG

The first stage of development in all insects is egg. Majority of insects are oviparous. Egg stage is inconspicuous, inexpensive and inactive. Yolk contained in the egg supports the embryonic development. Eggs are laid under conditions where the food is available for feeding of the future Youngones. Eggs are laid either individually or in groups. The outer protective shell of the egg is called chorion. Near the anterior end of the egg, there is a small opening called micropyle which allows the sperm entry for fertilization. Chorion may have a variety of textures. Size and shape of the insect eggs vary widely.

## Types of Eggs

(a) **SINGLY LAID**

1. **Sculptured egg :** Chorion with reticulate markings and ridges, e.g., Castor butterfly.

2. **Elongate egg :** Eggs are cigar shaped, e.g., Sorghum shoot fly.

3. **Rounded egg :** Eggs are either spherical or globular, e.g., Citrus butterfly

4. **Nit :** Egg of head louse is called nit. It is cemented to the base of the hair. There is an egg stigma at the posterior end, which assists in attachment. At the anterior end, there is an oval lid which is lifted at time of hatching.

5. **Egg with float :** Egg is boat shaped with a conspicuous float on either side. The lateral sides are expanded. The expansions serve as floats, e.g., Anopheles mosquito.

(b) **EGGS LAID IN GROUPS**

1. **Pedicellate eggs :** Eggs are laid in silken stalks of about 1.25 mm length in one groups on plants, e.g., Green lacewing fly.

2. **Barrel shaped eggs :** Eggs are barrel shaped. They look like miniature batteries. They are deposited in compactly arranged masses, e.g., Stink bug.

3. **Ootheca (Pl. Oothecae) :** Eggs are deposited by cockroach in a brown bean like chitinous capsule. Each ootheca consists of a double layered wrapper protecting two parallel rows of eggs. Each ootheca has 16 eggs arranged in two rows. Oothecae are carried for several days protruding from the abdomen of female prior to oviposition in a secluded spot. Along the top, there is a crest which has small pores which permit gaseous exchange without undue water loss. Chitinous egg case is produced out of the secretions of colleterial glands.

4. **Egg pod** : Grasshoppers secrete a frothy material that encases an egg mass which is deposited in the ground. The egg mass lacks a definite covering. On the top of the egg, the frothy substance hardens to form a plug which prevents the drying of eggs.

5. **Egg cass** : Mantids deposit their eggs on twigs in a foamy secretion called spumaline which eventually hardens to produce an egg case or ootheca. Inside the egg case, eggs are aligned in rows inside the egg chambers.

6. **Egg mass:** Moths lay eggs in groups in a mass of its body hairs. Anal tuft of hairs found at the end of the abdomen is mainly used for this purpose, e.g., Rice stem borer.

   Female silk worm moth under captivity lays eggs on egg card. Each egg mass is called a dfl (diseases free laying).

7. **Eff raft** : In Culex mosquitoes, the eggs are laid in a compact mass consisting of 200-300 eggs called egg raft in water.

## LARVAE

Larval stage is the active growing stage. It is the immature stage between the egg and pupal stage of an insect having complete metamorphosis. This stage differs radically from the adult.

## TYPES OF LARVAE

There are three main types of insect larvae namely oligopod, polypod and apodous.

1. **OLIGOPOD** : Thoracic legs are well developed. Abdominal legs are absent. There are two subtypes.

   (a) **Campodeiform** : They are so called from their resemblance to the dipluran genus Campodea. Body is elongate, depressed dorsoventrally and well sclerotised. Head is prognathous. Thoracic legs are long. A pair of abdominal cerci or caudal processes is usually present. Larvae are generally predators and are very active, e.g., grub of antlion or grub of lady brid beetle.

   (b) **Scarabaeiform** : Body is 'C' shaped, stout and subcylindrical. Head is well developed. Thoracic legs are short. Caudal processes are absent. Larva is sluggish, burrowing into wood or soil, e.g., grub of rhinoceros beetle.

2. **POLYPOD or ERUCIFORM** : The body consists of an elongate trunk with large sclerotised head capsule. Head bears a pair of

powerful mandibles which tear up vegetation. Two groups of single lensed eyes (Stemmata) found on either side of the head constitute the visual organs. The antenna is short. Three pairs of thoracic legs and upto five pairs of unjointed abdominal legs or prolegs are present. Thoracic legs are segmented and they end in claws which are used for holding on to the leaf. Bottom of the proleg is called planta which typically bears rows or circlet of short hooked spines or crochets which are useful in clinging to the exposed surface of vegetation and walking. Abdominal segments three to six and ten typically bear prolegs, e.g., Caterpillar (larvae of moths ad butterflies).

(a) **Hairy caterpillar** : The body hairs may be dense, sparse or arranged in tufts. Hairs may cause irritation, when touched, e.g., Red hairy caterpillar.

(b) **Slug caterpillar** : Larva is thick, short, stout and fleshy. Laval head is small and retractile. Thoracic legs are minute. Abdominal legs are absent. Abdominal segmentation is indistinct. Larva has poisonous spines called scoli distributed all over the body. Such larva is also called platyform larva.

(c) **Semilooper** : Either three or four pairs of prolegs are present. Prolegs are either wanting or rudimentary in either third or third and fourth abdominal segments, e.g., castor semilooper.

(d) **Looper** : They are also called measuring worm or earth measurer or inch worm. In this type, only two pairs of prolegs are present in sixth and tenth abdominal segments, e.g., Daincha looper.

3. **APODOUS:** They are larvae without appendages for locomotion. Based on the degree of development and sclerotization of head capsule there are three subtypes.

(a) **Eucepalous** : Larva with well developed head capsule with functional mandibles, maxillae, stemmata and antennae. Mandibles act transversely, e.g., Wriggler (larva of mosquito) and grub of red palm weevil.

(b) **Hemicephalous** : Head capsule is reduced and can be withdrawn into thorax. Mandibles act vertically, e.g., Larva of horse fly and robber fly.

(c) **Acephalous** : Head capsule is absent. Mouthparts consist of a pair of protrusible curved mouth hooks and associated internal sclerites. They are also called vermiform larvae, e.g., Maggot (larva of house fly).

## PUPA

It is the resting and inactive stage in all holometabolous insects. During this stage, the insect is incapable of feeding and is quiescent. During the transitional stage, the larval characters are destroyed and new adult characters are created. There are three main types of pupae.

1. **OBTECT** : Various appendages of the pupa viz., antennae, legs and wing pads are glued to the body by a secretion produced during the last larval moult. Exposed surfaces of the appendages are more heavily sclerotised than those adjacent to body, e.g., moth pupa.

    (a) *Chrysalis* : It is the naked obtect pupa of butterfly. It is angular and attractively coloured. The pupa is attached to the substratum by hooks present at the terminal end of the abdomen called cremaster. The middle part of the chrysalis is attached to the substratum by two strong silken threads called gridle.

    (b) *Tumbler* : Pupa of mosquito is called tumbler. It is an obtect type of pupa. It is comma shaped with rudimentary appendages. Breathing trumpets are present in the cephalic end and anal paddles are present at the end of the abdomen. Abdomen is capable of jerky movements which are produced by the anal paddles. The pupa is very active.

2. **EXARATE** : Various appendages viz., antennae, legs and wing pads are not glued to the body. They are free. All oligopod larvae will turn into exarate pupae. The pupa is soft and pale, e.g., Pupa of rhinoceros beetle.

3. **COARCTATE** : The pupal case is barrel shaped, smooth with no apparent appendages. The last larval skin is changed into case containing the exarate pupa. The hardened dark brown pupal case is called puparium, e.g., Fly pupa.

---

# INSECT TAXONOMY

---

## TAXONOMY

The term taxonomy is derived from the greek words taxis, arrangement and nomos, law and was first proposed in its French form by de Candolle in 1813 for the theory of plant classification. Later on it was accepted for zoological classification too. According to Simpson, "taxonomy is the theoretical study of classification, including its bases, principles, procedures and rules".

## SYSTEMATICS

The term systematics is derived from latinized Greek word systema as applied to the systems of classification developed by the early naturalists, notably Linnaeus. According to Simpson, "systematics is the scientific study of the kinds and diversity of organisms and of any and all relationships among them", or more simply, "systematics is the science of the diversity of organisms". The term 'relationship' is not used in a narrow phylogenetic sense, but is broadly conceived to include all biological relationships among organisms. This explains why such a broad area of common interest has developed between systematics, evolutionary biology, ecology and behavioral biology.

## CLASSIFICATION

According to Simpson, "zoological classification is the ordering of animals into groups or sets on the basis of their relationships, that is, of associations by contiguity, similarity or both."

## NOMENCLATURE

"It is the application of distinctive names to each of the groups recognized in classification."

Systematics is broader than taxonomy and includes it and classification and nomenclature, which are narrower than taxonomy and are, in a somewhat different way included in it. This relationship among systematics, taxonomy, classification and nomenclature can be represented as under :

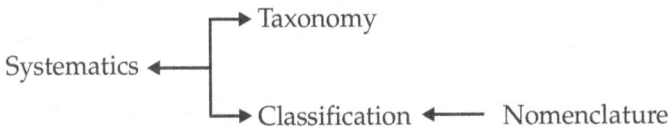

```
                        ┌──► Taxonomy
      Systematics ◄─────┤
                        └──► Classification ◄──── Nomenclature
```

**Stages of taxonomy:** The taxonomy of any group passes through several stages.

  (i) *Alpha taxonomy (α)* : It is concerned with naming and characterisation of species.

 (ii) *Beta taxonomy (β)* : Concerned with classification

(iii) *Gama taxonomy (γ)* : Concerned with evolutionary relations and phylogeny.

**Basis for classification** : Classification is the ordering of a large group of organisms based on certain characters into small groups. Classification is mainly based on evolutionary relationship and not based on superficial resemblance. Points considered while classifying are (i) external structure, (ii) internal characters, (iii) developmental history, (iv) physiological data and (v) cytogenetic data.

The biological system of classification is called hierarchial concept of classification. This was introduced by Carl Von Linnaeus (1758). A large group of organism is successively subdivided into small group. These groups are called taxa (taxon-singular). Each group is at a particular level in this system. This level is called the rank. Groups of the same rank are grouped together and that constitutes the taxonomic category, e.g., Class. Certain taxonomic categories are obligate, while others are optional. For describing and classifying any organism the basic taxonomic category is species. The lowermost category for classifying an organism is subspecies.

## STUDY OF INSECT ORDERS

## 1. THYSANURA

**Synonyms :** Ectognatha, Ectotrophi.

**Etymology :** Thysan - fringe; ura - tail.

**Common names :** Silverfish, Fire brat, Bristle tail.

**Characters:** Body is elongate and flattened. Body is glistening and clothed with scales. Compound eyes are present or absent. Antennae is long, filiform and multisegmented. Mouthparts are ecotognathous, biting type. They are primarily wingless insects. Abdomen is 11 segmented. Varying number of bilateral styli are present on abdominal sternites. Styli are belived to be reduced abdominal legs. Female has elongate jointed ovipositor. Abdomen at its tip carries a pair of elongate many segmented cerci and a median caudal filament. Insemination is indirect. Metamorphosis is absent. Moulting continues even after attaining sexual maturity.

**Importance:** It is often a pest in home and libraries. Ctenolepisma sp. is the common household silverfish. It feeds and destroys paper, book bindings and starched clothing. It can be collected from amongst old books, behind calendar, photo frames, etc.

## 2. COLLEMBOLA

**Synonyms :** Oligentoma, Oligoentomata

**Etymology :** coll-glue; embol - wedge or peg.

**Common names :** Spring tail, Snow flea

**Characters:** They are minute insects. Body is globose or tubular. Compound eyes are absent. One to several pairs of lateral ocelli form an eye patch. Antenna is four segmented. Mouthparts are entognathous biting type and found within a pouch. Tibia is fused with tarsus to form tibio-tarsus. They are primarily wingless. Abdomen is six segmented with there medially situated pregenital appendages.

- *Ventral tube or Collophore or Glue peg:* It is a bilobed adhesive organ found on the first abdominal sternite. It is beleived to be associated with respiration, adhesion and water absorption.

- *Hamula or Tenaculum or Retinaculum:* It is present on the third abdominal sternite. It consists of a fused basal piece, corpus and free distal part called rami. It holds the furcula.

- *Furcula or Springing organ:* It consists of a basal manubrium, paried dens and distal claws called mucro. It is held under tension beneath the abdomen by retinaculum when at rest.

Malpighian tubules, tracheal system and metamorphosis usually absent.

**Importance:** Sminthurus viridis is a pest on alfalfa. It can be collected from moist places in soil. They are also found in mushroom houses as a pest.

## 3. EPHEMEROPTERA

**Synonyms :** Ephemerida, Plectoptera.

**Etymology :** Ephemero-living for a day; ptera-wing.

**Common names :** Mayflies, Shadflies, Dayflies.

**Characters:** Small to medium sized soft bodied insects. Compound eyes are large. There are three ocelli. Antenna is short and setaceous. Mouthparts in adults are atrophied. Forewings are large and triangular. Hind wings are small and absent in some species. Numerous cross and intercalary veins are present. Wings are held vertically over the abdomen. Wing flexing mechanism is absent. Abdomen is slender with a pair of long cerci. Median caudal filament may be present or absent. Metamorphosis is incomplete with three stages viz., egg, naiad and adult. Naiad is aquatic with biting mouthparts. It breaths through bilateral abdominal gills. At tip of the abdomen a pair of long cerci and a median caudal filament are usually present. Immediately after the adult emergence body of the insect is covered with closely set fine hairs called pellicle and this stage is called as subimago. It is dull in colour with opaque wings and legs and cerci are not well developed. In imago wings are transparent. Legs and cerci are well developed. Body is shiny and not covered with pellicle. Adults are found near lakes and ponds and are also attracted by light.

**Importance:** Naiads are important fish food. Adults are shortlived and hence the name dayfly. When they emerge in large numbers the pose nuisance problem.

## 4. ODONATA

**Etymology :** Odon-tooth

**Common names :** Dragonflies and damselflies

**Characters:** Medium to large sized insects. They are attractively coloured. Head is globular and constricted behind into a petiolate neck. Compound eyes are large. Three ocelli are present. Mouthparts are adapted for biting. Mandibles are strongly toothed Lacinia and galea are fused to form mala which is also toothed.

Wings are either equal or subequal, membraneous; venation is net work like with many cross veins. Wings have a dark pterostigma towards the costal apex. Sub costa ends in nodus. Wing flexing mechanism is absent.

**Legs are anteroventrally placed :** They are suited for grasping, holding and conveying the prey to the mouth. Spinous femora and tibiae are useful for holding the prey. Forward shift of leg attachments allow easy transfer of prey items to mouth in flight. Legs are held in such a way that a basket is formed into which the food is scooped.

**Abdomen is long and slender :** In male gonopore is present on ninth abdominal segment. But the functional copulatory organ is present on the second abdominal sternite. Before mating sperms are transferred to the functional penis. Cercus is one segmented.

**Metamorphosis is incomplete with three life stages :** The naiad is aquatic. Labium is greatly elongated, jointed and bears two hooks at apex. It is called mask. It is useful to capture the prey.

**Importance :** Adults are aerial predators. They are able to catch, hold and devour the prey in flight. Naiads are aquatic predators. Dragonflies and damselflies can be collected with an aerial net near streams and ponds especially on a sunny day. Naiads can be collected from shallow fresh water ponds and rice fields.

**Classification :** There are two sub-orders. Dragonflies are classified under Ani-soptera and damselflies are grouped under Zygoptera.

## 5. ORTHOPTERA

**Synonyms :** Saltatoria, Saltatoptera.

**Etymology :** Ortho-straight; ptera-wings.

**Common names :** Grasshoppers, Locust, Katydid, Cricket, Mole cricket.

**Characters :** They are medium to large sized insects. Antenna is filiform. Mouthparts are mandibulate. Prothorax is large. Pronotum is curved, ventrally covering the pleural region. Hind legs are saltatorial. Forewings are leathery, thickened and known as tegmina. They are capable of bending without breaking. Hindwings are membranous with large anal area. They are folded by longitudinal pleats between veins and kept beneath the tegmina.

**Cerci are short and unsegmented :** Ovipositor is well developed in female. Metamorphosis is gradual. In many Orthopterans the newly hatched frist instar nymphs are covered by loose cuticle and are called pronymphs. Wing pads of nymphs undergo reversal during development. Specialized stridulatory (sound-producing) and auditory (hearing) organs are present.

## Classification

This order is sub divided into two suborders, viz., Caelifera and Ensifera.

I. **Suborder :** Caelifera

**Family :** Acrididae : Short horned grasshoppers and locusts.

II. **Suborder :** Ensifera

**Families :** 1. Tettigonidae: Long horned grasshoppers, Katydids and bush crickets.

2. Gryllidae : Crickets. 3. Gryllotalpidae : Mole crickets.

## 6. PHASMIDA

**Synonyms :** Phasmodea, Phasmatodea.

**Etymology :** Phasma - an apparition.

**Common names :** Stick insects, Leaf insects.

**Characters :** Body is stick - like or leaf - like. Head is prognathous. Mouthparts are chewing type. Prothorax is short. Meso and metathorax are long. Metathorax is closely associated with the first abdominal segment. Legs are widely separted. They are long and slender resembling twigs in stick insect. Tibia and femur shows lamellate expansion in leaf insects. A line of weakness is found between the tro-chanter andrest of the leg. The legs get broken easily at this region and such legs get regenerated subsequetly. Tarsus is five segmented. Wings may be present or absent. Forewings when present are small and modified into tegmina. In leaf insects the wing venation mimics leaf venation. Cerci are short and unsegmented. They show protective resemblance. They are herbivorous.

**Classification :** There are two families. Stick insects are grouped under Phasmidae and leaf insects are classified under Phyllidae.

## 7. DERMAPTERA

**Synonyms :** Euplexoptera, Euplecoptera.

**Etymology :** Derma - skin; ptera - wing.

**Common names :** Earwigs.

**Characters:** They are generally elongate insects. Head is with a distinct `Y' shaped epicranial suture. They have chewing mouthparts. Prothorax is large, well developed and mobile. Meso and Metathorax are fused with the first abdominal segment. Forewings are short, leathery and veinless. Both the wings meet along a mid dorsal line. They are called tegmina or elytra.

They are protective in function and are not used for flight. Hindwings are large, membranous, semicircular and ear like. The anal area of the wings is large with a number of branches of anal veins which are radially arranged. They are folded fan like, longitudinally and twice transversely and kept beneath the forewings at rest. Wings do not cover the abdomen fully. Cerci are found at the end of the abdomen. They are unsegmented enlarged, highly sclerotised and forceps like. They are large and bowed in male and nearly straight in female. They are useful in defence, folding and unfolding of wings, prey capture and copulation. Parental care is shown by female earwigs. It literally, roost on the eggs until hatching occurs and also cares for the nymphs.

**Importance :** Euborellia annulipes bores into groundnut pods and feeds on the kernel.

## 8. EMBIOPTERA

**Synonyms :** Embiodea, Embiidina.

**Etymology :** Embia-lively; ptera-wings.

**Common names :** Embiids, Webspinners.

**Characters:** They are small elongate soft bodied insects. Antenna is filiform. Mouthparts are chewing type. Basitarsus of the foreleg is greatly enlarged. Silk glands and spinnerets are found in the basitarsus. Hind femur is enlarged and helps in backward running. Male has well developed wings; while female is apterous. Wings are elongate, nearly equal, smoky brown with reduced wing venation. Radial vein is thick. Cerci are asymmetrical; left cercus is one segmented and it serves as clasper. Cerci are equal and two segmented in female. Embiids are gregarious and live inside tubular silken tunnels beneath stones, logs and bark of trees. Silken tunnels give protection against predators, prevent excessive water loss from the body and provide a humid atmosphere. Females show strong parental care and they nurse the eggs and nymphs. They feed on decaying plant matter.

## 9. DICTYOPTERA

**Synonyms :** Oothecaria, Blattiformia.

**Etymology :** Dictyon - net work; ptera - wings.

**Common names :** Cockroaches and preying mantids.

**Characters:** Head is hypognathous. Antenna is filiform. Mouthparts are chewing type. Tarsus is five segmented. Forewings are more on less thickened, leathery with a marginal costal vein. They are called tegmina. Hindwings are large, membranous and folded fanlike and kept beneath the forewings. Cerci are short and many segmented. Eggs are contained in an ootheca.

**Classification:** Dictyoptera is divided into two suborders viz., Blattaria (cock-roaches) and Mantodea (preying mantids). The important families are Blattidae and Mantidae.

## 10. ISOPTERA

**Synonyms :** Termitina, Termitida, Socialia.

**Etymology :** Iso - equal; ptera - wing.

**Common names :** Termites, White ants.

**Characters :** They are small greyish white, soft bodied insects. The body is pale yellow in colour because of weak sclerotization. Compound eyes are present in alate forms and usually absent in apterous forms. Antennae are short and moniliform. Mouthparts are adapted for biting and chewing. Two pairs of wings are present which are identical in size form and venation. Wings are membranous and semitransparent. Venation is not distinct. Veins near the costal and anal margin alone are distinct. Anterior veins are more sclerotised. Wings are flexed over the abdomen at rest. They are extended beyond the abdomen. Wings are present only in sexually mature forms during swarming season. Wing shedding takes place along the basal or humeral suture, after swarming. The remanant or the stump remaining behind is called `scale'. Abdomen is broadly jointed to the thorax without constriction. External genital organs are lacking in both the sexes. Cerci are short.

**Specialities :** They are ancient polymorphic, social insects living in colonies.

**Internal characters :** Salivary glands are well developed. Rectum is distended forming rectal pouch to accommodate large number of intestinal symbionts. Fat body development is extensive in male and female reproductives. Soil inhabiting termites construct earthern mounds called termitaria. They have evolved complex relationships with other organisms like bacteria, protozoa and fungi which help them in the digestion of wood. Incessant food sharing (trophallaxis) occurs between the members of the commounity by mouth-to-mouth and anus-to-mouth food transfer.

**Importance :** Termites are nature's scavengers. They convert logs, stumps, branches etc, to humus. Many are injurious to crops, furniture and wood works of buildings.

## 11. PSOCOPTERA

**Synonyms :** Corrodentia, Copeognatha.

**Etymology :** Psoco-rub small; ptera - wings; Psochos-dust like.

**Common names :** Book lice, Bark lice, Dust lice.

**Characters :** They are minute and soft bodied insects. Head has a distinct `Y' shaped epicranial suture. Clypeus is swollen. Mouthparts are biting and chewing type. Mandibles are with well developed molar and incisor areas. Lacinia is rodlike (`pick') which is partially sunken into the head capsule. Legs are slender. Wings may be present or absent. Forewings are larger than hind wings. Wings are held roof like over the abdomen. Cerci are absent. Psocids are frequently gregarious. Some psocids have the ability to spin silk. Dorsal pair of labial glands are modified into silk glands.

**Imporance :** The common book louse is Liposcelis sp. They feed on paper paste of book binding, fragments of animal and vegetable matter and stored products. They are also damage dry preserve insects and herbarium specimens.

---

# 12. MALLOPHAGA

**Synonyms :** Phthirapters.

**Etymology :** Mall-wool; phaga-eat.

**Common names :** Chewing lice, Biting lice, Bird lice.

**Characters :** They are minute insects. Body is dorsoventrally flattened. Head is large triangular and broder than thorax. Compound eyes are reduced. Mouthparts are biting type with large dentate mandibles. Mandibles are useful to clip off the host's skin debris or feather and cling to the host. Prothorax is invariably free and not fused with pterothorax. Meso and metathorax may be free or fused. Legs terminate in a pair of claws usually which are adapted for clinging to feathers. The tarsus is either unsegmented or two segmented. Wings are absent and are secondarily wingless. Eggs are called nits and are cemented to the feathers.

**Importance :** They are obligate parasites on birds and less frequently on mammals. They severely infest the poultry bird. Affected birds will become restless and peck at one another continuously, leading to loss of plumage. Louse infestation results in reduced body weight and decline in productivity. Bird lice feed on feathers, hairs, skin scales, scabs and possible blood clots around wounds. They cause irritation while feeding and crawling. In order to obtain relief, birds have dust bath.

For example, Menopon pallidum and M. gallinae are the two common lice assoicated with poultry.

---

# 13. SIPHUNCULATA

**Synonyms :** Anoplura.

**Etymology :** Siphunculus - a little tube.

**Common names :** Sucking lice.

**Characters :** They are minute insects. Body is dorsoventraly flattened. Head is small, conical and narrower than thorax. Mouth is surrounded by a row of hooks which are anchored in the host skin while feeding. Mouthparts are piercing and sucking type. There are three slender stylets which are withdrawn into a pouch in the head capsule at rest. This pouch is variously called stylet sac, buccal sac and trophic sac. Legs are scansorial, inwardly bent and adapted for clinging to mammalian hair. Tarsus is one segmented and ends in a single large claw which folds back on a thumb like projection of tibia forming an efficient organ of clinging. Wings are absent. They are secondarily wingless. Thoracic spiracles are dorsally located. Abdominal pleurites are highly sclerotised. Eggs are called nits and are strongly glued to the base of the hairs.

**Importance :** They are obligate blood sucking ecotoparasites on mammals. The prescence of lice lesions on the skin is known as pediculosis. Louse infestation causes itching and anaemia. The following lice are associated with man.

1. **Head louse:** Pediculus humanus captis. Eggs are glued to hairs. It is also called cootie.

2. **Body louse:** Pediculus humanus corporis. It infests neck, armpits and crotch. It transmits epidemic typhus, relapsing fever and trench fever, which are serious and often fatal diseases to humans. Eggs are attached to clothing.

3. **Crab louse:** Pthirus pubis. It infests armpits, pubic and perianal regions. Pthiriasis causes intense itching.

# 14. HEMIPTERA

**Synonym :** Rhynchota.

**Etymology :** Hemi-half; ptera-wing.

**Common name :** True bugs.

**General characters :** Head is opisthognathous. Mouthparts are piercing and sucking type. Two pairs of bristle like stylets which are the modified mandibles and maxillae are present. Stylets rest in the grooved labium or rostrum. Both labial palps and maxillary palps are atrophied. Mesothorax is represented dorsally by scutellum. Forewings are either uniformly thickened throughout or basally coriaceous and distally membranous. Cerci are always absent. Metamorphosis usually gradual; rarely complete. Alimentary canal is suitably modified (filter chamber) to handle liquid food. Salivary glands are universally present. Extra-oral digestion is apparently widespread. Abdominal ganglia fused with thoracic ganglia.

Classification : There are two suboders viz., Heteroptera and Homoptera.

## Important Families of Heteroptera

1. Gerridae: (Jesus bugs, Water striders, or Pond skater)
2. Reduviidae: (Assassin bugs, Kissing bugs or cone nose bugs)
3. Cimicidae: (Bed bugs)
4. Tingidae: (Lacewing bugs)
5. Miridae: (Plant bugs or Leaf bugs)
6. Lygaeidae: (Seed bugs or Chinch bugs)
7. Pyrrhocoridae: (Red bugs or Stainers)
8. Coreidae: (Squash bugs or leaf footed bugs)
9. Pentatomidae: (Stink bugs or Shield bugs)
10. Nepidae: (Water scorpions)
11. Belostomatidae: (Giant water bugs or electric light bugs)

## Important Families of Homoptera

1. Cicadidae: (Cicadas)
2. Membracidae: (Tree hoppers or Cowbugs)
3. Cicadellidae: (Leaf hoppers or Jassids)
4. Cercopidae: (Spittle bug or Cuckoo-spilt or Frog hopper)
5. Delphacidae: (Plant hoppers)
6. Lophopidae: (Aeroplane bugs)
7. Psyllidae: (Jumping plant lice)
8. Aleyrodidae: (Whiteflies)
9. Aphididae: (Aphids or Plant lice or Greenflies)
10. Coccidae: (Scale insects or Soft scales)
11. Diaspididae: (Armoured scale)
12. Kerridae: (Lac insect)
13. Pseudococcidae: (Mealy bug)

# 15. THYSANOPTERA

**Synonyms :** Physopoda.

**Etymology :** Thysano - fringe; ptera - wings.

**Common name :** Thrips.

**Characters :** They are minute, slender, soft bodied insects. Mouthparts are rasping and sucking. Mouth cone is formed by the labrum and labium together with basal segments of maxillae. There are three stylets derived from two maxillae and left mandibles. Right mandible is absent. Hence mouthparts are asymmetrical. Wings are either absent or long, narrow and fringed with hairs which increase the surface area. They are weak fliers and passive flight in wind is common. Tarsus is with one or two segments. At the apex of each tarsus a protrusible vesicle is present. Abdomen is often pointed. An appendicular ovipositor may be present or absent. Nymphal stage is followed by prepupal and pupal stages which are analogous to the pupae of endopterygote insects.

This order is subdivided into two suborders.

1. **Terebrantia:** Female with an appendicular ovipositor. Abdomen end is not tube like.

   Wing venation is present. Important family is Thripidae

2. **Tubulifera:** Ovipositor is absent. The abdomen end is tubular.

   Wing venation is absent.

**Importance:** Most of the thrips species belong to the family Thripidae and are phytophagous. They suck the plant sap. Some are vectors of plant diseases. Few are predators, e.g., Rice thrips: Stenchaetothrips biformis is a pest in rice nursery.

## 16. NEUROPTERA

**Etymology :** Neuro-nerve; ptera - wings.

**Common names :** Lace wings, Ant lions, Mantispidflies, Owlflies.

**Characters:** They are soft bodied insects. Antenna is filiform, with or without a terminal club. Mouthparts are chewing type in adults. Wings are equal, membranous with many cross veins. They are held in a roof-like manner over the abdomen. They are weak fliers Larva is campodeiform with mandibulosuctorial mouthparts. Pupa is exarate. Pupation takes place in a silken cocoon. Six out of eight Malpighian tubules are modified as silk glands. They spin the cocoons through anal spinnerets.

**Classification:** This order is subdivided into two suborders viz., Megaloptera and Planipennia.

**Suborder :** Planipennia:

1. Chrysopidae: (Green lacewings, Goldeneyes, Stinkflies, Aphid lions)
2. Mantispidae: (Mantispidflies).

3. Myrmeleontidae: (Ant lions)

4. Ascalaphidae: (Owlflies)

# 17. DIPTERA

**Etymology** : Di-two; ptera-wing.

**Common names** : True flies, Mosquitoes, Gnats, Midges.

**Characters** : They are small to medium sized, soft bodied insects. The body regions are distinct. Head is often hemispherical and attached to the thorax by a slender neck. Mouthparts are of sucking type, but may be modified. All thoracic segments are fused together. The thoracic mass is largely made up of mesothorax. A small lobe of the mesonotum (scutellum) overhangs the base of the abdomen. They have a single pair of wings. Forewings are larger, membranous and used for flight. Hind wings are highly reduced, knobbed at the end and are called haltere. They are rapidly vibrated during flight. They function as organs of equilibrium. Flies are the swiftest among all insects. Metamorphosis is complete. Larvae of more common forms are known as maggots. They are apodous and acephalous. Mouthparts are represented as mouth hooks which are attached to internal sclerites. Pupa is generally with free appendages, often enclosed in the hardened last larval skin called puparium. Pupa belongs to the coarctate type.

**Classification** : This order is sub divided in to three suborders.

**Nematocera (Thread-horn)** : Antenna is long and many segmented in adult. Larval head is well developed. Larval mandibles act horizontally. Pupa is weakly obtect. Adult emergence is through a straight split in the thoracic region.

**Brachycera (Short-horn)** : Antenna is short and few segmented in adult. Larval head is retractile into the thorax. Larval mandibles act vertically. Pupa is exarate. Adult emergence is through a straight split in the thoracic region.

**Cyclorrhapha (Circular-crack)** : Antenna is aristate in adult. Larval head is vestigial with mouth hooks. Larval mouth hooks act vertically. Pupa is coarctate. The coarctate pupa has a circular line of weakness along which the pupal case splits during the emergence of adult. The split results due to the pressure applied by an eversible bladder ptilinum in the head.

(i) **Nematocera**

   1. Culicidae: (Mosquitoes)

   2. Cecidomyiidae: (Gall midges)

(ii) **Brachycera**

   3. Asilidae: (Robber flies)

   4. Tabanidae: (Horse flies)

**(iii)  Cyclorrhapha**

    5.   Syrphidae: (Hover flies, Flower flies)

    6.   Tephritidae: (Fruit flies)

    7.   Drosophilidae: (Vinegar gnats, Pomace flies)

    8.   Tachinidae: (Tachinid flies)

    9.   Muscidae: (Housefly)

   10.   Hippoboscidae: (Dogfly)

---

# 18. COLEOPTERA

---

**Synonym** : Elytroptera.

**Etymology** : Coleo - Sheath; ptera-wing.

**Common names** : Beetles, Weevils.

**Characters** : They are minute to large sized insects. Antenna is usualy 11 segmented. Mouthparts are chewing type. Mandibles are short with blunt teeth at the mesal face in phytophagous group. In predators the mandibles are long, sharply pointed with blade like inner ridge. In pollen feeders teeth are absent and the mandibles are covered with stiff hairs. Prothorax is large, distinct and mobile. Mesothorax and metathorax are fused with the first abdominal segment.

Forewings are heavily sclerotised, veinless and hardened. They are called elytra. Forewings do not overlap and meet mid-dorsally to form a mid-dorsal line. It is not used for flight. They serve as a pair of convex shields to cover the hind wings and delicate tergites of abdomen. Hind wings are membranous with few veins and are useful in flight. At rest they are folded transversely and kept beneath the elytra. In some weevils and ground beetles the forewings are fused and hind wings are atrophied. A small part of the mesothorax known as scutellum remains exposed as a little triangle between the bases of elytra. Cerci and a distinct ovipositor are absent. Metamorphosis is complete. Larvae are often called grubs. Pupae are usually exarate and rarely found in cocoons.

**Importance :**It is the largest order. It includes predators, scavengers and many crop pests. They also damage stored products.

**Classification** : This order is divided into two suborders, viz., Adephaga (devourers) and Polyphaga (eaters of many things). Adephaga includes Cicindelidae, Carabidae and Dytiscidae. Other families listed out below come under Polyphaga.

## Families of Predators

    1.  Cicindelidae: (Tiger beetles)

2. Carabidae: (Ground beetles)

3. Dytiscidae: (True water beetles, Predaceous diving beetles)

4. Gyrinidae: (Whirligig beetles)

5. Coccinellidae: (Lady bird beetles)

6. Lampyridae: (Fireflies, Glow worms)

## Families of scavengers

1. Scarabaeidae: (Scarabs, Dung beetles)

2. Hydrophilidae: (Water scavenger beetles)

## Families of stored product pests

1. Anobiidae: (Wood worms, Wood borers)

2. Bostrychidae: (Grain borers)

3. Bruchidae: (Pulse beetles, Seed beetles)

4. Tenebrionidae: (Meal worms)

## Families of crop pests

1. Apionidae: (Ant like weevils)

2. Buprestidae: (Jewel beetles, Metallic wood borers)

3. Cassididae: (Tortoise beetles)

4. Cerambycidae: (Longicorn beetles)

5. Curculionidae: (Weevils, snout beetles)

6. Dynastidae: (Unicorn beetles, Rhinoceros beetles)

7. Elateridae: (Click beetles, Wire worms)

8. Galerucidae: (Pumpkin beetles)

9. Meloidae: (Blister beetles, Oil beetles)

10. Melolonthidae: (Chafer beetles, June beetles, White grubs)

## 19. HYMENOPTERA

**Etymology :** Hymen - membrane; ptera - wings.

Hymeno - god of marriage; ptera - wings, (Marriage/union of fore and hind wings by hamuli).

**Common names :** Ichneumonflies, Ants, Bees, Wasps, Parasitoids.

**Characters :** Mouthparts are primarily adapted for chewing. Mandibles are very well developed. In bees both labium and maxillae are integrated to form the lapping tongue. Thorax is modified for efficient flight. Pronotum is collar like. Mesothorax is enlarged. Metathorax is small. Both prothorax and metathorax are fused with mesothorax. Wings are stiff and membranous. Forewings are larger than hindwings. Wing venation is reduced. Both forwings and hindwings are coupled by a row of hooklets (hamuli) present on the leading edge of the hindwing.

**Abdomen is basally constricted :** The first abdominal segment is called propodeum. It is fused with metathorax. The first pair of abdominal spiracles is located in the propodeum. The second segment is known as pedicel which connects the thorax and abdomen. Abdomen beyond the pedicel is called gaster or metasoma. Ovipositor is always present in females. It is variously modified for oviposition or stinging or sawing or piercing plant tissue.

**Metamorphois is complete :** Often the grub is apodous and eucephalous. Larva is rarely eruciform. Pupa is exarate and frequently enclosed in a silken cocoon secreted from labial glands. Sex is determined by the fertilization of the eggs. Fertilized eggs develop into females and males are produced from unferti-lized eggs. Males are haploid and females diploid.

**Classification:** This order is subdivided into two suborders.

# Suborder : Symphyta
1. Tenthredinidae : (Sawflies)

# SUBORDER : APOCRITA
1. Ichneumonidae: (Ichneumonflies)
2. Braconidae : (Braconid wasps)
4. Bethylidae: (Bethylid wasps)
5. Chalcididae: (Chalcid wasps)
6. Eulophidae: (Pupal parasitoids)
7. Trichogrammatidae: (Egg parasitoids)
8. Evaniidae: (Ensign wasps)
9. Agaonidae: (Fig wasps)
10. Vespidae: (Yellow jackets, Hornets)
11. Sphecidae: ( Thread waisted wasp, Digger wasp, Mud dauber)
12. Formicidae: (Ants)
13. Apidae : (Honey bees)
14. Megachilidae: (Leaf cutter bees)
15. Xylocopidae: (Carpenter bees)

# 20. LEPIDOPTERA

**Synonym** : Glossata

**Etymology** : Lepido - scale; ptera - wings.

**Common names** : Moths, Butterflies, Skippers

**Characters** : Body, wings, appendages, are densely clothed with overlapping scales, which give colour, rigidity and strength. They insulate the body and smoothen air flow over the body. Mouthparts in adults are of siphoning type. Mandibles are absent. The galeae of maxillae are greatly elongated and are held together by interlocking hooks and spines. The suctorial proboscis is coiled up like a watch spring and kept beneath the head when not in use.

Wings are membranous and are covered with overlapping pigmented scales. Forewings are larger than hind wings. Cross veins are few. Wings are coupled by either frenate or amplexiform type of wing coupling.

**Larvae are polypod-eruciform type** : Mouthparts are adapted for chewing with strong mandibles. A group of lateral ocelli is found on either side of the head. The antenna is short and three segmented. There are three pairs of five segmented thoracic legs ending in claws. Two to five pairs of fleshy unsegmented prolegs are found in the abdomen. At the bottom of the proleg, crochets are present. Pupa is generally obtect. It is either naked or enclosed in a cocoon made out of soil, frass, silk or larval hairs.

**Classification** : Majority of Lepidopteran insects (97%) are grouped under the suborder Ditrysia in which the female insects have two pores. The copulatory pore is located in eighth abdominal sternite and the egg pore in ninth abdominal sternite. Ramaining insects are grouped under the suborder Monotrysia in which the female insects have one pore.

## BUTTERFLY FAMILIES

1. Nymphalidae: (Brush footed or four footed butterflies)
2. Lycaenidae: (Blues, Coppers, Hair streaks)
3. Papilionidae: (Swallow tails)
4. Pieridae: (whites and Sulphurs)
5. Satyridae: (Browns, Meadow - browns)

## MOTH FAMILIES

6. Arctiida : (Tigermoths)
7. Bombycidae: (Silkworm moths)

8. Cochlididae: (Slugcaterpillar)

9. Crambidae : (Grassmoths)

10. Gelechiidae: (Paddymoth)

11. Geometridae: (Loopers)

12. Lymantridae: (Tussockmoths)

13. Noctuidae: (Noctuamoths)

15. Pyraustidae: (Grassborers)

16. Saturniidae: (Moon months, giant silk wormmoths)

17. Sphingidae : (Hawk moths, Sphinx moths, Horn worms)

## SKIPPER FAMILY

18. Hesperiidae (Skipper)

# INTEGRATED PEST MANAGEMENT

## Definition IPM definition by FAO (1967)

Integrated Pest Management (IPM) is a system that, in the context of associated environment and population dynamics of the pest species, utilizes all suitable techniques and methods in as compatible a manner as possible and maintains pest populations at levels below those causing economic injury.

### IPM definition by Luckmann and Metcalf (1994)

IPM is defined as the intelligent selection and use of pest control tactics that will ensure favourable economical, ecological and sociological consequences

## Need for Pest Management or Why Pest Management

1. Development of resistance in insects against insecticides, e.g., OP and synthetic pyrethroid resistance in Helicoverpa armigera.

2. Out break of secondary pests, e.g., Whiteflies emerged as major pest when spraying insecticide against H. armigera.

3. Resurgence of target pests, e.g., BPH of rice increased when some OP chemicals are applied.

4. When number of application increases, profit decreases.

5. Environmental contamination and reduction in its quality.

6. Killing of non-target animals and natural enemies.

7. Human and animal health hazards.

### Stages in Crop Protection Leading to IPM

| | | |
|---|---|---|
| 1. | Subsistence phase | Only natural control, no insecticide use |
| 2. | Exploitation phase | Applying more pesticides, growing HY varieties and get more yield and returns |
| 3. | Crisis phase | Due over use pesticides, problem of resurgence, resistance, secondary pest outbreak, increase in production cost |
| 4. | Disaster phase | Due to increased pesticide use - No profit, high residue in soil - Collapse of control system |
| 5. | Integrated Management Phase | IPM integrates ecofriendly methods to optimize control rather than maximise it. |

## TOOLS OR COMPONENTS OF INTEGRATED PEST MANAGEMENT

1. **Cultural Methods :** Manipulation of cultural practices to the disadvantage of pests.

### I. Farm level practices.

| S.No. | Cropping Techniques | Pest Checked |
|---|---|---|
| 1. | Ploughing | Red hairy caterpillar |
| 2. | Puddling | Rice mealy bug |
| 3. | Trimming and plastering | Rice grass hopper |
| 4. | Pest free seed material | Potato tuber moth |
| 5. | High seed rate | Sorghum shotfly |
| 6. | Rogue space planting | Rice brown planthooper |
| 7. | Plant density | Rice brown planthooper |
| 8. | Earthing up | Surgarcane whitefly |
| 9. | Detrashing | Sugarcane whitefly |
| 10. | Destruction of weed hosts | Citrus fruit sucking moth |
| 11. | Destruction of alternate host | Cotton whitefly |
| 12. | Floding | Rice armyworm |
| 13. | Trash mulching | Sugarcane early shoot borer |
| 14. | Pruning/topping | Rice stem borer |
| 15. | Intercropping | Sorghum stem borer |
| 16. | Trap cropping | Diamond back moth |
| 17. | Water management | Brown planthopper |
| 18. | Judicious application of fertilizers | Rice leaf folder |
| 19. | Timely harvesting | Sweet potato weevil |

## II. Community level practices

1. **Synchronized sowing :** Dilution of pest infestation, e.g., Rice, Cotton.

2. **Crop rotation :** Breaks insect life cycle.

3. Crop sanitation.

    (a) Destruction of insect infested parts, e.g., Mealy bug in brinjal.

    (b) Removal of fallen plant parts, e.g., Cotton squares.

    (c) Crop residue destruction, e.g., Cotton stem weevil.

| | *Advantages* | *Disadvantages* |
|---|---|---|
| 1. | No extra skill | 1. No complete control |
| 2. | No costly inputs | 2. Prophylactic nature |
| 3. | No special equipments | 3. Timing decides success |
| 4. | Minimal cost | |
| 5. | Good component in IPM | |
| 6. | Ecologically sound | |

## Physical Methods

Modification of physical factors in the environment to minimise or prevent pest problems. Use of physical forces like temperature, moisture, etc. in managing the insect pests.

### A. Manipulation of temperature

1. Sun drying the seeds to kill the eggs of stored product pests.

2. Hot water treatment (50-55°C for 15 min) against rice white tip nematode.

3. Flame throwers against locusts.

4. Burning torch against hairy caterpillars.

5. Cold storage of fruits and vegetables to kill fruitflies (1-2°C for 12-20 days).

### B. Manipulation of moisture

1. Alternate drying and wetting rice fields against BPH.

2. Drying seeds (below 10% moisture level) affects insect development.

3. Flooding the field for the control of cutworms.

### C. Manipulation of light

1. Treating the grains for storage using IR light to kill all stages of insects, e.g., Infrared seed treatment unit

2. Providing light in storage go downs as the lighting reduces the fertility of Indian meal moth, Plodia.

3. Light trapping.

**D. Manipulation of air**

1. Increasing the $CO_2$ concentration in controlled atmosphere of stored grains to cause asphyxiation in stored product pests.

**E. Use of irradiation**

Gamma irradiation from $CO_{60}$ is used to sterilize the insects in laboratory which compete with the fertile males for mating when released in natural condition, e.g., cattle screw worm fly, Cochliomyia hominivorax control in Curacao Island by E.F. Knipling.

**F. Use of greasing material**

Treating the stored grains particularly pulses with vegetable oils to prevent the oviposition and the egg hatching, e.g., bruchid adults.

**G. Use of visible radiation :** Yellow colour preferred by aphids, cotton whitefly : yellow sticky traps.

**H. Use of Abrasive dusts**

1. **Red earth treatment to red gram :** Injury to the insect wax layer.

2. **Activated clay :** Injury to the wax layer resulting in loss of moisture leading to death. It is used against stored product pests.

3. **Drie-Die :** This is a porous finely divided silica gel used against storage insects.

**I. Mechanical Control**

Use of mechanical devices or manual forces for destruction or exclusion of pests.

**A. Mechanical destruction**

Life stages are killed by manual (or) mechanical force.

# Manual Force

1. Hand picking the caterpillars

2. Beating : Swatting housefly and mosquito

3. Sieving and winnowing : Red flour beetle (sieving) rice weevil (winnowing)

4. Shaking the plants : Passing rope across rice field to dislodge caseworm and shaking neem tree to dislodge June beetles

5. Hooking : Iron hook is used against adult rhinoceros beetle

6. Crushing : Bed bugs and lice
7. Combing : Delousing method for Head louse
8. Brushing : Woolen fabrics for clothes moth, carper beetle.

## MECHANICAL FORCE

1. **Entoletter :** Centrifugal force - breaks infested kernels - kill insect stages- whole grains unaffected - storage pests.
2. **Hopper dozer :** Kill nymphs of locusts by hording into trenches and filled with soil.
3. **Tillage implements :** Soil borne insects, red hairy caterpillar.
4. **Mechnical traps :** Rat traps of various shapes like box trap, back break trap, wonder trap, Tanjore bow trap.

### B. Mechanical exclusion

Mechanical barriers prevent access of pests to hosts.

1. **Wrapping the fruits :** Covering with polythene bag against pomegrante fruit borer.
2. **Banding :** Banding with grease or polythene sheets - Mango mealybug.
3. **Netting :** Mosquitoes, vector control in green house.
4. **Trenching :** Trapping marching larvae of red hairy catepiller.
5. **Sand barrier :** Protecting stored grains with a layer of sand on the top.
6. **Water barrier :** Ant pans for ant control.
7. **Tin barrier :** Coconut trees protected with tin band to prevent rat damage.
8. **Electric fencing :** Low voltage electric fences against rats.

| Advantage of Mechanical control | Disadvantage |
|---|---|
| 1. Home labour utilization | 1. Limited application |
| 2. Low equipment cost | 2. Rarely highly effective |
| 3. Ecologically safe | 3. Labour intensive |
| 4. High technical skill not required in adopting | |

## Appliances in Controlling the Pests

1. **Light traps :** Most adult insects are attracted towards light in night. This principle is used to attract the insect and trapped in a mechanical device.

(a) **Incandescent light trap :** They produce radiation by heating a tungsten filament. The spectrum of lamp include a small amount of ultraviolet, considerable visible especially rich in yellow and red. Example, Simple incandescent light trap, portable incandescent light trap. Place a pan of kerosenated water below the light source.

(b) **Mercury vapour lamp light trap :** They produce primarily ultraviolet, blue and green radiation with little red, e.g., Robinson trap. This trap is the basic model designed by Robinson in 1952. This is currently used towards a wide range of Noctuids and other nocturnal flying insects. A mercury lamp (125 W) is fixed at the top of a funnel shaped or trapezoid galvanized iron cone terminating in a collection jar containing dichlorvos soaked in cotton as insecticide to kill the insect.

(c) **Black light trap :** Black light is popular name for ultraviolet radiant energy with the range of wavelengths from 320-380 nm. Some commercial type like Pest-O-Flash, Keet-O-Flash are available in market. Flying insects are usually attracted and when they come in contact with electric grids, they become elctrocuted and killed.

2. **Pheromone trap :** Synthetic sex pheromones are placed in traps to attract males. The rubberised septa, containing the pheromone lure are kept in traps designed specially for this purpose and used in insect monitoring/ mass trapping programmes. Sticky trap, water pan trap and funnel type models are available for use in pheromone based insect control programmes.

3. **Yellow sticky trap :** Cotton whitefly, aphids, thrips prefer yellow colour. Yellow colour is painted on tin boxes and sticky material like castor oil/ vaseline is smeared on the surface. These insects are attracted to yellow colour and trapped on the sticky material.

4. **Bait trap :** Attractants placed in traps are used to attract the insect and kill them. For example, Fishmeal trap: This trap is used against sorghum shootfly. Moistened fish meal is kept in polythene bag or plastic container inside the tin along with cotton soaked with insecticide (DDVP) to kill the attracted flies.

5. Pitfall trap helps to trap insects moving about on the soil surface, such as ground beetles, collembola, spiders. These can be made by sinking glass jars or metal cans into the soil. It consists of a plastic funnel, opening into a plastic beaker containing kerosene supported inside a plastic jar.

6. **Probe trap :** Probe trap is used by keeping them under grain surface to trap stored product insect.

7. **Emergence trap :** The adults of many insects which pupate in the soil can be trapped by using suitable covers over the ground. A

wooden frame covered with wire mesh covering and shaped like a house roof is placed on soil surface. Emerging insects are collected in a plastic beaker fixed at the top of the frame.

8. **Indicator device for pulse beetle detection :** A new cup shaped indicator device has been recently designed to predict timely occurrence of pulse beetle Callosobruchus spp. This will help the farmers to know the correct time of emergence of pulse beetle. This will help them in timely sun drying which can kill all the eggs.

# BIOLOGICAL CONTROL

## Definition

The study and utilization of parasitoids, predators and pathogens for the regulation of pest population densities. Biological control can also be defined as the utilization of natural enemies to reduce the damage caused by noxious organisms to tolerable levels.

Biological control is often shortened to biocontrol.

## *Advantages*

1. Protection and encouragement of natural enemies.
2. Introduction, artificial increase and colonizing specific parasitoids and predators.
3. Pathogens on insects like virus, bacteria, fungi and protozoa.
4. Use of botanicals like neem, pongam, etc.

Parasitoids and predators may be used in Agriculture and IPM in three ways. They are

1. **Conservation and encouragement of indigenous natural enemies**

   Defined as actions that preserve and increase natural enemies by environmental manipulation. For example, use of selective insecticides, provide alternate host and refugia for NE.

2. **Importation or Introduction**

   Importing or introducing natural enemies into a new locality (mainly to control introduced pests).

3. **Augmentation**

   Propagation (mass culturing) and release of natural enemies to increase its population. Two types,

   (i) **Inoculative release:** Control expected from the progeny and subsequent generations only.

(ii)    **Inundative release:** Natural enemies mass cultured and released to suppress pest directly, e.g., Trichogramma sp. egg parasitoid, Chrysoperla carnia predator.

## Merits

- Since biological control is safe to environment, it should be adopted as an important component of IPM.
- Biological control method can be integrated well with other methods namely cultural, chemical methods and host plant resistance (except use of broad spectrum insecticides)
- Biological control is self propagating and self perpetuating
- Pest resistance to NE is not known
- No harmful effects on humans, livestock and other organisms
- Biological control is virtually permanent
- Biological agents search and kill the target pest

### MICROBIAL CONTROL

- It is a branch of biological control
- Defined as control of pests by use of microorganisms like viruses, bacteria, protozoa, fungi, rickettsia and nematodes.

## I. VIRUSES

Viruses coming under family Baculoviridae cause disease in Lepidoptera larvae. Two types of viruses are common.

NPV (Nucleopolyhedro virus), e.g., HaNPV, SlNPV

GV (Granulovirus), e.g., CiGV

## Symptoms

Lepidopteran larva become sluggish, pinkish in colour, lose appetite, body becomes fragile and rupture to release polyhedra (virus occlusion bodies). Dead larva hang from top of plant with prolegs attached (Tree top disease or "Wipfelkrankeit").

## II. BACTERIA

Spore forming (Facultative - Crystalliferous)

2 types of bacteria Spore forming (Obligate)

Non-spore forming

(i) Spore forming (Facultative, Crystelliferous)

The produce spores and also toxin (endotoxin). The endotoxin paralyses gut when ingested, e.g., Bacillus thuringiensis effective against lepidopteran. Commercial products - Delfin, Dipel, Thuricide

(ii) Spore-forming (Obligate)

For example, Bacillus popilliae attacking beetles, produce 'milky disease' Commercial product - 'Doom' against 'white grubs'

(iii) Non-spore forming

For example, Serratia entomophila on grubs

## III. FUNGI

(i) Green muscardine fungus - Metarhizium anisopliae attack coconut rhinoceros beetle

(ii) White muscardine fungus - Beaveria bassiana against lepidopteran larvae

(iii) White halo fungus - Verticillium lecanii on coffee green scale.

**Other Microbs: Protoza, Nematodes**

**Limitations of biocontrol technique**

- Complete control not achieved - Slow process

- Subsequent pesticide use restricted

- Expensive to culture many NE

- Requires trained man power

## 5. Chemical methods

**Chemical Control:** Management of insect pests using chemical pesticides is termed as chemical control.

## 1. Attractants

Chemicals that cause insects to make oriented movements towards their source are called insect attractants. They influence both gustatory (taste) and olfactory (smell) receptors.

### Types of Attractants

1. **Pheromones :** Pheromones are chemicals secreted into the external environment by an animal which elicit a specific reaction in a receiving individual of the same species.

2. **Food lures :** Chemical present in plants that attract insect for feeding. They stimulate olfactory receptors.

3. **Oviposition lures :** These are chemicals that govern the selection of suitable sites for oviposition by insects. For example extracts of corn attracts Helicoverpa armigera for egg laying on any treated surface.

## Use of Attractants in IPM

Insect attractants are used in 3 ways in pest management.

(a) Sampling and monitoring pest population.

(b) Luring pests to insecticide coated traps or poison baits.

### *Examples of poison baits*

- **For biting insects:** Moistened Bran + molasses) + insecticides
- **For sucking insects:** Sugar solution + insecticide
- **For fruitflies:** Trimedlure/ Cuelure/ Methyl eugenol + insecticides
- **For cockroaches:** Sweet syrup + white or yellow phosphorus
- **For sweet-loving ants:** Thallous sulphste + sugar + honey + glycerine + water
- **For meat loving ants:** Thallous sulphate + peanut butter

(c) In distracting insects from normal mating, aggregation, feeding or oviposition The female insects if lured to wrong plants for egg laying, the emerging larva will starve to death Advantage of using attractants is that they are specific to target insects and NE not affected. But they cannot be relied as the sole method of control and can only be included in IPM as a component.

## 2. Repellents

Chemicals that induce avoiding (oriented) movements in insects away from their source are called repellents. They prevent insect damage to plants or animals by rendering them unattractive, unpalatable or offensive.

## TYPES OF REPELLENTS

## 1. Physical repellents

Produce repellence by physical means

(a) **Contact stimuli repellents:** Substances like wax or oil when applied on leaf surface changes physical texture of leaf which are disagreeable to insects.

(b) **Auditory repellents:** Amplified sound is helpful in repelling mosquitoes.

(c) **Barrier repellents:** Tar bands on trees and mosquito nets are examples.

(d) **Visual repellents:** Yellow light acts as visual repellents to some insects.

(e) **Feeding repellents:** Antifeedants are feeding repellents. They inhibit feeding.

## 2. Chemical repellents

(a) **Repellents of Plant origin:** Essentials oils of Citronella, Camphor and cedarwood act as repellents. Commercial mosquito repellent 'Odomos' uses citronella oil extracted from lemongrass, Andrpogon pardus as repellent. Pyrethrum extracted form Chrysanthemum is a good repellent and has been used against tsetse fly, Glossina morsitans.

(b) **Synthetic repellents:** Repellents synthetically produced.

**Uses of repellents:**

- They can be applied on body to ward off insects.
- Used as fumigants in enclosed area.
- Used as sprays on domestic animals.
- To drive away insects from their breeding place.

## 3. Insecticides

OC, OP, carbamates, pyrethroids, etc.

**Groups of pesticides :** The pesticides are generally classified into various groups based on pest organism against which the compounds are used, their chemical nature, mode of entry and mode of action.

1. **Based on organisms**

   (a) **Insecticides :** Chemicals used to kill or control insects, e.g., endosulfan, malathion

   (b) **Rodenticides :** Chemicals exclusively used to control rats, e.g., Zinc phosphide

   (c) **Acaricides :** Chemicals used to control mites on crops / animals, e.g., Dicofol

   (d) **Avicides :** Chemicals used to repel the birds, e.g., Anthraquionone

(e) **Molluscicides :** Chemicals used to kill the snails and slugs, e.g., Metaldehyde

(f) **Nematicides :** Chemicals used to control nematodes, e.g., Ethylene dibromide

(g) **Fungicides :** Chemicals used to control plant diseases caused by fungi, e.g., Copper oxy cholirde

(h) **Bactericide :** Chemicals used to control the plant diseases caused by bacteria, e.g., Streptomycin sulphate

(i) **Herbicide :** Chemicals used to control weeds, e.g., 2, 4, –D

2. **Based on mode of entry**

(a) **Stomach poison :** The insecticide applied in the leaves and other parts of the plant when ingested, act in the digestive system of the insect and bring about kill, e.g., Malathion.

(b) **Contact Poison :** The toxicant which brings about death of the pest species by means of contact, e.g., Fenvalerate.

(c) **Fumigant :** Toxicant enter in vapour form into the tracheal system (respiratory poison) through spiracles, e.g., Aluminium phosphide.

(d) **Systemic poison :** Chemicals when applied to plant or soil are absorbed by foliage or roots and translocated through vascular system and cause death of insect feeding on plant, e.g., Dimethoate.

3. **Based on mode of action**

(a) **Physical poison :** Toxicant which brings about kill of one insect by exerting a physical effect, e.g., Activated clay.

(b) **Protoplasmic poison :** Toxicant responsible for precipitation of protein, e.g., Arsenicals.

(c) **Respiratory poison :** Chemicals which inactivate respiratory enzymes, e.g., hydrogen cyanide.

(d) **Nerve poison :** Chemicals inhibit impulse conduction, e.g., Malathion.

(e) **Chitin inhibition :** Chemicals inhibit chitin synthesis, e.g., Diflubenzuron.

4. **Based on chemical nature**

(i) Inorganic pesticides

Inorganic chemicals used as insecticides. For example, Arsenic, Fluorine, Sulphur, lime sulphur (Insecticides) zinc phosphide (Rodenticide)

(ii) Organic pesticides

Organic compounds (constituted by C, H, O and N mainly)

(a) Hydrocarbon oil or Petroleum oil : For example, Coal tar oil, kerosine etc.,

(b) Animal origin insecticides : For example, Nereistoxin extracted from marine annelids - commercially available as cartap, padan.

(c) Plant origin insecticides : Nicotine from tobacco plants, pyrethrum from Chrysanthemum flowers, Rotenoids from roots of Derris and Lonchocarpus Neem - azadirachtin, Pongamia glabra, Garlic etc.,

d. Synthetic organic compounds : These organic chemicals are synthetically produced in laboratory.

(i) Chlorinated hydrocarbon or organochlorines. For example, Endosulfan, Lindane, Dicofol

(ii) Cyclodienes. For example, Chlordane, Heptachlor (Banned chemicals)

(iii) Organophosphates : (Esters of phosphoric acid). For example, Dichlorvos, Methyl parathion, Dimethoate, Malathion, Acephate, Chlorpyriphos

(iv) Carbamates: (Derivatives of carbamic acid). For example, Carbaryl, Carbofuran, Carbosulfan

(v) Synthetic pyrethroids; (Synthetic analogues of pyrethrum). For example, Allethrin, Cypermethrin, Fenvalerate

(vi) Miscellaneous compounds

a. Neonicotinoids (Analogues of nicotine). For example, Imidacloprid

b. Spinosyns (Isolated from actinomycetes). For example, Spinosad

c. Avermectins (Isolated from bacteria). For example, Avermectin, Vertimec

d. Fumigants, For example, Aluminium phosphide, Hydrogen cyanide, EDCT

## NEWER INSECTICIDES / COMPOUNDS

### I. Naturalytes

A. **Avermectins :** They are discovered from Streptomyces avermetilis by Merck & Co. The analogue Avermectin $B_1$ (Commercially available as Abamectin) is insecticidally most active (systemic).

B. **Spinosyns :** In 1994 Dow Elango - announced a new class of insect control active molecules called 'spinosyns'. They are naturally derived from a new species of Actinomycetes,

Saccharopolyspora spinosa. Commercially available as spinosad. It shows both contact and stomach activity against different types of insects. Spinosad causes persistent activation of Ach receptors in the insect nervous system.

C. **Cartap hydrochloride :** It is extracted from a marine annelid, Lumbriconereis heteropoda. It has systemic, contact and stomach poisons. It is effective against chewing and sucking pests. Commercially available as Caldan 50SP.

II.  **Neo nicotinoids**

(a)  Chlornicotynyl compounds.

The chemical Imidocloprid (Bayer) is available as Goucho 70WS for seed treatment and Confidor 200SL for spray application.

(b)  Thionictoynyl compounds.

The chemical Thiomethozam (Syngenta) is available as Cruiser 70WS for seed treatment and Actara 25WG for spray application.

(c)  Thionicotoynyl compounds : Chemical is yet to come in this group.

**MOA :** Neo nicotinoids bind the receptor portion of synapse

## HAZARDS CAUSED BY PESTICIDES

1.  In Kerala, in 1953, 108 people died due to parathion poisoning.

2.  'Bhopal Gas Tradedy' in 1984 at Bhopal where the gas called Methyl isocyanate (MIC) (an intermediate involved in manufacture of carbaryl) leaked killing 5000 people and disabling 50,000 people. Totally 2,00,000 persons were affected. Long term effects like mutagenic and carinogenic effects are felt by survivors.

3.  Cases of Blindness, Cancer, Liver and Nervous system diseases in cotton growing areas of Maharashtra where pesticides are used in high quantity.

4.  Psychological symptoms like anxiety, sleep disturbance, depression, severe head ache in workers involved in spraying DDT, malathion regularly.

5.  Endosulfan - causing problem due to aerial spraying in cashew in Kerala - recent controversy - yet to be studied in detail.

**Antidotes**

| S.No. | Antidote/Medicine | Used in poisoning due to |
|---|---|---|
| 1. | Common salt (Sodium chloride) | Stomach poison in general |
| 2. | Activated charcoal (7g) in warm Magnesium oxide (3.5 g) water Tannic acid (3.5 g) | Stomach poison in general |
| 3. | Gelatin (18 g in water) or Flour or milk power (or) Sodium thiosulphate | Stomach poison in general |
| 4. | Calcium gluconate | Chlorinated insecticide, Carbon tetra-chloride, ethylene dichloride, Mercurial compound |
| 5. | Phenobarbital or Pentobarbital intravenous administration | Stomach poison of chlorinated hydrocarbon insecticides |
| 6. | Sodium bicarbonate | Stomach poison of organophosphate compounds |
| 7. | Atropine sulphate (2-4 mg intramuscular/intravenous administration) or PAM (Pyridine-Z aldoxime-N-methliodide) | Organophosphate Compounds |
| 8. | Atropine sulphate (2-4 mg intra-muscular/intravenous administration) | Carbamates |
| 9. | Phenobarbital | Synthetic pyrethoid |
| 10. | Potassium permanganate | Nicotine, Zinc phosphide |
| 11. | Vitamin $K_1$ and $K_2$ | Warfarin, Zinc phosphide |
| 13. | Epinephrine | Methlyl bromide |
| 14. | Methyl nitrite ampule | Cyanides |

## 4. INSECT GROWTH INHIBITORS

Insect Growth Regulators (IGRs) are compounds which interfere with the growth, development and metamorphosis of insects. IGRs include synthetic analogues of insect hormones such as ecdysoids and juvenoids and non-hormonal compounds such as precocenes (Anti-JH) and chitin synthesis inhibitors.

Natural hormones of insects which play a role in growth and development are:

1. **Brain hormone:** The are also called activation hormone (AH). AH is secreted by neuro secretory cells (NSC) which are neurons of central nervous system (CNS). It's role is to activate the corpora allata to produce juvenile hormone (JH).

2. **Juvenile hormone (JH):** Also called neotinin. It is secreted by corpora allata which are paired glands present behind insect brain. Their

role is to keep the larva in juvenile condition. JH I, JH II, JH III and JH IV have been identified in different groups of insects. The concentration of JH decreases as the larva grows and reaches pupal stage. JH I, II and IV are found in larva while JH III is found in adult insects and are important for development of ovary in adult females.

3. **Ecdysone:** Also called Moulting hormone (MH). Ecdysone is a steroid and is secreted by Prothoracic Glands (PTG) present near prothoracic spiracles. Moulting in insects is brought about only in the presence of ecdysone. Ecdysone level decreases and is altogether absent in adult insects.

## IGRs used in Pest Management

(a) **Ecdysoids:** These compunds are synthetic analogues of natural ecdysone. When applied in insects, kill them by formation of defective cuticle. The development processes are accelerated bypassing several normal events resulting in integument lacking scales or wax layer.

(b) **Juvenoids (JH mimics) :** They are synthetic analogues of Juvenile Hormone (JH). They are most promising as hormonal insecticides. JH mimics were first identified by Williams and Slama in the year 1966. They found that the paper towel kept in a glass jar used for rearing a Pyrrhocoris bug caused the bug to die before reaching adult stage. They named the factor from the paper as 'paper factor' or 'juvabione'. They found that the paper was manufactured from the wood pulp of balsam fir tree (Abies balsamea) which contained the JH mimic.

Juvenoids have anti-metamorphic effect on immature stages of insect. They retain status quo in insects (larva remains larva) and extra (super numerary) moultings take place producing super larva, larval-pupal and pupal-adult intermediates which cause death of insects. Juvenoids are larvicidal and ovicidal in action and they disrupt diapause and inhibit embryogenesis in insects. Methoprene is a JH mimic and is useful in the control of larva of hornfly, stored tobacco pests, green house homopterans, red ants, leaf mining flies of vegetables and flowers

(c) **Anti-JH or Precocenes:** They act by destroying corpora allata and preventing JH synthesis. When treated on immature stages of insect, they skip one or two larval instars and turn into tiny precocious adults. They can neither mate, nor oviposit and die soon. For example, EMD, FMev, and PB (Piperonyl Butoxide)

(d) **Chitin Synthesis inhibitors:** Benzoyl phenyl ureas have been found to have the ability of inhibiting chitin synthesis in vivo by blocking the activity of the enzyme chitin synthetase.

1. **Diflubenzuron (Dimilin) :** Inhibits chitin synthesis and thus affects the moulting effectice agaisnt Lepidoptean and coleopteran. Insects.

2. **Buprofezin (Applaud) :** Mould inhibitors effective agaisnt sucking pests (BPH)

3. **Lufenuron :** Available as Match 5 EC or 'Rimon' 10 EC (Especially for Helocverpa and Diamond Black moth)

The effects they produce on insects include

- Disruption of moulting
- Displacement of mandibles and labrum
- Adult fails to escape from pupal skin and dies
- Ovicidal effect.

**IGRS from Neem :** Leaf and seed extracts of neem which contains azadirachtin as the active ingredient, when applied topically causes growth inhibition, malformation, mortality and reduced fecundity in insects.

**Hormone mimics from other living organisms :** Ecdysoids from plants (Phytoecdysones) have been reported from plants like mulberry, ferns and conifers. Juvenoids have been reported from yeast, fungi, bacteria, protozoans, higher animals and plants.

## Advantages of Using IGRs

- Effective in minute quantities and so are economical
- Target specific and so safe to natural enemies
- Bio-degradable, non-persistent and non-polluting
- Non-toxic to humans, animals and plants

## Disadvantages

- Kills only certain stages of pest
- Slow mode of action
- Since they are chemicals possibility of build-up of resistance
- Unstable in the environment

1. **Chemosterilants :** Any chemical which interfere with the reproductive capacity of an insect.

    (a) *Alkylating agents :* They inhibit nucleic acid synthesis inhibit gonad development produce mutagenic effect, e.g., TEPA, Chloro ethylamine.

    (b) *Antimetabolites :* Chemicals having structural similarity to biologically active substances. They interfere with nucleic acid synthesis, e.g., 5-Fluororacil, Amithopterin

2. **Antifeedants :** Antifeedants are chemicals that inhibit feeding in insects when applied on the foliage (food) without impairing their appetite and gustatory receptors or driving (repelling) them away from the food. They are also called gustatory repellents, feeding deterrents and rejectants. Since do not feed on trated surface they die due to starvation.

## Groups of Antifeedants

1. **Triazenes:** AC 24055 has been the most widely used triazene which is a oduorless, tasteless, non-toxic chemical which inhibit feeding in chewing insects like caterpillars, cockroaches and beetles.

2. **Organotins:** They are compounds containing tin. Triphenyl tin acetate is an important antifeedants in this group effective against cotton leaf worm, Colarado potato beetle, caterpillars and grass hoppers.

3. **Carbamates:** At sublethal doses thiocarbamates and phenyl carbamates act as antifeedants of leaf feeding insects like caterpillars and Colarado potato beetle. Baygon is a systemic antifeedants against cotton boll weevil.

7. **Botanicals:** Antifeedants from non-host plants of the pest can be used for their control The following antifeedants are produced from plants.

   (a) **Pyrethrum:** Extracted from flowers of Chrysanthemum cinerarifolium acts as antifeedants at low doses against biting fly, Glossina sp.

   (b) **Neem:** Extracted from leaves and fruits of neem (Azadirachta indica) is an antifeedant against many chewing pests and desert locust in particular

   (c) **Apple factor:** Phlorizin is extracted from apple which is effective against non-apple feeding aphids.

   (d) **Solanum alkaloids:** Leptine, tomatine and solanine are alkaloids extracted from Solanum plants and are antifeedants to leaf hoppers.

## BEHAVIOURAL METHODS

1. Pheromones
2. Allelochemics

Semiochemicals are chemical substances that mediate communication between organisms. Semiochemicals may be classified into Pheromones (intraspecific semiochemicals) and Allelochemics (interspecific semiochemicals). Pheromones are chemicals secreted into the external environment by an animal which elicit a specific reaction in a receiving individual of the same species. Pheromones are volatile in nature and they aid in communication among insects. Pheromones are exocrine in origin (i.e., secreted outside the body). Hence they were earlier called as ectohormones. In 1959, German chemists Karlson and Butenandt isolated and identified the first pheromone, a sex attractant from silkworm moths. They coined the term pheromone. Since this first report, hundreds of pheromones have been identified in many organisms. The advancement made in analytical chemistry aided pheromone research.

Based on the responses elicited pheromones can be classified into 2 groups

(a) **Primer pheromones:** They trigger off a chain of physiological changes in the recipient without any immediate change in the behaviour. They act through gustatory (taste) sensilla. For example, Caste determination and reproduction in social insects like ants, bees, wasps, and termites are mediated by primer pheromones. These pheromones are not of much practical value in IPM.

(b) **Releaser pheromones:** These pheromones produce an immediate change in the behaviour of the recipient. Releaser pheromones may be further subdivided based on their biological activity into

1. **Pheromone:** Semiochemical used for intraspecific communication which is an exocrine secretion that causes specific reaction in the receiving individuals of the same species.

2. **Sex pheromone:** Female produce to attract males, e.g., Bombyco (Bombyx mori) Cyplure (gypsy moth) and Gossyplure (Pink boll worm) ( In American boll weevil males produce)

3. **Alarm pheromone:** Semiochemicals used to warm other fellow individuals from mandibular glands or anal glands. For example, honey bees (E) B. Farnesene aphids.

4. **Trailmarking pheromone:** Semiochemicals used in route perception. For example, Ants, termites.

5. Aggregation pheromone- Semiochemicals which attract other fellow members to a particular spot. For example, Ferrolure of red palm weevil.

Releaser pheromones act through olfactory (smell) sensilla and directly act on the central nervous system of the recipient and modify their behaviour. They can be successfully used in pest management programmes.

(1) Sex pheromones are released by one sex only and trigger behaviour patterns in the other sex that facilitate in mating. They are most commonly released by females but may be released by males also. In over 150 species of insects, females have been found to release sex pheromones and about 50 species males produce.

Aphrodisiacs are substances that aid in courtship of the insects after the two sexes are brought together. In many cases males produce aphrodisiacs.

Major differences between male and female produced pheromones are listed.

## Insect orders producing sex pheromones

Lepidoptera, Orthoptera, Dictyoptera, Diptera, Coleoptera, Hymenoptera, Hemiptera, Neuroptera and mecoptera. In Lepidoptera, sex pheromonal system is highly evolved.

## Pheromone producing glands

In Lepidoptera they are produced by eversible glands at the tip of the abdomen of the females. The posture shown during pheromone release is called 'calling position'. Aphrodisiac glands of male insects are present as scent brushes (or hair pencils) at the tip of the abdomen, (e.g., Male butterfly of Danaus sp.).

Andraconia are glandular scales on wings of male moths producing aphrodisiacs.

The following are some of the female sex pheromones identified in insects.

| S. No. | Name of the Insect | Pheromone |
|--------|--------------------|-----------|
| 1. | Silkworm, Bombyx mori | Bombykol |
| 2. | Gypsy moth, Porthesia dispar | Gyplure, disparlure |
| 3. | Pink bollworm, Pectinophora gossypiella | Gossyplure |
| 4. | Cabbage looper, Trichoplusia ni | Looplure |
| 5. | Tobacco cutworm, Spodoptera litura | Spodolure, litlure |
| 6. | Gram pod borer, Helicoverpa armigera | Helilure |
| 7. | Honey bee queen, Apis sp. | Queen's substance |

## Examples of male sex pheromones

Cotton boll weevil, Anthonomas grandis, Coleoptera

Cabbage looper, Trichoplusia ni, Lepidoptera

Mediterranean fruitfly, Ceratitis capitata, Diptera.

## Pest Management With Sex Pheromones

Synthetic analogues of sex pheromones of quite large No. of pests are now available for use in Pest management. Sex pheromones are being used in pest management in three different ways.

(a)  In sampling and detection (Monitoring)

(b)  To attract and kill (Mass trapping)

(c)  To disrupt mating (Confusion or Decoy method)

### (a) In sampling and detection (Monitoring) :

Pheromones can be used for monitoring pest incidence/ outbreak in the following ways.

The Lures are available for following pests

1.  Helicoverpa armigera        - Heli lure

2.  Tobacco caterpillar S. litura  - Spode lure/Pherodin SL

3.  Pink boll worm Pectinphora gossypiella - Pectinolure/Gossyplure

4.  Rhionceros beetle Orycetes rhinoceros - Sime RB or Rhinolure

5.  Red palm weevil - Rhynchophorus ferrungineus - Ferrolure

6.  Spoted boll worm Eavis - Erin lure.

7.  ? Farnescene (EBF) has been identified as alarm pheromone of aphids - Aphis gossypii.

The number of traps required for monitoring is 12 /ha.

Set up at 1-2' inch above the crop level

## Types of pheromone trap

1.  Funnel trap for mamy insects

2.  Sticky trap / delta trap for pink boll worm

3.  Bucket trap - red palm weevil & rhinoceros beetle

## Allelochemicals

**Definition :** Chemicals that deliver behavioural messages which act either interspecial or intraspecifically.

1.  **Allomone -** Interspecific semiochemical that favours the produces, e.g., Repellents, Deterrents (feeding and ovipositional)

2.  **Kairomone -** Interspecific semiochemical that favours the receiver, e.g., attractants "Food love".

3.  **Synamone -** Interspecific semiochemical that favours both the producer and receiver, e.g., Plant odours attracting natural enemies of pests.

4.  **Apneumone -** Chemical from non-living materials eliciting behavioural response. For example, Fish meal attracting sorghum shoot fly.

### Regulatory/legal method

- Plant/animal quarantine
- Eradication and suppression programme

### Genetic/biotechnology method

- Release of genetically incompatible/sterile pests
- Transgenic plant

## HOST PLANT RESISTANCE IN IPM (HPR)

- HPR is a very important component of IPM.

- Selection and growing of a resistant variety minimise cost on all other pest management activities.

## Definition

"Those characters that enable a plant to avoid, tolerate or recover from attacks of insects under conditions that would cause greater injury to other plants of the same species" (Painter, R.H., 1951).

"Those heritable characteristics possessed by the plant which influence the ultimate degree of damage done by the insect" (Maxwell, F.G., 1972).

## Types of Resistance

**Ecological Resistance or Pseudo resistance :** Apparent resistance resulting from transitory characters in potentially susceptible host plants due to environmental conditions.

Pseudoresistance may be classified into 3 categories

(a) **Host evasion**

Host may pass through the most susceptible stage quickly or at a time when insects are less or evade injury by early maturing. This pertains to the whole population of host plant.

(b) **Induced Resistance**

Increase in resistance temporarily as a result of some changed conditions of plants or environment such as change in the amount of water or nutrient status of soil.

(c) **Escape**

Absence of infestation or injury to host plant due to transitory process like incomplete infestation. This pertains to few individuals of host.

# GENETIC RESISTANCE

A. **Based on number of genes**
- **Monogenic resistance:** Controlled by single gene
  Easy to incorporate into plants by breeding
  Easy to break also
- **Oligogenic resistance:** Controlled by few genes
- **Polygenic resistance:** Controlled by many genes
- **Major gene resistance:** Controlled by one or few major genes (vertical resistance)
- **Minor gene resistance:** Controlled by many minor genes. The cumulative effect of minor genes is called adult resistance or mature resistance or field resistance. Also called horizontal resistance.

B. **Based on biotype reaction**
- **Vertical resistance:** Effective against specific biotypes (specific resistance).
- **Horizontal resistance:** Effective against all the known biotypes (Non- specific resistance).

C. **Based on population/Line concept**
- **Pureline resistance:** Exhibited by liens which are phenotypically and genetically similar.
- **Multiline resistance:** Exhibited by lines which are phenotypically similar but genotypically dissimilar.

D. **Miscellaneous categories**
- **Cross resistance:** Variety with resistance incorporated against a primary pest, confers resistance to another insect.
- **Multiple resistance:** Resistance incorporated in a variety against different environmental stresses like insects, diseases, nematodes, heat, drought, cold, etc.

E. **Based on evolutionary concept**
- **Sympatric resistance:** Acquired by coevolution of plant and insect (gene for gene) Governed by major genes.
- **Allopatric resistance:** Not by co-evolution of plant and insect. Governed by many genes.

# MECHANISMS OF RESISTANCE

The three important mechanisms of resistance are

- Antixenosis (Non-preference)

- Antibiosis
- Tolerance

(a) **Antixenosis :** Host plant characters responsible for non-preference of the insects for shelter, oviposition, feeding, etc. It denotes presence of morphological or chemcial factor which alter insect behaviour resulting in poor establishment of the insect, e.g., Trichomes in cotton - resistant to whitefly, Wax bloom on carucifer leaves - deter feeding by DBM, Plant shape and colour also play a role in non-preference, Open panicle of sorghum - Supports less Helicoverpa.

(b) **Antibiosis :** Adverse effect of the host plant on the biology (survival, development and reproduction) of the insects and their progeny due to the biochemical and biophysical factors present in it.

Manifested by larval death, abnormal larval growth, etc.

Antibiosis may be due to :

- Presence of toxic substances
- Absence of sufficient amount of essential nutrients
- Nutrient imbalance/improper utilization of nutrients

Chemical factors in Antibiosis - Examples

| *Chemicals present in plants* | *Imparts resistance against* |
|---|---|
| 1. DIMBOA (Dihydroxy methyl benzoxazin) | Against European corn borer, Ostrinia nubilalis |
| 2. Gossypol (Polyphenol) | Helicoverpa armigera (American bollworm) |
| 3. Sinigrin | Aphids, Myzus persicae |
| 4. Cucurbitacin | Cucurbit fruit flies |
| 5. Salicylic acid | Rice stem borer |

(c) **Tolerance**

Ability to grow and yield despite pest attack. It is generally attributable to plant vigour, regrowth of damaged tissue, to produce additional branches, compensation by growth of neighbouring plants.

## Compatibility of HPR in IPM

(a) Compatability with chemical control
- HPR enhances efficacy of insecticides
- Higher mortality of leaf hoppers and plant hoppers in resistant variety compared to susceptible variety
- Lower concentration of insecticide is sufficient to control insects on resistant variety

(b) Compatibility with biological control

- Resistant varieties reduce pest numbers - thus shifting pest: Predatory (or parasitoid) ratio favourable for biological control, e.g., Predatory activity of mirid bug Cyrtorhinus lividipennis on BPH was more on a resistant rice variety IR 36 than susceptible variety IR 8

- Insects feeding on resistant varieties are more susceptible to virus disease (NPV)

(c) Compatibility with cultural method

- Cultural practices can help in better utilization of resistant varieties., e.g., use of short duration, pest resistant plants effective against cotton boll weevil in USA.

## Advantages of HPR as a Component in IPM

1. **Specificity:** Specific to the target pest. Natural enemies unaffected.

2. **Cumulative effect:** Lasts for many successive generations.

3. **Eco-friendly:** No pollution. No effect on man and animals.

4. **Easily adoptable:** High yielding insect resistant variety easily accepted and adopted by farmers. Less cost.

5. **Effectiveness:** Res. variety increases efficacy of insecticides and natural enemies.

6. **Compatability:** HPR can be combined with all other components of IPM.

7. **Decreased pesticide application:** Resistant varieties requires less frequent and low doses of insecticides.

8. **Persistence:** Some varieties have durable resistance for long periods.

9. **Unique situations:** HPR effective where other control measures are less effective, for example,

   (a) When timing of application is critical.

   (b) Crop of low economic value.

   (c) Pest is continuously present and is a single limiting factor.

## Disadvantages of HPR

1. **Time consuming:** Requires from 3-10 years by traditional breeding programmes to develop a res. variety.

2. **Biotype development:** A biotype is a new population capable of damaging and surviving on plants previously resistant to other population of same species.

3. **Genetic limiation:** Absence of resistance genes among available germination.

## Plant Quarantine

Legal restriction of movement of plant materials between countries and between states within the country to prevent or limit introduction and spread of pests and diseases in areas where they do not exist.

## Exotic Pests - Pests Accidentally Introduced into India

1. Pink bollworm - Pectinophora gossypiella
2. Cotton cushion scale - Icerya purchasi
3. Wooly aphid of apple - Aphelinus mali
4. SanJose scale - Quadraspidiotus perniciosus
5. Potato tuber moth - Gnorimoschima operculella
6. Cyst (Golden) nematode of potato - Globodera sp.
7. Giant african snail - Acatina fullica
8. Subabul psyllid - Heteropsylla cubana
9. Bunchytop disease of banana
10. Spinalling whitefly - Aleyrodicus dispersus

## Foreign Pests From Which India is Free

1. Mediterranean fruitfly - Ceratitis capitata
2. Grapeavine phylloxera
3. Cotton boll weevil - Anthonomos grandis
4. Codling moth of apple - Lasperysia pomonella

## RESISTANCE

Resistance may be defined as 'a heritable change in the sensitivity of a pest population that is reflected in the repeated failure of a product to achieve the expected level of control when used according to the label recommendation for that pest species'. Cross-resistance occurs when resistance to one insecticide confers resistance to another insecticide, even where the insect has not been exposed to the latter product. Clearly, because pest insect populations are usually large in size and they breed quickly, there is always a risk that insecticide resistance may evolve, especially when insecticides are misused or over-used.

## INSECTICIDE RESISTANCE

Insecticide resistance is the development of an ability to tolerate a dose of insecticide, which would prove lethal (kill) to majority of the individuals of the same species.

This ability is due to the genetic change in pest population in response to pesticide application.

Pesticide application can artificially select for resistant pests. In this diagram, the first generation happens to have an insect with a heightened resistance to a pesticide (red). After pesticide application, its descendants represent a larger proportion of the population because sensitive pests (white) have been selectively killed. After repeated applications, resistant pests may comprise the majority of the population.

Pesticide resistance is the adaptation of pest species targeted by a pesticide resulting in decreased susceptibility to that chemical. In other words, pests develop a resistance to a chemical through selection; after they are exposed to a pesticide for a prolonged period it no longer kills them as effectively. The most resistant organisms are the ones to survive and pass on their genetic traits to their offspring.

More specific definitions of pesticide resistance often apply to particular classes of pesticides. Manufacturers of pesticides tend to prefer a definition that is dependent on failure of a product in a real situation, sometimes called field resistance. For example, the Insecticide Resistance Action Committee (IRAC) definition of insecticide resistance is 'a heritable change in the sensitivity of a pest population that is reflected in the repeated failure of a product to achieve the expected level of control when used according to the label recommendation for that pest species'.

Resistance to insecticides was first documented by A.L. Melander in 1914 when scale insects demonstrated resistance to an inorganic insecticide. Between 1914 and 1946, 11 additional cases of resistance to inorganic insecticides were recorded. The development of organic insecticides, such as DDT, gave hope that insecticide resistance was an issue of the past. Unfortunately, by 1947 housefly resistance to DDT was documented. With the introduction of every new insecticide class - cyclodienes, carbamates, formamidines, organophosphates, pyrethroids, even Bacillus thuringiensis - cases of resistance surfaced within two to 20 years.

- In the US, studies have shown that fruit flies that infest orange groves were becoming resistant to malathion, a pesticide used to kill them.

- In Hawaii and Japan, the diamondback moth developed a resistance to Bacillus thuringiensis about three years after it began to be used heavily.

- In England, rats in certain areas have developed such a strong resistance to rat poison that they can consume up to five times as much of it as normal rats without dying.

- DDT is no longer effective in preventing malaria in some places, a fact which contributed to a resurgence of the disease.

- In the southern United States, the weed Amaranthus palmeri, which interferes with production of cotton, has developed widespread resistance to the herbicide Roundup.

- Rachel Carson predicted the phenomenon in her 1962 book Silent Spring.

## Insecticide Resistance in Insect Pests in India

| Name of pest | Common name | Insecticides to which resistant |
| --- | --- | --- |
| 1. *Aphis craccivora* | Aphid | Carbamates, OP, Cypermethrin, Endosulfan, Monocrotophos |
| 2. *Bemesia tabaci* | Whitefly | |
| 3. *Helicoverpa armigera* | Cotton boll worm | OP, Synthetic pyrethroid, Bacillus thuringiensis |
| 4. *Plutella xylostella* | Diamond back moth on cabbage, cauliflower | Abamectin, Bt, OP compounds |

**Simple resistance :** Insect develops resistance only against the insecticide to which it is exposed.

Multiple resistance is the phenomenon in which a pest is resistant to more than one class of pesticides. This can happen if one pesticide is used until pests display a resistance and then another is used until they are resistant to that one, and so on.

Cross resistance, a related phenomenon, occurs when the genetic mutation that made the pest resistant to one pesticide also makes it resistant to other pesticides, especially ones with similar mechanisms of action or ones in the same class.

## PEST RESURGENCE

Tremendous increase in pest population brought about by insecticides despite good initial reduction in pest population at the time of treatment.

Insecticides lead to pest resurgence in two ways.

(i)  After initial decline, resistant population increase in large numbers

(ii)  Killing of natural enemies of pest, cause pest increase, e.g., Quinalphos, phorate - Cause resurgence of BPH in rice, Carbofuran - Leaf folder in rice.

## IMPACT OF GLOBAL WARMING ON PESTS

### Effect of global warming on world and agriculture

-  Increase in overall temperature on earth, e.g., Earth's surface temperature has increased 1.4oF in lst one century (Forecast: 5°F rise in next century)
-  Change in climate tremendously
-  Melting of ice in Polar region
-  Increase in seas level and submerging of coastal areas
-  Flooding and intense down pours
-  Drought in warmer regions

### Impact of global warming on pest status

Temperature playes a key role in the life cycle of insects and global warming also has a definite effect of insect-pests.

1.  Due to change in climate, there will be overall effect on survival and distribution on insect-pests.
2.  Introdution and expansion of exotic/invasive/alien insects.
3.  Native species will be displaced or killed by the invaders.
4.  Life cycle of the insects will be shortened and there will be more generations per year leading to more insects and more loss.
5.  Fecundity rate willl also increase.
6.  Insects will eat more leading to more loss.
7.  There will be migration of insect pests and behavioural change in insects.

### PEST MONITORING, SURVER AND SURVILLENCE

### Pest Monitoring

Monitoring phytophagous insects and their natural enemies is a fundamental tool in IPM - for taking management decision. Monitoring -

estimation of changes in insect distribution and abundance - information about insects, life history - influence of biotic and abiotic factors on pest population.

## Pest Surveillance

Refers to the constant watch on the population dynamics of pests, its incidence and damage on each crop at fixed intervals to forewarn the farmers to take up timely crop protection measures.

## Objectives of Pest Surveillance

1. To know existing and new pest species.
2. To assess pest population and damage at different growth stage of crop.
3. To study the influence of weather parameters on pest.
4. To study changing pest status (Minor to major).
5. To assess natural enemies and their influence on pests effect of new cropping pattern and varieties on pest.

## Three basic components of pest surveillance

Determination of

(a) the level of incidence of the pest species
(b) the loss caused by the incidence
(c) the economic benefits, the control will provide

## PEST FORECASTING

Forecasting of pest incidence or outbreak based on information obtained from pest surveillance.

## Uses

- Predicting pest outbreak which needs control measure
- Suitable stage at which control measure gives maximum protection

## Two types of pest forecasting

(a) Short term forecasting - Based on 1 or 2 seasons
(b) Long term forecasting - Based on affect of weather parameters on pest

## SURVEY

Conducted to study the abundance of a pest species.

## Two types of survey - Roving survey and fixed plot survey

## Roving Survey
- Assessment of pest population/damage from randomly selected spots representing larger area
- Large area surveyed in short period
- Provides information on pest level over large area

## Fixed Plot Survey

Assessment of pest population/damage from a fixed plot selected in a field.

The data on pest population/damage recorded periodic from sowing till harvest, e.g., sq.m. plots randomly selected from 5 spots in one acre of crop area in case of rice. From each plot 10 plant selected at random. Total tillers and tillers affected by stem borer in these 10 plants counted. Total leaves and number affected by leaf folder observed. Damage expressed as per cent damaged tillers or leaves. Population of BPH from all tillers in 10 plants observed and expressed as number/tiller.

Qualitative survey - Useful for detection of pest

Quantitative survey - Useful for enumeration of pest

## Sampling Techniques

Absolute sampling - To count all the pests occurring in a plot

Relative sampling - To measure pest in terms of some values which can be compared over time and space, e.g., Light trap catch, Pheromone trap.

## Methods of Sampling
(a) *In situ* **counts** : Visual observation on number of insects on plant canopy (either entire plot or randomly selected plot)
(b) **Knock down** : Collecting insects from an area by removing from crop and (Sudden trap) counting (Jarring)
(c) **Netting** : Use of sweep net for hoppers, odonates, grasshopper
(d) **Norcotised collection** : Quick moving insects anaesthesised and counter
(e) **Trapping - Light trap** : Phototropic insects
    *Pheromone trap - Species specific*
    *Sticky trap - Sucking insects*
    *Bait trap - Sorghum shootfly - Fishmeal trap*
    Emergence trap - For soil insects
(f) **Crop samples** : *Plant parts removed and pest counted, e.g., Bollworms.*

## Stage of Sampling

- Usually most injurious stage counted
- Sometimes egg masses counted - Practical considerations
- Hoppers - Nymphs and adult counted

## Sample Size

- Differs with nature of pest and crop
- Parger sample size gives accurate results

## Decision Making

- Population or damage assessed from the crop
- Compared with ETL and EIL
- When pest level crosses ETL, control measure has to be taken to prevent pest from reducing EIL.

# BIOTECHNOLOGICAL APPLICATIONS IN PEST MANAGEMENT

## Definition of Biotechnology

Biotechnology is the term derived through the fusion of "Biology" and "Technology".

Biotechnology may be defined as "The controlled use of biological agents, such as microorganisms, or cellular components for beneficial use" ( U.S. National Science Foundation).

Biotechnology is the integrated use of bio-chemistry, microbiology and engineering sciences in order to achieve technological and Industrial application of the capabilities of microorganisms, cultured tissue cells and parts there of (European Federation of Biotcchnology).

Any technological application that uses biological systems, living organisms or derivatives thereof, to make or modify products or processes for specific use (UN Convention on Biological Diversity, 1992).

## APPLICATION OF BIOTECHNOLOGY IN PEST MANAGEMENT

Biotechnology has enormous potential to provide a wave of new products for pest management. Biotechnology in fact provides several innovative, safe and effective tools that may essentially resolve pest management problems. There are several biotechnological strategies, which are being used to develop to minimize the losses caused by insect pest. These include.

1. Production of transgenic crops
2. Genetic manipulation of entomopathogens such as viruses and fungi
3. Genetic modification or engineering of natural enemies
4. Genetic modification or engineering of insects
5. Development of new insecticide molecules

## BIOAGENTS FOR ORGANIC HORTICULTURE

The organic crop management with biological devices could satisfy the actual needs of important components of biointensive integrated pest management (BIPM). Cultural control, mechanical control, NSE, parasites, predators and microbial pesticides are highly potential components of organic plant protection. Biological control is utilization parasitoids, predators and pathogens of crop pests below economic threshold level. The first sparkling success of biological control by use of 129 Vedalia, ladybird beetles (Rodolia cardinalis) against cottony cushion scale (Icerya purchasi) is well documented to safeguard the interest of citrus growers in California in 1888. In India after 1975 researches have amply proved that about 70-80% major pests of major crops could be suppressed by biocontrol. It is the ecofriendly, economic and effective method of pest suppression. The classical biological control rarely finds place in fast reviving plant protection system.

**Potent Bioagents and Biopesicides for Organic Vegetables & Fruit Crops**

| S.No. | Bioagents | Bioagent Group | Doses | Target Pest |
|---|---|---|---|---|
| **I. Tomato/Okra/Brinjal/Cauliflower** | | | | |
| 1. | *Trichogramma chilonis* Trichogrammatidae: Hymenoptera | Egg parasitoid | 5 Trichocards (Each 1cc Card) | Shoot & fruit borers |
| 2. | *Trichogramma pretisosum* Trichogrammatidae: Hymenoptera | Egg par | 5 Trichocards (Each 1cc Card) | DBM, Spodoptera |
| 3. | *Cotesia putellae* Encyrtidae Hymenoptera Telenomus remus | Larval pupal par Egg-Parasite | 50,000 adults 10,000 adults | DBM cabbage butterfly Spodoptera |
| 4. | (i) *Copidosoma koehleri* Encyrtidae Hymenoptera (ii) *Trichogramma chilonis* Trichogrammatidae Hymenoptera | Egg-Larval parasitoid Egg parasitoid | 50,000 adults 5 Trichocards (Each 1cc Card) | Potato tuber moth Cut worms. |
| **II. Coconut :** | | | | |
| 1. | *Goniozus nephantidis* Bethylidae Hymenoptera | Larval parasitoid | 500 adults | Coconut black headed caterpillar. |
| 2. | *Brcon brevicornis* Braconidae Hymenoptera | Larval parasitoid | 500 adults | Coconut black headed caterpillar. |
| 3 | *Platymaris lavuecollis* | Predator | 500 Adults | Rhinoceros Beetle |

*Contd...*

| S.No. | Bioagents | Bioagent Group | Doses | Target Pest |
|---|---|---|---|---|
| **III. Fruit crops and Polyhouse crops (Rose & Orna)** | | | | |
| 1. *Rodelia cardinalis*(Vedalia beetle) Coccinellidae : Coleoptera | Predator | Predator 150 beetles | Cottony coushion scale, Iceria purchasi |
| 2. *Cryptolaemus montrouzieri* Mulsant Coccinellidae Coleoptera | Predator | 1500 grubs or adults, 3-4 grubs/vine | Mealy bugs on fruit crops. |
| 3. *Trichogramma chilonis* Trichogrammatidae: Hymenoptera | Egg parasitoid | 5 Trichocards (Each 1cc Card) | Lepidopteran Shoot, fruit, seed borers |
| 4. *Encarsia formosa* | Parasitoid | 1000/100 sq m | Greenhouse white flies |

### Potent Biopesicides for Organic Suppression of Crop Pests and Weeds

| S.No. | Bioagents | Bioagent Group | Doses | Target Pest |
|---|---|---|---|---|
| **I. Tomato/Okra/Brinjal/Cauliflower** | | | | |
| 1. HaNPV Baculloviridae: Bacullovirus | Insect virus | 250 LE | Tomato fruit borer, H. armigera. |
| 2. *Bacillus thuringiensis* Bacillaceae: Eubacterials | Insect pathogen | 0.5-1.0 kg. | L. Arbonalrs, H. Armigera, DBM |
| 3. *Spodoptera litura* NPV (SlNPV) Baculloviridae Bacullovirus | Virus | $1 \times 10^9$ POB's /ml: 500 ml | Spodoptera on vegetables. |
| 4. *Hirsutella thompsoni* Moniliaceae: Moniliales | Fungal pathogen | 2 kg | Mites |
| **II. Coconut** | | | | |
| 1. *Metarrhyzium anisopliae* Moniliaceae Moniliales | Fungal pathogen | 80 ml/cubic meter manure | Rhinoceros beetle Oryctes rhinocerous |
| 2. *Baculovirus oryctes* Baculovidae: Bacculovirus | Insect virus | $1 \times 10^9$ POB's/ ml: 500 ml | White grub. |
| **III. Fruit crops and Polyhouse crops (Rose & Orna)** | | | | |
| 1. Verticillium lecanii Moniliaceae: Moniliales | Fungal pathogen | $1 \times 10^8$ cfu/g 2-4 kg/ha | Scale insects, mealy bugs, white flies, thrips, mites, aphids |
| 2. *Ampelomyces quasaqalis* Moniliaceae: Moniliales | Fungal pathogen | $1 \times 10^8$ cfu/g 2-4 kg/ha | Fungal diseases of folliage |
| 3. (i) *Paeclomyces lilicacinus* (ii) *Trichoderma* spp, $1 \times 10^8$ cfu/g Moniliaceae:Moniliales (iii) *Pseudomonas fluroscence* Bacillaceae: Eubacterials | Nematode trapping fungi Bacterial pathogen | 5-10 g/kg seed or 5 kg/ha in soil as 0.5% suspension | Root-knot nematodes in various crops. Fungal & Bacterial diseases |
| 4. *Trichoderma viridae*/hargianum+ *Paeclomyces lilicacinus* $1 \times 10^8$ cfu/g | Biofungicide Cum nematicide | 5-10 g/kg seed or 5 kg/ha in soil as 0.5% suspension | Plant wilts caused by root rot fungi and root-knot mema-todes in various crops |
| 5. HaNPV Baculloviridae: Bacullovirus | Insect virus | 250 LE | Grape bunch feeder, H. armigera. |

## Conclusions and Measures to Popularize Biocontrol

1. Biological control is inevitable component of organic plant protection and 70-80% major pests of major crops are manageable by biointensive IPM.

2. Biopesticides are emerging as major driving force to chemical pesticides. Bt, fungal and viral biopesticides availability is increasing.

3. The quality control is more concern than registration of biopesticides. Establishment of 10 referral labs( one lab/3-4 districts) in SAUs in M.S.is required to test quality of biopesticides samples collected from open market by pesticide inspector with regional monitoring team charging quality control charges of Rs.200/q/annum to SSI by the State Government for production capacity of the SSI.

4. Biopesticide requirement of Maharashtra is 12000 tons against present production capacity of 3000 tons.

5. Considering past experience that biopesticides are almost safe and produced by small scale industries; set up of State Biopesticide Board and Registration Committee (SBB/RC) is necessary to speed up the registration process and to sustain mission of organic plant protection.

6. For promoting NSE use "Mahila Bachat Gats" in villages (> 5000 population) could be encouraged to establish neem seed collection, powdering and sale units providing 90% subsidy of Rs. 1.80 lakhs.

7. Distributing pamphlets/leaflets through 'Panchayats' to villagers for propaganda to conserve and augment naturally available bioagents. Planting shrubs and trees along field bunds, few rows of cowpea, maize, mustard, setaria in crops; and need based use of systemic insecticide as seed treatments and soil application.

# 6

# REASONING

## How Many Insect Species Exist?

We have compiled a lot of different insect facts and interesting information into some of the most frequently asked questions (FAQS) about insects that you will find anywhere! There are many more kinds of insects on earth than there are of any other kind of living creature! Insects have the greatest number of individuals living of at any one time. It is estimated that there are approximately 10 quintillion (10,000,000,000,000,000,000) individual insects alive. There are 1,017,018 species of insects in the world with some experts estimating that there just might be as many as 10 million species out there. That means you could spend your whole life looking at different kinds of insects and you'll never be able to see them all. It's hard to imagine, but 95% of all the animal species on the earth are insects! There are so many insects with different insect characteristics we had to limit our insect facts to the top 100 most interesting insect facts. To give you an idea just how many insects there are consider that on average there are millions of insects living on every single acre of land! Over one million species have been discovered by scientists and entomologists as they think there might be more than ten times as many insects than we currently know about which haven't even been named yet!

## Insect Orders, Families, and Species

Insects are arthropods (a type of invertebrate, animals that lack a backbone). These insect creatures are divided up into 32 orders, or groups. The largest insect order is the beetles (Coleoptera) with 125 different families and around 500,000 different species. In fact, one out of every four animals on earth is some type of beetle bug.

## Insects in Terms of Biomass Returns Surprising Results

Scientists estimate that 10% of the animal biomass of the world is composed of ants, and another 10% of termites. This means that 'social insects' could possibly make up an incredible 20% of the total animal biomass on this planet! There isn't anywhere on land where you can go and not find insects. Even on the frozen extremes of the Arctic and Antarctica you can find insects alive and active during the warmer months. Insects are ubiquitous, they are in the soil beneath your feet, in the air above your head, on and in the bodies plants and animals around you, as well as on and in you. They are incredibly adaptable creatures and have evolved to live successfully in most environments existing on earth. The only place where insects are not commonly found is in the oceans. Insects eat more plants than all other creatures on earth combined! They are also tremendously important in the breakdown of plant and animal matter. Without them, we would have a world covered in dead plants and animals! In addition to all of this insects are a major food source for a wide range of animal species.

## How Far can Insects Fly?

Painted Lady butterflies (*Vanessa cardui*) that migrate from Africa to England must travel over 3,000 miles to accomplish their task. It is said that the Painted Lady butterfly may be the most widespread butterfly in the entire world. If you are looking to have Painted Lady butterflies fill your home or office permanently then try our electronic moving butterflies for a longer lasting effect. They are so realistic that everyone will think your indoor display is made up of live butterflies! We also offer Painted Lady butterflies (Vanessa Cardui) available for butterfly releases at weddings and special events. Please make your reservations ASAP.

## Insect Longest Migration

The longest migration for an insect goes to the Desert Locust (*Schistocerca gregaria*) who travels about 2800 miles yearly.

Monarch butterflies (Danaus plexippus) migrate over 2000 miles as they fly from Canada to central Mexico every fall. Monarch butterflies (Danaus plexippus) can flyover 1000 km without stopping. Monarchs are very popular as collector items and are now available framed or as electronic moving butterflies. They are also perfect for butterfly releases! Shine a new light on any wedding and special occasion and create a memory that will last a lifetime!

## Oldest Insect Fossil

The oldest identifiable insect fossil is a 390-million year old bristletail, a primitive wingless insect from the order Thysanura found in Quebec, Canada. The insect fossil is remarkably similar to modern-day silverfish. The first flights on earth were undoubtedly made by insects!

# What Type of Insect has Been on Earth the Longest?

The oldest group of insects on Earth are the cockroaches, particularly those in the Blattaria order, as this species dates back 300 million years.

## Largest Living Insect

The largest living insects are Giant Walking Sticks from the (order Phasmida). This bug can grow to up to thirteen inches in length making it surely one of the longest insects in the world. The record holder for the longest insect goes to a specimen commonly found in Western Malaysia and Singapore called the West Malaysian Pharnacia Serratipes and it reaches an amazing length of twenty-two inches. There are approximately 3,000 tropical species of walking sticks, with only 10 species living in North America.

Although principally tropical and Asian in distribution, walking sticks are also found in the temperate regions of Europe and North America. One species, Megaphasma dentricus, is the longest insect in the United States, attaining 7 inches in length. The Malaysian Walking Leaf (Heteropteryx dilatata) is another very large insect to add to that list. Females of this species can be very large, up to 6.5 inches long which makes them half an inch short of being the longest insect. Framed walking stick insects are available in a variety of species for the discriminate insect collector.

## What was the Biggest Insect Ever?

Fossils of a prehistoric dragonfly (*Meganeura monyi*) in the order Protodonata makes it the largest insect ever to be found with a full wingspan of 30 inches and a body length of 18 inches. The Meganeura dragonfly lived about 250 million years ago until it became extinct at the end of the Paleozoic Era.

## Largest Insect Egg

The Malaysian stick insect (*Heteropteryx dilitata*) lays the largest insect egg with the egg measuring .05 inches.

## Heaviest Insect

The Goliath Beetle (*Goliathus cacicus*) from Africa weighs 1/4 pound or 100 grams and is 5 inches long. The heaviest insects belong to the same family of scarab beetles as the Eastern Hercules Beetle, which is about 2 inches long including the horn. African Goliath beetles from the family Scarabaeidae are an important part of nature's clean-up crew as they recycle "used material" keeping the earth clean. Another competitor for the title is the extemely rare the South American longhorn beetle (Tytanus giganteus) that measures 25 cm!

## Strongest Animal

Rhinoceros Beetles are the strongest animal for their relative size. The Hercules beetle (*Dynastes hercules*) can grow to a length of 6 inches and has tremendous strength. An average man can pull about 0.86 times his own weight, a Leaf Beetle (Donacia), on the other hand, can pull 42.7 times its own weight. Horses can pull half their weight, and ants can pull 52 times their weight (comparable to a human pulling 4.5 tons). If you would like a Rhinoceros Beetle of your own to hang on the wall in your home, or office, we sell them framed and mounted in elegant shadow boxes.

## Animal has the Most Legs

Millipedes have the most legs of any animal, specifically the the Illacme plenipes species from California, with its over 375 pairs of legs! Millipede means "thousand legs", but they don't have 1,000 legs. In reality millipedes have between 47 to 197 pairs of legs, depending on their species. Millipedes are invertebrates; they have a hard exoskeleton and many jointed legs. Millipedes and centipedes are believed to be more closely related to insects than to spiders. For a very unique display in your home or office use framed millipedes and framed centipedes.

## Biggest Group of Insects

The biggest group of insects are the beetles, numbering 330,000 species, making them the largest order of insects in the world. You can now start collecting framed beetles including the rhino beetle, dung beetle and many more species. The metallic-coloured wing covers of some beetles are used for jewelry. If you are a entomologist, zoologist or just a bug collector we have just the right of framed beetles and insect jewellry for you to choose from!

## Firefly

Fireflies (*Pyractomena borealis*) are also known as lightning bugs and glowworms. Fireflies are not really "flies" as entomologists know them, but instead are beetles from the family Lampyridae. There are over 2000 species of fireflies living in the tropical and temperate regions of this world. They range in size from 1/5 of an inch to 1 inch in length. Most fireflies in the United States are about an inch or less in length. If you happen to live in the United States West of Kansas, you are not likely to have flashing fireflies in your area. Although some isolated sightings of luminous fireflies have been reported from time to time throughout the regions of the Western United States. The reason for the regional distribution of this insect phenomenon is not known. Fireflies inhabit warm humid areas throughout the world, especially tropical Asia, and Central and South America. Some species may be found in very arid regions and are usually spotted following rain. Fireflies live in areas of abundant water, such as ponds, streams, marshes or even depressions, ditches, etc., that may retain moisture longer than other surrounding areas.

Flashing fireflies or Lightning Bugs are glowing to try to attract a mate. Amongst most, but not all, species of North American Lightning Bugs, it is the males that fly about flashing their luminescence bodies while the females will perch on vegetation, usually nearer to the ground. When a female sees a flasher she likes, she responds by flashing right after the male's last flash. A short flash dialogue takes place as the male gets closer and closer, and then, if all goes well, they will mate. Some mated females use aggressive mimicry to trick another organism into thinking it is another insect. These females belong to several species in the genus Photuris where they mimic the female responses of other fireflies in the same area in order to attract the males of the mimicked species to them. When the males are tricked into landing near these mimics to mate, they are pounced upon and eaten! Check out our electronic firefly light effects kit to see a decoration that looks like real fireflies flying around in the dark!

## How do Fireflies Produce Light?

Luminescence Firefly light, called bioluminescence, is known as "cold light" because it emits almost no heat. Fireflies (or lightning bugs) produce light via a chemical reaction when oxygen, breathed in through their abdominal trachea, combines with a substance called luciferin in the presence of the enzyme luciferase, and ATP (adenosine triphosphate) in special cells called podocytes. When these components are added, light is produced. Relatively few insects have this ability and most firefly species are bioluminescent only as adults. However, all known firefly larvae are bioluminescent, as are firefly eggs. Also not all bioluminescent beetles are fireflies. Related beetle families which have bioluminescent members include some click beetles (family Elateridae), phengodid beetles (family Phengodidae) and several others. Some Fireflies, or Lightning Bugs, don't flash at all!

All known Firefly larvae have photic organs that produce light. The Firefly larvae uses their luminescence as a warning signal (aposematism) to communicates with potential predators that they taste bad because of the defensive chemicals in their bodies. Larvae also increase both the intensity and frequency of their glow when disturbed. All known firefly larvae that are wingless and live mostly on the ground or under bark produce light. If you see only a faint glow on the ground, it can be tricky to decide whether you're seeing a firefly larva, a glow-worm, or another luminescent insect.

Almost 100% of a Firefly's luminescence light is given off as light. By comparison, a normal electric light bulb gives off only 10% of its energy as light, while 90% is wasted entirely as heat.

We now have electronic and realistic moving Fireflies that you can use to decorate your yard, patio, and business for any special occasion! Just place them in any dark area to create a display that simulates the true amazing look of Fireflies!

## Other Real Insects Glow

The Owlet moth or Noctuids (Noctuidae) gets its name from the fact that its eyes will glow in the dark when you shine a light on it. The larval stage of the Owlet Moth is called Armyworms, or Cutworms, because of their eating habits are destructive on plants. The Lantern Fly is one of a number of mainly tropical fly species, most of which have a narrow head that is drawn out into a long process. In some species this process looks like a lantern. Many lantern flies are large, with a wing span up to 6 inches.

Besides the insects that produce their own light, all insects seem to be attracted to light. Most scientists think that bright lights confuse the insects' guidance systems so they can't fly straight any more. We offer Lantern Flys as part of our framed insect collection and now we have glow-in-the-dark electronic insects and realistic moving firefly lights too! There are in fact a few living insects that have natural glowing or reflective traits on their bodies.

## How can a Real Butterfly Glow?

The African butterfly (*Bicyclus anynana*) is known for having natural reflective wing patterns. As if that wasn't enough, American scientists and researchers have created the first glowing butterfly by inserting jellyfish marker genes, which have bioluminescence traits making them fluorescent, into the DNA of the African butterfly making it glow luminous green in the pitch dark. Scientists say this experiment to create these mutant butterflies was done to understand how wing patterns emerged in butterflies.

## Are there Electronic Glowing Insects?

Wings in Motion glow-in-the-dark electronic insect products demonstrate the amazing scientific principle of luminescence (the principle of returning light previously absorbed thereafter producing a low-temperature emission of light). Electronic moving insect wings create an enchanting, cheery, eerie glow every time! Since they move like a real butterfly, dragonfly and moth species they are a top decorative choice when you need that nature look to be incorporated into your next interior decorative project. The Luna moth (*Actius luna*) has natural fluorescent green wings and we now have Luna or Lunar moths available as a realistic glow-in-the-dark electronic moving insect!

## Insect Mimicry

Mimicry in the animal kingdom is a phenomenon where one organism tricks another organism into thinking it is something its not. This is done to fool predators into associating it with a similar appearing poisonous, or bad tasting, species. The Monarch butterfly (*Danaus plexippus*) is sometimes called the "milkweed butterfly" because its larvae eat the plant and therefore absorbs the poisonous toxins which it contains.

The Viceroy Butterfly (*Limenitis archippus*) has evolved to mimic and look like the poisonous monarch butterfly so that predators will avoid it! There electronic moving butterflies, dragonflies, and moths are also very realistic and mimic the real insect species! Once you put an electronic Monarch butterfly in your home your family and friends with be fooled into thinking a real butterfly has landed!

# REFERENCES

## PART A–BOOKS / JOURNALS / WEBSITES

### 1. Books

1. Alford, D.V. 1999. *A Textbook of Agricultural Entomology*. Blackwell Science, London.

2. Atwal, AS, Dhaliwal, G.S. & David, B.V. 2001. *Elements of Economic Entomology*. Popular Book Depot, Chennai.

3. Boucias DG & Pendland JC. 1998. *Principles of Insect Pathology*. Kluwer Academic Publisher, Norwel.

4. Burges, H.D. & Hussey, N.W. (Eds.) 1971. *Microbial Control of Insects and Mites*. Academic Press, London.

5. Busvine, J.R. 1971. *A Critical Review on the Techniques for Testing Insecticides*. CABI, London.

6. Chapman JL & Reiss MJ. 2006. *Ecology : Principles & Applications*. 2nd Ed. Cambridge Univ. Press, Cambridge.

7. Chapman RF. 1998. *The Insects : Structure and Function*. Cambridge Univ. Press, Cambridge.

8. Chattpadhyay, S.B. 1995. *Principles and Procedures of Plant Protection*. Oxford & IBH, New Delhi.

9. Chhillar, B.S., Gulati, R. & Bhatnagar, P. 2007. *Agricultural Acarology*. Daya Publ. House, New Delhi.

10. Coppel, H.C. & James, W.M. 1977. *Biological Insect Pest Suppression*. Springer Verlag. Berlin.

11. Crampton, J.M. & Eggleston, P. 1992. *Insect Molecular Science*. Academic Press, London.

12. Dakeshott, J. & Whitten, M.A. 1994. *Molecular Approaches to Fundamental and Applied Entomology*. Springer - Verlag, Berlin.

13. David BV & Ananthkrishnan TN. 2004. *General and Applied Entomology*. Tata- McGraw Hill, New Delhi.

14. De Bach, P. 1964. *Biological Control of Insect Pests and Weeds*. Chapman & Hall, New York.

15. Dhaliwal, G.S. & Arora, R. 2001. *Integrated Pest Management : Concepts and Approaches*. Kalyani Publ., New Delhi.

16. Dhaliwal, G.S. & Koul, O. 2007. *Biopesticides and Pest Management*. Kalyani Publ., New Delhi.

17. Dhaliwal, G.S., Singh, R. & Chhillar, B.S. 2006. *Essentials of Agricultural Entomology*. Kalyani Publ., New Delhi.

18. Dhaliwal, G.S. & Singh, R. (Eds.) 2004. *Host Plant Resistance to Insects. Concepts and Applications*. Panima Publ., New Delhi.

19. Dunston, A.P. 2007. *The Insects : Beneficial and Harmful Aspects*. Kalyani Publ., New Delhi.

20. Duntson PA. 2004. *The Insects : Structure. Function and Biodiversity*. Kalyani Publ., New Delhi.

21. Evans JW. 2004. *Outlines of Agricultural Entomology*. Asiatic Publ., New Delhi.

22. Evans, J.W. 2005. *Insect Pests and their Control*. Asiatic Publ., New Delhi.

23. Gotelli NJ & Ellison AM. 2004. *A Primer of Ecological Statistics*. Sinauer Associates, Inc., Sunderland, MA.

24. Gupta RK. 2004. *Advances in Insect Biodiversity*. Agrobios, Jodhpur.

25. Gupta, H.C.L. 1999. *Insecticides: Toxicology and uses*. Agrotech Publ., Udaipur.

26. Gupta, S.K. 1985. *Handbook of Plant Mites of India*. Zoological Survey of India, Calcutta.

27. Gwilyn, O. & Evans, G.O. 1998. *Principles of Acarology*. CABI, London.

28. Hayes, W.J. and Laws, E.R. 1991. *Handbook of Pesticide Toxicology*. Academic Press, New York.

29. Hoy, M.A. 2003. *Insect Molecular Genetics: An Introduction to Principles and Applications*. 2nd Ed. Academic Press, New York.

30. Huffaker, C.B. & Messenger, P.S. 1976. *Theory and Practices of Biological Control*. Academic Press, London.

31. Ignacimuthu, S.S. & Jayaraj, S. 2003. *Biological Control of Insect Pests*. Phoenix Publ., New Delhi.

32. Ishaaya, I. & Degheele (Eds.). 1998. *Insecticides with Novel Modes of Action*. Narosa Publ. House, New Delhi.

33. Jeppson, L.R., Keifer, H.H. & Baker, E.W. 1975. *Mites Injurious to Economic Plants*. University of California press, Berkeley.

34. Kerkut GA & Gilbert LI. 1985. *Comprehensive Insect Physiology, Biochemistry adn Pharmacology*. Vols. I-XIII. Pergamon Press, New York.

35. Khare, B.P. 1994. *Stored Grains Pests and Their Management*. Kalyani Publ. New Delhi.

36. Magurran AE. 1988. *Ecological Diversity and its Measurement*. Princeton Univ. Press, Princeton.

37. Matsumura, F. 1985. *Toxicology of Insecticides*. Plenum Press, New York.

38. Maxwell, F.G. & Jennings, P.R. (Eds.) 1980. *Breeding Plants Resistant to Insects*. John Wiley & Sons, New York.

39. Mayr, E. & Ashlock, P.D. 1991. *Principles of Systematics Zoology*. 2nd Ed. McGraw Hill, New York.

40. Mayr, E. 1969. *Principles of Systematics Zoology*. McGraw Hill, New York.

41. O'Brien, R.D. 1974. *Insecticides Action and Metabolism*. Academic Press, New York.

42. Painter, R.H. 1951. *Insect Resistance in Crop Plants*. MacMillan. London.

43. Panda, N. & Khush, G.S. 1995. *Plant Resistance to Insects*. CABI, London.

44. Panda, N. 1979. *Principles of Host Plant Resistance to Insects*. Allenheld, Osum ad Co., New York.

45. Patnaik BD. 2002. *Physiology of Insects*. Dominant, New Delhi.

46. Perry, A.S., Yamamoto, I., Ishaaya, I. and Perry, R. 1998. *Insecticides in Agriculture and Environment*. Narosa Publ. House, New Delhi.

47. Prakash, A. and Rao, J. 1997. *Botanical Pesticides in Agriculture*. Lewis Publ., New York.

48. Prakash, I. & Mathur, R.P. 1987. *Management of Rodent Pests*. ICAR, New Delhi.

49. Price PW. 1997. *Insect Ecology*. 3rd Ed. John Wiley, New York.

50. Quicke, D.L.J. 1993. *Principles and Techniques of Contemporary Taxonomy*. Blackie Academic and Professional, London.

51. Richards OW & Davies RG. 1977. *Imm's General Textbook of Entomology*. 10th Ed. Chapman & Hall, London.

52. Ross H. 1974. *Biological Systematics*. Addison Wesley Publ. Co.

53. Sadasivam, S. and Thayumanavan, B. 2003. *Molecular Host Plant Resistance to Pests*. Marcel Dekker, New York.

54. Saxena RC & Srivastava RC. 2007. *Entomology: At a Glance*. Agrotech Publ. Academy, Jodhpur.

55. Smith, C.M. 2005. *Plant Ressistance to Arthropods - Molecular and Conventional Approaches*. Springer, Berlin.

56. Smith, C.M., and Khan, Z.R. and Pathak, M.D. 1994. *Techniques for Evaluating Insect Resistance in Crop Plants*. CRC Press, Boca Raton, Florida.

57. Snodgross R.E. 1993. *Principles of Insect Morphology*. Cornell Univ. Press, Ithaca.

58. Steinhaus E.A. 1984. *Principles of Insect Pathology*. Academic Press, London.

59. Triplehorn CA & Johnson NF. 1998. *Borror and DeLong's Introduction to the Study of Insects*. 7th Ed. Thomson/Brooks/Cole, USA/Australia.

60. Van Driesche, & Bellows, T.S. Jr. 1996. *Biological Control*. Chapman & Hall, New York.

61. Walter, D.E. & Proctor, H.C. 1999. *Mites-Ecology. Evaluation and behaviour*. CABI, London.

62. Wigglesworth VB. 1984. *Insect Physiology*. 8th Ed. Chapman & Hall, New York.

63. Wratten SD & Fry GLA. 1980. *Field and Laboratory Exercises in Ecology*. Arnold, London.

# Appendix I :
# International Entomological Journals

## Journals

1. Name: European Journal of Entomology
   Editor: Svacha P
   Publisher: Institute of Entomology Czech republic Periodicity: Monthly
2. Name: European Journal of Plant Pathology
   Editor: B.M. Cooke
   Publisher: European Foundation of Plant Pathology
   Periodicity: Monthly
3. Name: International Journal of Pest Management
   Publisher: Taylor & Francis
   Periodicity: Monthly
4. Name: International Journal of Dipterological Research
   Editor: Gergey. Y.U.
   Publisher: Russian Academy of Sciences, Russia
   Periodicity: 3 months
5. Name: Journal of Applied Entomology
   Editor: Reinhard Schopt
   Publisher: Journal Publisher Manager, Germany Volume Number: 130/2006
   Periodicity: 9 times per year
6. Name: Journal of Economic Entomology
   Editor: John T. Trumble
   Publisher: University of California, Department of Entomology
   Periodicity: Monthly
7. Name: Journal of Entomology
   Editor: Dr. Robin L. Cooper U.S.A.
   Publisher: Journal Publisher Manager, U.S.A.
   Periodicity: 4 times per year

8. Name: Journal of Plant Sciences
   Editor: Dr. Robin L. Cooper U.S.A.
   Publisher: Journal Publisher Manager, U.S.A.
   Periodicity: 4 times per year

9. Name: Journal of Microbiology
   Editor: J.A. Laursen
   Publisher: J.A.Laursen Publishing editor, U.S.A.
   Periodicity: 4 times per year

10. Name: Journal of Parasitology
    Editor: Dr. Robin L. Cooper U.S.A.
    Publisher: Journal Publisher Manager, U.S.A.
    Periodicity: 4 times per year

11. Name: Journal of Hymenoptera Research
    Editor: Dr.Zdenek, U.S.A.
    Publisher: International Society of Hymenoptera Research, U.S.A.
    Periodicity: 4 times per year

12. Name: Journal of Insect behaviour
    Editor: D.D.Paine
    Publisher: College of Food and natural Resources, Colombia, U.S.A.
    Periodicity: 4 times per year

13. Name: Journal of Pest Science
    Editor: J.Gross
    Publisher: Springer Berling, Federal Biological Research Centre, Germany
    Periodicity: 4 times per year

14. Name: Journal of Phyto Pathology
    Editor: Alan A. Brund
    Publisher: College of Food and natural Resources, Colombia, U.S.A.
    Periodicity: 10 times per year

15. Name: Journal of Biological Control
    Publisher: Society for Bio Control Advancement
    Periodicity: Monthly

16. Name: Journal of Pestology
    Publisher: Scientific Publications Pvt. Ltd.
    Periodicity: Monthly

17. Name: Indian Journal of Plant Protection
    Publisher: Plant Protection Association of India
    Periodicity: Monthly

18. Name: Journal of Pesticides
    Publisher. Colour Publications Pvt. Ltd.
    Periodicity: Monthly

19. Name: Journal of Pesticide Information
    Publisher: Pesticide Association of India
    Periodicity: Quarterly

20. Name: Journal of Plant Protection Bulletin
    Publisher: Plant Protection Advisor to the Govt. of India, New Delhi
    Periodicity: Quarterly

# Appendix II :
# Websites on Pesticide Toxicology

1. Extoxnet

   http://ace.orst.eduJinfo/extoxnet/

2. National Pesticide Telecommunications Network (NPTN)

   http:// ace.orst.eduJinfo/nptn/

3. Agricultural and Environmental News

   http://www2.tricity. wsu.edu/aenews

4. California Department of Pesticide Regulation (DPR):

   http://www.cdpr.ca.gov/

5. California Environmental Protection Agency (CalEP A)

   http://www.calepa.ca. gov/

6. Cooperative State Research, Education, and Extension Service (CSREES):

   http://www.reeusda.gov/

7. Food Animal Residue Avoidance Databank (FARAD)

   http://www.FARAD.org

8. www.pesticide-residues.org

9. http://www. pesticides.gov. uk/prc.asp?id=65 8

10. http://www.atsdr . cdc. gov /tfacts3 5 .html

11. http://www. pesticideinfo.org/

12. http://www.aenews. wsu.edu

13. http://depts.wasltington.edu/pnaslt/news/webmedia.lztm 22.ltttp://eltp.nielts.nilt.gov/

14. http://npi. corst. edu

15. http://www.cdc.gov/llasd

# Appendix III :
## Biological Control Websites

www.nematode.unl.edu/wormepns.htm nysaes. cornell. edu/entlbiocontro

www.oznet.ksu.edu/library/ENTML2/MF2222en.wikipedia.org

www.entomoloqy.wisc.edu/mbcn/misc412.html

www.entomoloqy.wisc.edu/mbcn/fea101.html

www.marchbiolo~ical.com/

www.marchbiological.com/

www.umext.maine .edu/on I inepubs/PD F pu bs/7144.

www.ipm.ucdavis.edu/IPMPROJECT/ADS/cdbiocontrol.html

www.ent.iastate.edu/listJdirectory/108

www.scitechresources.gov/Results/show result.php?rec=1138 -nematode.unl.edu

www.cbit.uq.edu.au/biocontrol/biocontrol/biolo~icalcontrolofinsects.htm

www. thebeneficialwww. ippc. orst .edu/cicp/tactics/biocontrol. htm

www.aUra.orq/aUra-pub/farmscape.html

www.cbit.uq.edu.au/biocontrol/biocontrol/biocontrolmain.htm

www.ncera125.ent.msu.edu/

www.atl.cfs.mcan.qc.ca/index-e/what-e/science-e/ecoloQyecosystem se/biocontrol-
.html

www.ias.ac.in/currsci/iuI10/articles11.html

www.publish.csiro.au/pid/3195.html

www.state.ni.us/aqriculture/planUbiolab.htmlmtwow.org/knapweed-insectary. html

www.csiro.au/csiro/contenUstandard/ps3r ... html

www-rci .rutgers. edu/-insects/biopesticides.html

www-rci .rutgers. edu/-insects/biopesticides.html

www.aqr.kyushu-u.ac.jp/enqlish/bio/index.html

www.ento.vt.edul-idlab/idlist.html

www.pk.uni-bonn.de/ppiqb/ipm.html

www.cdpr.ca.qov/docs/ipminov/bensupp/selrefs.html

www.cplbookshop.com/contents/C100.html

www.librarv.uiuc.edu/envi/beneficialinsects.html

www.center-biolooisk-bekaempelse.dk/insectsuk.html

www.ipmalmanac.com/basics/beneficial.asp

www.earthlife.neUinsects/pub/iliinois.html

www. uri. Edu/ce/factsheets/sheets/biocontrol turf.html

www.eureka.be/altrearmethods

www. bio-control.com/7f.asp

# Appendix IV :
# Pest and Pest Management Websites

- www.idph.state.il.us/eDvhealth/entpestfshts.htm
- www.msstate.edu/Entomology/EntHome.html-
- www.biilmakers.com/entomol o~y/qlossa ry.a sp
  www.ipm.montana.edu/
- www.une.edu.au/agronomy/insects/
- www.apgea . army. mill ento/
- www.entm.purdue.edu/entomoloQv/urban/home.html
- www.amazon.com/exec/obidos/tq/detail/-I
- www.vonl.com/CHIPS/slagpm.htm-ipm.uiuc.edul-rec.ifas.ufl.edu/lsol
  owww.uky.edu/Ag/Entomology/entdepUeg4.htm -
- www.uky.edu/Ag/Entomology/entfacts/fruiUef218.htm -
- www.entomolo~v.umd.edu/Research/aqpestmqmt.html
- www.nysaes.comell.edu/pp/resourceguide/ -agrifor.ac.uk/browse/cabi/
- www.apgea.army.mil/entolosites.htm
- www.mediarelations.kstate.edu/WEB/News/NewsReleases.html-
- www.vonl.col..1/chips/entopes5.htm
- www.ento.vt.edu/Main/Extensionlnfo.html
- www.cabipublishinq.orq
- www.nysaes.comell.edu/pubs/press/2004/060124
- www.uvm.edu/-entlab/international.html-
- www.wvu.edu/-aqexten/ipm/others.htm
- www.crcpress.co.uk/shoppingcart/categories/categoriesprod ucts.asp? parent id
- www.emr.ac.uk/entomology.htm -
- www.i m.iastate.edu  m/icm/1996/11-11-1996/internet.html
- www.ndsu.edu/gradschool/bulletin/d_entomology.html-
- www.cplbookshop.com/contents/C2508.htm
- www.buqwood.orq/entomoloQv.html

- www.aqric.wa.qov.au/pls/portaI30/docs/
- www.entiastate.edu/list/ -
- www.sqrl.csiro.au/news/default.html
- www.oardc.ohio-state.edu/entomology/degree.asp -
- www.krishiworld.com/html/insect_pest_crops1.html-
- www.ars.usda.qov/research/proiects/proiects.htm

## Important Indian websites

www.icar.org

www.irac.org.in

www.cibrc.nic.in

www.plantquarantineindia.org

www.fao.org

www.ipcc.int

# Appendix V :
# Pesticides and Formulations Registered for use in the Country under the Insecticides Act, 1968

(Tentative list subject to corrections - As on 3, August 2010)

| Sr. No. | Name of the Pesticides | Formulation registered |
|---|---|---|
| 1. | 2 ,4-Dichlorophenoxy Acetic Acid (2,4-D Sodium Amine and Ester Salt | (a) 2,4-D Sodium Salt used as Tech a.i. 80% w/w min. |
| | | (b)2,4-D Amine Salt 58% SL 22.5% SL |
| | | (c)2,4-D Ethyl Ester 38% EC, 4.5% Gr., 20%WP, |
| 2. | Acetamiprid | 20 SP |
| 3. | Acephate | 75% SP |
| 4. | Alachlor | 50%EC, 10% Gr |
| 5. | Allethrin | 0.5% Coil, 4% Mat, 0.5% Aer., 3.6% L, 0.2% & 0.02% Coil |
| 6. | Alphacypermethrin | 10% EC, 5% WP, 0.5% Chalk, 10% SC, 0.1% RTU |
| 7. | Alphanaphthyl Acetic Acid | 4.5% Sol. |
| 8. | Aluminium phosphide *(R) | 56% Tab', 56% P |
| 9. | Anilofos | 30% EC, 18% EC |
| 10. | Atrazine | 50% WP |
| 11. | Aureofungin | 46.15% SP |
| 12. | Azadirachtin (neem products) | 25%, 10%, 0.03% EC 0.1 EC, 0.15 EC, 5 EC,0.3% 15% extract concentrate, 1% EC |
| 13. | Bacillus thuringiensis (b.t.) | Liquid & WP formulations, 5% AS |
| 14. | Barium Carbonate | 1% P |
| 15. | Beta cyfluthrin | 2.45% SC |
| 16. | Beauveria bassiana | 1.15% WP |
| 17. | Bendiocarb | 80% WP |
| 18. | Benfuracarb | 40%EC, 3.0% GR |

*Contd...*

| Sr. No. | Name of the Pesticides | Formulation registered |
|---------|------------------------|------------------------|
| 19. | Benomyl | 50% WP |
| 20. | Bitertanol | 25% WP |
| 21. | Bifenthrin | 10% EC, 2.5%EC |
| 22. | Bromadiolone | 0.25% CB, 0.005% RB & 0.005% RB cake |
| 23. | Buprofenzin | |
| 24. | Butachlor | 50% EC, 5% Gr., 50% EW, |
| 25. | Captan | 50% WP, 75% WP, 50% WDG |
| 26. | Carbaryl | 5% DP, 10% DP, 50% WP, 85% WP, 4% Gr., 40% LV, 42% Flow |
| 27. | Carbendazim | 25% DS, 50% WP, 46.27% SC |
| 28. | Carbofuran | 3% CG, 50% SP for Government use |
| 29. | Carbosulfan | 25% DS, 25% EC, 6% Gr. |
| 30. | Carboxin | 75% WP |
| 31. | Carpropamid | 27.8% SC |
| 32. | Cartap Hydrochloride | 4% Gr., 50% SP |
| 33. | Chlorfenapyr | 10% SC (FI) |
| 34. | Chlorimuron ethyl | 25% WP, |
| 35. | Chlormequat Chloride | 50% Sol. |
| 36. | Chlorofenviphos | 10% Gr. |
| 37. | Chlorothalonil | 75% WP |
| 38. | Chlorpyriphos | 20% EC, 10% Gr,., 1.5% DP, 50% EC, 2% RTU |
| 39. | Chlorpyriphos Methyl | 40% EC |
| 40. | Cinmethylene | 10% EC |
| 41. | Clodinafop-propinyl (Pyroxofop-propinyl) | 15% WP |
| 42. | Clomazone | 50% EC |
| 43. | Copper Oxychloride | 50% WP, 40% Paste, 5% DP, 50 WG |
| 44. | Copper Hydroxide | 77% WP |
| 45. | Copper Sulphate | Used as Tech. 8%, 25% w/w min., 2.62% SC |
| 46. | Coumachlor | 0.5% CB, 0.025% RB |
| 47. | Coumatetralyl | 0.75% TP, 0.037% Bait. |
| 48. | Cuprous Oxide | 4% DP |
| 49. | Cyfluthrin | 10% WP, 5% EW,Cyfluthrin + Propoxur (0.5%) (0.015%) |
| 50. | Cyhalofop-butyl | 10% EC |

*Contd...*

| Sr. No. | Name of the Pesticides | Formulation registered |
|---------|------------------------|------------------------|
| 51. | Cymoxonil | 80% WP |
| 52. | Cypermethrin | 10% EC, 25% EC, 1% Chalk, 0.1% Aquous (HH), 0.25 DP, 3% Smoke Generator |
| 53. | Cyphenothrin | 5% EC, 0.15% in combination as Aer. |
| 54. | Dazomet | Dazomet Technical GR |
| 55. | Decamethrin (Deltamethrin) | 2.5% Flow, 2.5% WP, 2.8% EC, 0.5% Chalk, 1.25% ULV, 25% Tab., 11% EC, 0.5% Tablet bait |
| 56. | Diafenthiuron | 50% WP |
| 57. | Diazinon | 20% EC, 10% Gr, 2% DP, 40% WP, 5% Gr., 25% Micro Encapsulation |
| 58. | Dichloro Diphenyl Trichloroethane (DDT) | 50% WP, 75% WP |
| 59. | Dichloropropene and Dichloropropanes mixture (DD Mixture) *(R) | 1:1 |
| 60. | Diclorvos (DDVP) | 76% EC |
| 61. | Diclofop-methyl | 28% EC |
| 62. | Dicofol | 18.5% EC |
| 63. | Difenoconazole | 25% WP |
| 64. | Diflubenzuron | 25% WP |
| 65. | Dimethoate | 30% EC |
| 66. | Dimethomorph | 50% WP |
| 67. | Dinocap | 48% EC |
| 68. | Dithianon | 75% WP |
| 69. | Diuron | 80% WP |
| 70. | Dodine | 65% WP, 50% flow |
| 71. | D-trans allethrin | 2% Mat, 0.1% coil, 0.1% coil (12 hr.) |
| 72. | Edifenphos | 50% EC |
| 73. | Endosulfan | 2% DP, 4% DP, 35% EC, 4% Gr. |
| 74. | Ethephon | 39% SL, 10% Paste |
| 75. | Ethion | 50% EC |
| 76. | Ethofenprox (Etofenprox) | 10% EC |
| 77. | Ethoxysulfuron | 10% EC |
| 78. | Ethylene Dibromide and Carbon Tetrachloride mixture (EDCT mixture 3: 1) | 3:1 |
| 79. | Fenarimol | 12% EC |

*Contd...*

| Sr. No. | Name of the Pesticides | Formulation registered |
|---|---|---|
| 80. | Fenazaquin | 10 EC |
| 81. | Fenitrothion | 5% DP, 40% WP, 50% EC, 82.5% EC, 2% Spray, 20% OL |
| 82. | Fenobucarb (BPMC) | 50% EC |
| 83. | Fenoxaprop-p-ethyl | 10% EC, 9.3% EC one time import |
| 84. | Fenpropathrin | 10% EC, 30% EC |
| 85. | Fenthion | 82.5% EC, 2% Gr., 2% Spray |
| 86. | Fenvalerate | 0.4% DP, 20%EC |
| 87. | Fipronil | 0.3% Gr., 5% SC, 0.05% Gel (Import) |
| 88. | Fluchloralin | 45% EC |
| 89. | Flusilazole | 40% EC |
| 90. | Flufenacet | 60% WP |
| 91. | Flufenoxuron | 10% DC |
| 92. | Fluvalinate | 25% EC |
| 93. | Fosetyl-Al | 80% WP |
| 94. | Gibberellic Acid | Tech. P, 0.186% SP, 0.001% W/W |
| 95. | Glufosinate Ammonium | 13.5% SL |
| 96. | Glyphosate | 41% SL, 20.2% SL, 5%SL |
| 97. | Hexaconazole | 5% EC, 5% SC, 2% SC |
| 98. | Hydrogen cyanamid | 49% age, 50% SC |
| 99. | Imazethapyr | 10% EC, Imazethapyr+Pendimethalin 2% + 30% |
| 100. | Imidacloprid | 17.8% SL, 70% WS, 48% FS, 30.5% SC, 2.5% Gel |
| 101. | Iprobenfos (Kitazin) | 48% EC, 17% Gr. |
| 102. | Imiprothrin | 50% MUP (Imiprothrin 0.1% + Cyfenothrin 0.15%) |
| 103. | Indoxacarb | 14.5% SC |
| 104. | Iprodione | 50% WP |
| 105. | Isoprothiolane | 40% EC |
| 106. | Isoproturon | 50% WP, 75% WP, 50% Flow |
| 107. | Kasugamycin | 3% SL |
| 108. | Lambdacyhalothrim | 5% EC, 10% WP, 2.5% EC, 0.5% Chalk |
| 109. | Lime Sulphur | 22% SC *Contd...* |
| 110. | Lindane *(R) | 0.65% DP, 1.3% DP, 6.5% WP, 20% EC, 6% Gr. |
| 111. | Linuron | 50% WP |
| 112. | Lufenuron | 5.4% EC |

*Contd...*

| Sr. No. | Name of the Pesticides | Formulation registered |
|---|---|---|
| 113. | Malathion | 5% DP, 25% WP, 50% EC, 0.25% Spray and 96% ULV, 2% Spray, 5% Spray |
| 114. | Mancozeb | 75% WP, 35% SC (suspension concentrate), 75% WG |
| 115. | Mepiquat Chloride | 5% AS, 50% TK |
| 116. | Milbemectin | 1% EC |
| 117. | Metalaxyl | 35% WS, 40% WS |
| 118. | Metaldehyde | 2.5% DP |
| 119. | Metiram | 70% WG |
| 120. | Methomyl | 40% SP |
| 121. | Methabenzthiazuron | 70% WP |
| 122. | Methoxy ethyl mercury chloride *( R) | 3% FS, 6% FS |
| 123. | Methyl bromide *(R) | 99% L, 98% L |
| 124. | Methyl chlorophenoxy acetic acid | 40% SL or 40% As |
| 125. | Methyl Parathion *( R) | 2% DP, 50% EC |
| 126. | Metasulfuron methyl | 20% WD |
| 127. | Metolachlor | 50% EC |
| 128. | Metoxuron | 80% WP |
| 129. | Metribuzin | 70% WP |
| 130. | Monocrotophos | 36% SL |
| 131. | Myclobutanil | 36% SL |
| 132. | Novaluron | 10% EC (FI) |
| 133. | Oxadiazon | 25% EC |
| 134. | Oxadiargyl | 80% WP, 6% EC |
| 135. | Oxycarboxin | 20% EC |
| 136. | Oxydemeton-methyl | 25% EC |
| 137. | Oxyfluorfen | 23.5% EC, 0.35% Gr. |
| 138. | Paclobutrazol | 23% SC |
| 139. | Paraquat dichloride | 24% SL |
| 140. | Penconazole | 10% EC |
| 141. | Pendimethalin | 30% EC, 5% Gr. |
| 142. | Permethrin | 25% EC, 5% SG. |
| 143. | Phenthoate | 2% DP, 50% EC |
| 144. | Phorate | 10% CG |
| 145. | Phosalone | 4% DP, 35% EC |

*Contd...*

| Sr. No. | Name of the Pesticides | Formulation registered |
|---|---|---|
| 146. | Phosphamidon | 40% SL, |
| 147. | Primiphos-methyl | 25% WP, 50% EC, 1% Spray |
| 148. | Prallethrin | 0.8% mat for 12 hours, 1% Mat, 0.8% L, 1.6% L, 0.5% mosquito coil, 0.04% Mosquito coil, 1.2% mat, 19% w/w VP, 0.6% mat |
| 149. | Pretilachlor | 50% EC, 30.7% w/w EC |
| 150. | Profenophos | 50% EC |
| 151. | Propanil | 35% EC |
| 152. | Propergite | 57% EC |
| 153. | Propetamphos | 20% EC, 1% Spray |
| 154. | Propiconazole | 25% EC |
| 155. | Propineb | 70% WP |
| 156. | Propoxur | 20% EC, 1% Aer., 2% Aer. 1% HH Spray, 2% Bait |
| 157. | Pyrethrins (Pyrethrum) | 0.2% DP, 2.5% EC, 0.05% Spray, 0.2% PH, 2.0% EC |
| 158. | Quinalphos | 1.5% DP, 25% EC, 20% AF |
| 159. | Quizalofop ethyl | 5% EC (FI) |
| 160. | S-Bioallethrin | 2.4% mat |
| 161. | Sirmate | 38 WP, 4 Gr. |
| 162. | Sodium Cyanide *( R) | Used as Tech., 96% a.i. min |
| 163. | Spinosad | 45% SC, 2.5% SC |
| 164. | Streptomycin + Tetracycline | 90: 10 SP |
| 165. | Sulfosulfuron | 75% WG |
| 166. | Sulphur | 85% DP, 80% WP, 40% SC, 80% WG/WDG, 55.16 SC (800 gm / L) |
| 167. | Tebuconazole | 2.5% DS, 2% DS |
| 168. | Temephos | 50% EC, 1% Sand Granules |
| 169. | Thiobencarb(Benthiocarb) | 50% EC, 10% Gr. |
| 170. | Thiodicarb | 75% WP |
| 171. | Thiomethoxiam | 25% WG, 70% WS, 30%FS |
| 172. | Thiometon | 25% EC |
| 173. | Thiophanate-methyl | 70% WP |
| 174. | Thiacloprid | 21.7% SC |
| 175. | Thiram | 80% WP |
| 176. | Transfluthrin | 0.88% Liquid Vaporiser, 0.03% Mos. Coil, 20% MV Gel (30 days mat tray) |

*Contd...*

| Sr. No. | Name of the Pesticides | Formulation registered |
|---|---|---|
| 177. | Triadimefon | 25% WP |
| 178. | Triallate | 50% EC |
| 179. | Triazophos | 40% EC, 20% EC |
| 180. | Trichlorfon | 5% DP, 50% EC, 5% Gr. |
| 181. | Tricontanol | 0.05% EC, 0.1% EW, 0.05% GR |
| 182. | Tricyclazole | 75% WP |
| 183. | Tridemorph | 80% EC |
| 184. | Trifluralin | 48% EC |
| 185. | Validamycin | 3% L |
| 186. | Zinc Phosphide | 2% RB |
| 187. | Zineb | 80% WP, 27% Colloidal Suspension |
| 188. | Ziram | 80 WP, 27% CS |

## 2. APPROVED FORMULATION OF COMBINATION PESTICIDES

| Sr. No. | Combination Product | Company |
|---|---|---|
| **A. Insecticides** | | |
| 1. | Carbaryl 4% + Gamma BHC 4% Gr. | M/s Avantis Crop Science India Ltd.,Mumbai |
| 2. | Deltamethrin 1% + Triazophos 35% EC | -do- |
| 3. | Profenofos 40% + Cypermethrin 4% EC | M/s Novartis Crop Protection Ltd., Mumbai |
| 4. | Chlorpyriphos 50% + Cypermethrin 5% EC | M/s De-Nocil, Mumbai |
| 5. | Cypermethrin 3% + Quinalphos 20% EC | M/s United Phosphorus Ltd., Mumbai |
| 6. | Chloropyriphos 16% + Alphacypermethrin 1% EC | M/s Acco Industries Ltd., Mumbai |
| 7. | Acephate 25% + Fenvalerate 3% EC | M/s Rallis India Ltd., Bangalore |
| 8. | Phosalone 24% + Cypermethrin 5% EC | M/s Aventis Cropscience Ltd. |
| 9. | Ethion 40% + Cypermethrin 5% EC | M/s Rallis India Ltd., Bangalore. |
| 10. | Deltamethrin 0.75% +Endosulfan 29.75% EC | |
| 11. | Cymoxanil 8% + Mencozeb 64% WP | EI Dupont |
| 12. | Methyl bromide 98% + chlorpicrin 2% | |
| 13. | Propoxur 0.25% + cyfluthrin 0.025% Aerosal | M/s Bayer India |
| 14. | Cyfluthrin 0.025% + Tranfluthrin 0.04% | M/s Bayer India |
| 15. | Imiprothrin 0.1% + Cyphendthrin 0.15% | |
| 16. | Propoxur 0.5% + Cyfluthrin 0.025% Spray, Propoxur 0.5% + Cyfluthrin 0.015% Spray | |
| 17. | Acephate 5% + Imidacloprid 1.1% | M/s UPL Ltd |

# Appendix VI :
# List of Pesticide/Pesticide Formulations Banned in India

## A. Pesticides Banned for manufacture, import and use (27 Nos.)

| | | | | | |
|---|---|---|---|---|---|
| 1 | Aldrin | 10 | Heptachlor | 19 | Toxafen |
| 2 | Benzene Hexachloride | 11 | Menazone | 20 | Aldicarb |
| 3 | Calcium Cyanide | 12 | Nitrogen | 21 | Chlorobenzilate |
| 4 | Chlordane | 13 | Paraquat Dimethyl Sulphate | 22 | Dieldrine |
| 5 | Copper Acetoarsenite | 14 | Pentachloro Nitrobenzene | 23 | Maleic Hydrazide |
| 6 | CIbromochloropropane | 15 | Pentachlorophenol | 24 | Ethylene Dibromide |
| 7 | Endrin | 16 | Phenyl Mercury Acetate | 25 | TCA (Trichloro acetic acid) |
| 8 | Ethyl Mercury Chloride | 17 | Sodium Methane Arsonate | 26 | Metoxuron |
| 9 | Ethyl Parathion | 18 | Tetradifon | 27 | Chlorofenvinphos |

## B. Pesticide/Pesticide formulations banned for use but their manufacture is allowed for export (2 Nos.)

| | | | |
|---|---|---|---|
| 1 | Nicotin Sulfate | 2 | Captafol 80% Powder |

## C. Pesticide formulations banned for import, manufacture and Use (4 Nos.)

| | | | |
|---|---|---|---|
| 1 | Methomyl 24% L | 3 | Phosphamidon 85% SL |
| 2 | Methomyl 12.5% L | 4 | Carbofuron 50% SP |

## D. Pesticide Withdrawn (7 Nos.)

| | | | | | |
|---|---|---|---|---|---|
| 1 | Dalapon | 2 | Nickel Chloride | 3 | Simazine |
| 4 | Ferbam | 5 | Paradichlorobenzene (PDCB) | 6 | Warfarin |
| 7 | Formothion | | | | |

## LIST OF PESTICIDES REFUSED REGISTRATION

| | | |
|---|---|---|
| 1 Calcium Arsonate | 7 Carbophenothion | 13 Thiodemeton / Disulfoton |
| 2 EPM | 8 Mephosfolan | 14 Fentin Acetate |
| 3 Azinphos Methyl | 9 Vamidothion | 15 Fentin Hydroxide |
| 4 Lead Arsonate | 10 Azinphos Ethyl | 16 Chinomethionate (Morestan) |
| 5 Mevinphos (Phosdrin) | 11 Binapacryl | 17 Ammonium Sulphamate |
| 6 2,4, 5-T | 12 Dicrotophos | 18 Leptophos (Phosvel) |

## PESTICIDES RESTRICTED FOR USE IN INDIA

| | | |
|---|---|---|
| 1 Aluminium Phosphide | 6 Sodium Cyanide | 11 Fenthion |
| 2 DDT | 7 Monocrotophos | 12 Dazomet |
| 3 Lindane | 8 Endosulfan | 13 Methoxy Ethyl Mercuric Chloride (MEMC) |
| 4 Methyl Bromide | 9 Fenitrothion | |
| 5 Methyl Parathion | 10 Diazinon | |

## PART B

# KEY NOTES ON
# PLANT PATHOLOGY
# AND MICROBIOLOGY

# 1

# DISCOVERIES

| Year | Name of the Scientist | Discovery/Contribution |
|------|----------------------|------------------------|
| - | S.N. Winogradasky | Discovered the autotrophic mode of life amongst bacteria  Established the microbial transformation of nitrogen and sulphur. Developed the enrichment culture techniques involving the successive transfer of microorganisms. |
| - | V.P. Bhide | Started Agril. Microbiology subject for first time in India. Worked on fungal and Bacterial diseases as well as nitrogen fixing bacteria in the soil. |
| - | M.J. Thirumalchar | Developed "Aerofurgin" a broad spectrum antifungal antibiotics. |
| - | Nandi | Active in building a school of antibiotics and biofertilizer. |
| - | Robert Hartig | Father of forest pathology |
| - | M.N. Kamat and G. Rangaswami | Book on Plant Pathology and Mycology |
| - | M.K. Patel | Pioneer worker on Bacterial plant pathogenis in India. |
| 1675 | Antony van Leeuwenhoek | Developed first microscope |
| 1753 | Carlous Linnaeus | Established Latin Bionomial System of Nomenclature of plants and animal in his book, 'Species Plantarum'. |
| 1773 | Needham | *Anguine tritici* was first plant parasitic nematode |
| 1807 | Prevost | $CUSO_4$ as seed treatment for Bunt of wheat. |
| 1821 | Roberson | First used sulphur as a fungiside. |
| 1831-88 | Anton de Bary | Father of Plant Pathology |
| 1846-48 | M.J. Berkley | Came out with the parasitic theory of plant diseases. He published several papers on diseases on vegetable's cercals and other crops. |
| 1847 | Berkeley | Discribed oidium as a cause of powdery mildew of canes. |
| 1853 | Anton De-bary | Shows the fungus origin of *Phytophthora*. Late blight of potatoes. |
| 1854-1915 | Paul Ehrlich | Father of Chemotherapy, Salvarsan, Pioneer of modern chemotherapy |

| Year | Name of the Scientist | Discovery/Contribution |
|------|----------------------|------------------------|
| 1876 | Robert Koach | Proved Bacterial nature of Anthrax disease in animals and developed a plate method of isolation of Bacteria and fungi. |
| 1880 | Burrill T.J. | First to show that bacteria cause plant disease. |
| 1882 | P.A. Millardet | Discovered the use of Bordeaux mixture for the control of downy mildew of grapes. |
| 1884 | Gram, H.C. | Gram staining. |
| 1885 | Frank | Coined the term mycorrhizae. |
| 1886 | Adolf Mayer | Showed infectious nature of tobacco mosaic virus by infecting the juice from diseased plant into healthy plant. |
| 1887 | Jensen | Developed hot water treatment of seed's for loose smut of wheat. |
| 1888 | M.W. Beijerink | Discovery of bacteria (now named *Rhizobium)* is responsible for symbiotic nitrogen fixation in legumes. |
| 1892 | Iwanoski | Filterable nature of viruses. |
| 1894 | Eriksson | Studied on Biological Races in Cereal rust. He also put forth the mycoplasma theory. |
| 1902 | Takami | Show insect transmittion in rice stunt virus. |
| 1905 | Beijerink | Discovery of *Azotobacter.* |
| 1907 | | Establishment of American Phytopathological Society. |
| 1917 | E. C. Stakeman | Demonstrated physiologic races in stem (From Minnesota) rust of wheat. |
| 1925 | Beijerink | Discovery of *Azospirillum.* |
| 1925 | Coons, G.H. and Kotila J.C. | Isolation of bactriophases of *Bacillus carotovorus.* |
| 1931 | J.C. Luthra | Solar heat treatment for loose smut of wheat. |
| 1932 | Fred | Introduction of YEMA medium for Rhizobium. |
| 1933-38 | Ruska, E. and Von Borries | Developed electron microscope. |
| 1935 | W.M. Stanley | Established crystalline nature of virus particle. |
| 1936-37 | Bowden and Pirie | Showed that viruses are Nuclear proteins. |
| 1939 | P.K. Dey | Discovery of BGA as nitrogen fixation in paddy field. |
| 1944 | S.A. Waksman | Discovered the antibiotic streptomycin. |
| 1945 | Alexander Fleming | Nobel prize for discovery of penicillin. |
| 1945 | Florey, H.W. and Chain E.B. | Nobel prize for demonstrating chemotherapeutic use of  penicillin. |
| 1952 | Lederberg. J. | Coined the term plasmid. |
| 1952 | Zinder and Leder- | Discovery of transduction-berg.J. |

| Year | Name of the Scientist | Discovery/Contribution |
|------|----------------------|------------------------|
| 1960 | P.K. Dey and R. Bhattacharyya | Isolation of new non-symbiotic N- fixing organism *Derxia* gummosa |
| 1962 | Stolp, H. | Discovery of *Bdellovibrio.* |
| 1964 | Klement, Z et al | Discovery of hypersensitive response. |
| 1967 | Doi, Y. *et.al.* | Discovery of Mycoplasma like Organisms (MLO's). |
| 1970 | Subba Rao N.S. | A pioneer worker in field of BNF. He worked extensively on all types of $N_2$ fixing systems. He also published several books as BNF. |
| 1973 | Boyer H. and Cohen, S. | Cloned the first DNA using plasmid. |
| 1980 | Dye, D.W. *et.al.* | Introduction of pathovar system of taxonomy of plant pathogenic bacteria. |
| 1986 | Hellriegel and Wilfarth | Discovery of symbiotic nitrogen fixation. |

# 2

# TERMINOLOGY

| Term | Terminology |
|------|-------------|
| **Abiogenesis** | Belief that life arose from vital forces in non living or decomposing matter. |
| **Abiotic** | Nonliving, or caused by a nonliving agent; e.g., abiotic disease. |
| **Acervulus** | A sub epidermal, saucer-shaped, asexual fruiting body producing conidia on short conidio phores. |
| **Acquired resistance** | Plant resistance to disease aestivated after inoculation of the plant with certain microorganisms or treatment with certain chemical compounds. |
| **Actinomycetes** | Formerly, a group of bacteria forming branching filamentous. |
| | A group of organisms intermediate between bacteria and fungi. |
| **Active defense** | Defenses induced in the plant after attack by a pathogen. |
| **Aecium** | A cup-shaped fruiting body of rust fungi that produces aeciospores. |
| **Aerobe** | An organism which grows in the presence of oxygen. |
| **Aerobic** | A microorganism that lives or a process that occurs, in the presence of molecular oxygen. |
| **Aerobiology** | The study of micro-organisms carried in the air. |
| **Aflatoxin** | A mycotoxin produced by the fungus *Aspergillus flavus* and by some other fungi. |
| **Agar** | A gelatin-like material obtained from seaweed and used to prepare culture media on which microor-ganisms are grown and studied. |

| Term | Terminology |
| --- | --- |
| **Agglutination** | A serological test in which viruses or bacteria suspended in a liquid collect into clumps whenever the suspension is treated with antiserum containing antibodies specific against viruses or bacteria. |
| **Agroterrorism** | Terrorism caused by scaring con-sumers away from buying certain agricultural products such as vegetables, milk, and meat, by contaminating them on the farm or in the market with human pathogens. Also, scaring people for future shortages of food by spreading plant pathogens on crops so that terrorists reduce the amount of food produced. |
| **Alarm signal** | A chemical compound presumably produced by a host plant, in response to infection, and sent out to host cell proteins and genes that the plant activates to produce substances inhibitory to the pathogen. |
| **Allele** | One of two or more alternate forms of a gene occupying the same locus on a chromosome. |
| **Allozyme** | An enzyme with slightly altered properties produced by an allele of the original gene. |
| **Alternate host** | One of two kinds of plants on which a parasitic fungus (e.g., rust) must develop to complete its life cycle. |
| **Ammensalism** | One species is suppressed while the second is not affected, typically the result of toxin production. |
| **Anaerobe** | An organism which grows in the absence of oxygen. |
| **Anaerobic** | A microorganism that lives or a process that occurs, in the absence of molecular oxygen. |
| **Anamorph** | The imperfect or asexual stage of a fungus. |
| **Anastomosis** | The union of a hypha with another, resulting in intercommunication of their genetic material. |
| **Antagonist** | An organism or substance which can inhibit the growth of or interferes with the activity of another due to the neutralizing effect of their antibiotic substance. |

| Term | Terminology |
|---|---|
| **Antheridium** | The male sexual organ found in some fungi. |
| **Anthracnose** | A disease that appears as black, sunken, leaf, stem, or fruit lesions, caused by fungi that produce their asexual spores in an acervulus. |
| **Antibiosis** | Antagonistic association between two organisms in which one is adversely affected. |
| **Antibiotic** | A natural, semisynthetic or wholly synthetic antimicrobial compound which is effective at low concentrations. |
| | A chemical compound produced by one microorganism that inhibits or kills other microorganisms. |
| **Antibody** | One of a class substances (protein) produced by an animal in response to the introduction of an antigen. |
| **Antigen** | A substance which when introduced into an animal body, stimulates the production of specific entities (antibodies) that react or unite with the substance introduced (antigen). |
| **Antiseptic** | It is a substance that opposes sepsis, putrefaction or decay and referred especially to chemical substances applied to living tissues. |
| **Antiserum** | The blood serum of a warm–blooded animal that contains antibodies. |
| **Antiserum titer** | The highest dilution of an antiserum that will react with its homologues virus. |
| **Aseptic** | Free of microorganisms capable of causing infection or contamination. |
| **Asexual reproduction** | Any type of reproduction not involving the union of gametes. |
| **Autotroph** | Organism able to utilize carbon dioxide as sole source of carbon. |
| **Auxotroph** | Microorganisms which lacks the ability to synthesize one or more essential growth factors. |
| **Bacterial viruses** | Viruses that multiply in bacteria. |

| Term | Terminology |
|------|-------------|
| **Bacteriocide** | Any physical or chemical agent which is able to kill some type of bacteria. |
| **Bacteriocin** | A protein or a peptide with antibiotic activity normally against narrows range of closely related bacteria; usually it is plasmid borne. |
| **Bacteriology** | The scientific study of bacteria. |
| **Bacteriophase** | A virus that infects bacteria and usually kills them. |
| **Biofertilizers** | Carrier based inoculants containing live or latent cells of efficient strains of N fixing, P solubilizing or cellulolytic microorganisms used for application to seed or soil. |
| **Biotechnology** | The application of microbes, genetic and biochemical technology for the exploitation of biological systems and processes for economic use. |
| **Biotrophs** | These are those organisms which, regardless of the ease with which they can be rutted in nature, obtain their food from the living tissues on which they complete their life cycles. Some typical examples are rusts, smuts and mildews. |
| **Botulism** | Food poisoning due to the toxin of *Clostrodium bolulinum* |
| **Chelating agent** | A chemical such as sodium EDTA which will bind with bivalent and trivalent cations to assist in virus purification |
| **Chemoautotrophs** | Using chemical energy as source and $CO_2$ as principle carbon source. Example, Nitrosomonas and Nitrobacter. |
| **Chemoheterotrophs** | Using chemical energy as source and an organic substrate as principle carbon source, e.g., protozoa, fungi, metazoan animals, and bacteria. |
| **Chemolithotroph** | An organism which obtains its energy from endogenous, light independent chemical reactions and oxidation of inorganic compounds/substrates. |
| **Chemotherapy** | The treatment of disease by the use of chemicals. |
| **Chemotrophs** | The organisms that is dependent on a chemical energy source. |

| Term | Terminology |
|------|-------------|
| **Circulative Virus** | A virus which is transmitted by an insect in a persistent manner and which circulates from the insects digestive tract, through the haemolymph to the salivary glands, before being transmitted in the saliva as the insect feeds. |
| **Cloning** | A term used to express the construction of recombinant DNA molecules. |
| **Coding** | Sequence of nucleotide within a certain areas of RNA determines the sequence of amino acids in the synthesis of particular proteins. |
| **Codon** | The coding unit, consisting of three adjacent nucleotides which codes for specific amino acids. |
| **Colony** | A macroscopically visible growth of micro-organisms on a solid culture medium. |
| **Commensalism** | Living together of two organisms when one is beneficial by the association while the other is apparently neither benefited nor harmed. |
| **Competition** | A condition in which there is a suppression of one organism for limiting quantities of nutrients, $O_2$ or other common requirements. |
| **Complementary DNA** | Synthesized by reverse transcript age from RNA. |
| **Complementation** | When virus is assisted by another virus to replicate. |
| **Conjugation** | Transfer of genetic material from one cell to another through cell contact. |
| **Constrictive resistance** | Genetically controlled, inherited resistance. |
| **Contamination** | Entry of undesirable organisms into some material or object. |
| **Cork** | An external, secondary tissue impermeable to water and gases. It is often formed in response to wounding or infection. |
| **Cross protection** | The phenomenon in which plant tissues infected with one strain of a virus are protected from infection by other, more severe, strains of the same virus. |
| **Culture** | A population of microorganisms cultivated in a medium. |

| Term | Terminology |
|------|-------------|
| | To artificially grow microorganisms or plant tissue on a prepared food material; a colony of microorganisms or plant cells artificially maintained on such food material. |
| **Cuticle** | A thin, way layer on the outer wall of epidermal cells consisting primarily of wax and cutin. |
| **Cutin** | A waxy substance comprising the inner layer of the cuticle. |
| **Cyst** | An encysted zoospore (fungi); in nematodes, the carcass of dead adult females of the genus *Heterodera* or *Globodera,* which may contain eggs |
| **Cytokinins** | A group of plant growth-regulating substances that regulate cell division. |
| **Cytoplasmic resistance** | Resistance controlled by genetic material present in the cell cytoplasm |
| **Dalton** | A unit of mass equaling the atomic weight of a hydrogen atom. |
| **Damping-off** | Destruction of seedlings near the soil hue, resulting in the seedlings falling over on the ground. |
| **Defense activators** | Synthetic chemicals that, when applied to plants as sprays, injections, root treatments, etc., induce systemic acquired resistance in them to several types of pathogens. |
| **Defensins** | A group of defense-related, cysteine-rich, antimicrobial peptides present in the plasma membrane of most plant species that provide resistance to different pathogens. |
| **Denatured protein** | Protein whose properties have been altered by treatment with physical or chemical agents. |
| **Density-gradient centrifugation** | A method of centrifugation in which particles are separated in layers according to their density. |
| **Dependent transmission** | Transmission of a virus (by aphids) that only occurs when the vector feeds on a source plant that is jointly infected by a second virus. The second virus is referred to as helper virus, and the virus that is not transmissible on its own is called as dependent virus. |

| Term | Terminology |
| --- | --- |
| **Dieback** | Progressive death of shoots, branches, and roots, generally starting at the tip. |
| **Differential centrifugation** | Cycles of low and high speed  clarification and sedimentation used in the purification of a virus. |
| **Differential host** | A plant which gives distinctive symptoms when infected with a specific virus, allowing the virus to be distinguished from others. |
| **Differential media** | Media which distinguish between colonies of one desire organism from another. |
| **Dikaryotic** | Mycelium OT spores containing two sexually compatible nuclei per cell. Common in the basidiomycetes. |
| **Dilution end point** | The lowest dilution in a serial dilution of a virus preparation that will infect a mechanically inoculated plant. |
| **Disease** | Any malfunctioning of host cells and tissues those results from continuous irritation by a pathogenic agent or environmental factor and leads run development of symptoms. |
| **Disease cycle** | The chain of events involved in disease development, including the stages of development of the pathogen and the effect of the disease on the host. |
| **Disease gradient** | The change in incidence of a disease with increasing distance from the source of infection. |
| **Disinfectant** | It is used to free a substance from infection, and is usually a chemical agent which destroys disease germs or other harmful microorganisms or inactivates viruses. |
| | A physical or chemical agent that frees a plant, organ, or tissue from infection. |
| **Disinfectant** | An agent that kills or inactivates pathogens in the environment or on the surface (II) plant or plant organ before infection takes place. |
| **Dominant gene** | A gene that is fully expressed in the phenotype of the heterozygote. |

| Term | *Terminology* |
|------|---------------|
| **Double antibody sandwich** | A method in enzyme linked immunosorbent assays (ELISA) in which the reactants are added to the test plate in the order of antibody, virus, and antibody enzyme complex. |
| **Downy mildew** | A plant disease in which the sporangiophores and spores of a fungus appear as a downy growth on the lower surface of leaves and stems, fruit, etc., caused by fungi in the family Peronosporaceae. |
| **Durable resistance** | Used to describe resistance that is long lasting. |
| **Ecdysis** | Moulting of integuments of an insect that occurs between each of its growth stages. |
| **Ectoparasite** | A parasite feeding on a host from the exterior. |
| **Effector protein** | A protein coded by a bacterial pathogenicity/virulence gene that is exported into (hl) plant and interacts with an R-gene protein. |
| **Egg** | A female gamete. In nematodes, the first stage (I) the life cycle containing a zygote or a juvenile. |
| **Electrophoresis** | Separation of DNA/RNA or protein on acrylamide/agarose gel by application of an electric field. |
| **Elicitors** | Molecules produced by a pathogen tholt induce a defense response by the host. |
| **ELLSA** | A serological test in which one antibody carries with it an enzyme that releases a coloured compound. |
| **Enation** | Tissue malformation or overgrowth, induced by certain virus infections. |
| **Encapsulation** | The enclosure of virus nucleic acid genome within a protein shell. |
| **Endoparasite** | A parasite that enters a host and feeds from within. |
| **Endospore** | A thick walled spore formed in the bacteria cell. |
| **Enzyme** | A protein produced by living cells that can catalyze a specific organic reaction. |
| **Epidemic** | A disease increase in a population; usually a widespread and severe outbreak of a disease. |
| **Epidemic rate** | The amount of increase of disease per unit or time in a plant population. |

| Term | Terminology |
|------|-------------|
| **Epidemiology** | The study of factors affecting the out-break and spread of infectious diseases. |
| **Epidermis** | The superficial layer of cells occurring on all plant parts. |
| **Epiphytically** | Existing on the surface of a plant or plant organ without causing infection. |
| **Epiphytotic** | A widespread and destructive outbreak of a disease of plants; epidemic. |
| **Eradicant** | A chemical substance that destroys a pathogen at its source. |
| **Eradication** | Control of plant disease by eliminating the pathogen after it is established or by eliminating the plants that carry the pathogen. |
| **Etiology of disease** | The determination and study of the cause of a disease. |
| **Expressed sequence tag (EST)** | Molecular landmarks that provide a profile of mRNAs and allow cloning of a large number of genes being expressed in a cell population. |
| **Facultative anaerobe** | Aerobic bacteria which can also grow both in the absence of oxygen. |
| **Facultative parasite** | Having the ability to be a parasite. |
| **Fermentation** | Anaerobic oxidation of compounds by enzyme action of microorganisms; gaseous oxygen is not involved in this energy-yielding process. |
| | Oxidation of certain organic substances in the absence of molecular oxygen. |
| **Fertilization** | The sexual union of two nuclei, resulting in doubling of chromosome numbers. |
| **Filamentous** | Thread like; filiform. |
| **Fission** | Transverse splitting in two of bacterial cells; asexual reproduction. |
| **Fitness** | The ability of a pathogen to survive and reproduce. |
| **Flagellin** | A receptor system for general elicitors very similar and common to plants and animals. |
| **Flagellum** | A whip-like structure projecting from a bacterium or zoospore and functioning as an organ of locomotion; also called a *cilium*. |

| Term | Terminology |
| --- | --- |
| **Focus** | The site of initial disease infection from which secondary spread may occur. |
| **Forma specialist (f.sp.)** | A group of races and biotypes of a pathogen species that can infect only plants within a certain host genus or species. |
| **Forma specialist (f.sp.)** | A group of races and biotypes of a pathogen species that can infect only plants within a certain host genus or species. |
| **Free-living** | Of a microorganism that lives freely, unattached, or a pathogen living in the soil, outside its host. |
| **Frevind's incomplete adjuvant** | A substance containing an emulsifier and mineral oil which is mixedwith a virus before it is injected into the muscles of an animal to produce antiserum. The adjuvant allows slow release of the virus following injection. |
| **Fructification** | Production of spores by fungi; also, a fruiting body. |
| **Fruiting body** | A complex fungal structure containing spores. |
| **Fumigant** | A toxic gas or volatile substance that is used to disinfest soil or certain areas from various pests. |
| **Fumigation** | The application of a fumigant for disinfestations of an area or soil. |
| | Application of fungicides to foliage or roots through the irrigation system. |
| **Functional genomics** | Genetic studies focusing on the functions and interactions of genes or groups of genes that may belong to plants, pathogens, or both. |
| **Fungicide** | A compound toxic to fungi. |
| **Fungistatic** | A compound that prevents fungus growth without killing the fungus. |
| **Gall** | A swelling or overgrowth produced on a plant as a result of infection by certain pathogens. |
| **Gametangium** | A cell containing gametes or nuclei that acts as gametes. |
| **Gamete** | A male or female reproductive cell or nuclei within a gametangium. |

| Term | Terminology |
|------|-------------|
| **Gel chromatography** | A molecular sieving procedure, by which viruses bare separated from different sized molecules when passed through the pores of gel beads such as agarose. Used for virus purification. |
| **Gel double diffusion** | A serological test in which the antibody and antigen reactants diffuse towards each other in gel and react to form a visible precipitation line. |
| **Gene** | A linear portion of the chromosome that determines or conditions one or more hereditary characters; the smallest functioning unit of the genetic material. |
| **Gene cloning** | The isolation and multiplication of an individual gene sequence by its insertion into a bacterium, which can multiply the gene as it multiplies itself. |
| **Gene flow** | The process by which certain genes move from one population to another geographically separated one. |
| **Gene for gene** | The concept that for each gene for virulence in a pathogen there is a corresponding gene for resistance in the host toward that pathogen. |
| **Gene knockout** | The disruption of a target gene by transformation or mutation and characterization of the function of the gene by assessing the phenotype of the resulting mutant. |
| **Gene silencing** | The interruption or suppression of the activity of a targeted gene that prevents it from coordinating the production of specific proteins. |
| **Generation time** | The time from completion of the first bacterial division till the completion of next generation. |
| **Genetic drift** | The occurrence of random effects (mutations, etc.) in individuals of a population that affect the survival of various genetic traits in subsequent generations. |
| **Genetic engineering** | The use of *in vitro* techniques in the isolation, manipulation, recombination and expression of DNA. |
| | Alteration of the genetic composition of a cell or organism by various procedures (transformation, protoplast fusion, etc.). |

| Term | Terminology |
|------|-------------|
| **Genetic load or drag** | Accumulation of excess genes for any characteristic, even for virulence, that imposes a fitness penalty to the organism. |
| **Genome sequencing** | The orderly reading of all the millions of nucleotides constituting the total DNA of a living organism. |
| **Genomics** | Studies focusing on the analysis of whole genomes of organisms. |
| **Genotype** | The genetic constitution of an organism. |
| **Genotype flow** | Transfer of entire genotypes of asexually only reproducing microorganisms from one population to another. |
| **Germ theory** | The proposal that infectious and contagious diseases are caused by germs (microorganisms). |
| **Germ tube** | The early growth of mycelium produced by a germinating fungus spore. |
| **Gibberellins** | A group of plant growth-regulating substances with a variety of functions. |
| **G-proteins** | A subset of the GTPase super family of proteins that is concerned with the accuracy of recognition or interaction of activated receptor sites. |
| **Grafting** | A method of plant propagation by transplantation of a bud or a scion of a plant on another plant; also the joining of cut surfaces of two plants so as to form a living union. |
| **Growth regulator** | A natural substance that regulates the enlargement, division, or activation of plant cells. |
| **Gum** | Complex polysaccharidal substances formed by cells in reaction to wounding or infection. |
| **Gummosis** | Production of gum by or in a plant tissue. |
| **Guttation** | Exudation of water from plants, particularly along the leaf margin. |
| **Habitat** | The natural place of occurrence of an organism. |
| **Haploid** | A cell or an organism whose nuclei have a single complete set of chromosomes. |

| Term | Terminology |
|---|---|
| **Harpins or pilins** | Proteins coded by *hrp* (hypersensitive response and pathogenicity) genes that are used to make type III protein secretion systems. |
| **Haustorium** | A simple or branched projection of hyphae into host cells that acts as an absorbing organ. |
| **Hectare** | An area of land equal to 2.5 acres. |
| **Hemibiotrophic** | An organism that lives part of its life as a parasite on another organism and the other part as a saprophyte. |
| **Hemibiotrophs** | These are organisms which attack living tissues in the same way as biotrophs but continue to develop and sporulate after the tissue is dead. Typical examples of these are leaf-spotting fungi. |
| **Herbaceous plant** | A higher plant that does not develop woody tissues. |
| **Hermaphrodite** | An individual bearing both functional male and female reproductive organs. |
| **Heteroecious** | Requiring two different kinds of hosts to complete its life cycle, pertaining particularly to rust fungi. |
| **Heteroecism** | Some rust fungi requires two distinct species of hosts for the completion of the full life cycle such rust are called heteroecious and the phenomenon as Heteroecism. |
| **Heterokaryosis** | The condition in which a mycelium contains two genetically different nuclei per cell. |
| **Heterologus reaction** | A serological reaction in which an antiserum is reacted against and antigen other than the one used in its preparation. |
| **Heteroploid** | A cell, tissue, or organism that contains more or fewer chromosomes per nucleus than the normal IN or 2N for that organism. |
| **Heterothallic fungi** | Fungi producing compatible male and female gametes on physiologically distinct mycelia. |
| **Heterothallism** | The spermatization in rusts takes place between +ve and −ve spermatia of different origin i.e. the fungus show the phenomenon of heterothallism. |

| Term | Terminology |
|------|-------------|
| **Heterotrophs** | The organisms dependent on an organic carbon source. |
| **Homologous reaction** | A serological reaction in which an antiserum is reacted against the antigen used for its preparation. |
| **Homothallic fungus** | A fungus producing compatible male and female gametes on the same mycelium. |
| **Horizontal resistance** | Partial resistance equally effective against all races of a pathogen. |
| **Hormone** | A growth regulator, frequently referring particularly to auxins. |
| **Host** | An organism that harbours or supports the activities of a parasite is known as the host. |
| | A plant that is invaded by a parasite and from which the parasite obtains its nutrients. |
| **Host range** | The various kinds of host plants that may be attacked by a parasite. |
| **Hyaline** | Colorless; transparent. |
| **Hybrid** | The offspring of two individuals differing in one or more heritable characteristics. |
| **Hybridization** | The crossing of two individuals differing in one or more heritable characteristics. |
| **Hybridoma** | A hybrid animal cell produced by the fusion of a spleen cell and a cancer cell and able to produce monoclonal antibodies and to multiply. |
| **Hydathodes** | Structures with one or more openings that discharge water from the interior of a leaf to its surface. |
| **Hydrolysis** | The enzymatic breakdown of a compound through the addition of water. |
| **Hyper parasite** | A parasite parasitic on another parasite. |
| **Hyperplasia** | A plant overgrowth due to increased cell division. |
| **Hypersensitivity** | Excessive sensitivity of plant tissues to certain pathogens. Affecter cells are killed quickly, blocking the advance of obligate parasites. |
| **Hypertrophy** | A plant overgrowth due to abnormal cell enlargement. |

| Term | Terminology |
| --- | --- |
| **Hypha** | A single branch of a mycelium. |
| **Hypo virulence** | Reduced virulence of a pathogen strain as a result of the presence of transmissible double-stranded RNA. |
| **Immune** | Cannot be infected by a given pathogen. |
| **Immunology** | The study of acquired immunity in animals and man against infectious disease. |
| **Imperfect fungus** | A fungus that is not known to produce sexual spores; also known as a deuteromycete or a mitosporic fungus. |
| **Imperfect stage** | The part of the lifecycle of a fungus in which no sexual spores are produced; the anamorphic stage. |
| **In breeding depression** | Loss in vigour due to continuous breeding. |
| ***In vivo*** | In culture, outside the host. |
| | In the host. |
| **Inclusion bodies** | Crystalline or amorphous structures in virus-infected plant cells that are produced by and consist largely of viruses and are visible under a compound microscope. |
| **Inclusion body** | Virus induced structures that may occur in the cytoplasm or nucleus of the infected plants. |
| **Incomplete dominance** | Occurs when a dominant gene is only partially expressed in the phenotype of the heterozygote. |
| **Incubation period** | The period between the time infection occurs and the appearance of the first symptoms. |
| **Incubation period** | The period of time between penetration of a host by a pathogen and the first appearance of symptoms on the host. |
| **Indexing** | A procedure to determine whether a given plant is infected by a virus or a xylem or phloem infecting fastidious bacterium. It involves the transfer of a bud, scion, sap, etc. from one plant to one or more kinds of (indicator) plants that are sensitive to the virus or other pathogen. |

| Term | Terminology |
|---|---|
| **Indicator** | A plant that reacts to certain viruses or environmental factors with production of specific symptoms and is used for detection and identification of these factors. |
| **Induced systemic resistance** | A systemic resistance in plants that is triggered by certain strains of nonpath-ogenic root-colonizing bacteria; its signaling requires jasmonic acid and ethylene. |
| **Inducible or induced** | A substance, usually an enzyme, whose production has been or may be stimulated by another compound, often a substrate or a structurally related compound called an inducer. |
| **Infection** | It implies the establishment of the pathogen inside the host following penetration in which a parasitic relationship between the two organisms is established. |
| | The establishment of a parasite within a host plant. |
| **Infectious disease** | A disease that is caused by a pathogen that can spread from a diseased to a healthy plant. |
| **Infectivity assay** | A bioassay using mechanical sap transmission to quantitatively determine the amount of infectious virus. |
| **Infested** | Containing great numbers of insects, mites, nematodes, etc. as applied to an area or field. Also applied to a plant surface, soil, container, or tool con-taminated with bacteria, fungi, etc. |
| **Injectosome** | In gram-positive bacteria, injectosome is a complex consisting of the Exportable, a novel organelle that organizes the general secretary machinery (Sec), the Sec pathway for protein export, and a pore-forming cytolysin, and functions to inject signal transduction proteins into host cells. |
| **Injury** | Damage of a plant by an animal, physical, or chemical agent. |
| **Inoculate** | To bring a pathogen into contact with a host plant or plant organ. |
| **Inoculation** | Artificial introduction of micro-organisms on or into a medium living system. |

| Term | Terminology |
| --- | --- |
| **Inoculation** | The arrival or transfer of a pathogen onto a host. |
| **Inoculation feeding period** | The length of time a vector feeds on a test plant during transmission experiments. |
| **Inoculation threshold period** | The minimum feeding period a vector needs on a test plant to transmit a virus. |
| **Inoculum** | The pathogen or its parts that can cause infection; that portion of individual pathogens that are brought into contact with the host. |
| **Instars** | A growth phase between moults in an insect's. |
| **Integrated control** | An approach that attempts to use all available methods of control of a disease or of all the diseases and pests of a crop plant for best control results but with the least cost and the least damage to the environment. |
| **Integrated pest management** | The attempt to prevent pathogens, insects, and weeds from causing economic crop losses by using a variety of management methods that are cost effective and cause the least damage to the environment. |
| **Intercalary** | Formed along and within the mycelium, not at the hypha I tips. |
| **Intercellular** | Between cells. |
| **Intracellular** | Within or through the cells. |
| **Introns** | Sections of 70-140 nucleotide noncoding pre-messenger RNA that exist between exons and are spliced during the processing of mRNA. |
| **Invasion** | The spread of a pathogen into the host. |
| **IPM** | The attempts to prevent pathogens, insects, and weeds from causing economic crop losses by using a variety of management methods that are cost effective and cause the east damage to the environment. |
| **Isoelectric point** | The pH at which a virus particle has a zero net charge. |
| **Isolate** | A single spore or culture and the subcultures derived from it. Also used to indicate collections of a pathogen made at different times. |

| Term | Terminology |
|------|-------------|
| **Isolation** | The separation of a pathogen from its host and its culture on a nutritive medium. |
| **Isometric** | Used to describe virus particles that are approximately spherical in shape. |
| **Isozymes** | The different forms of an enzyme that carry out the same enzymatic reaction but require different conditions (pH, temperature, etc.) for optimum activity. |
| **Juvenile** | The life stages of a nematode between the embryo and the adult; an immature nematode. |
| **Kilobase** | One thousand continuous bases (nucleotides) of single-stranded RNA or DNA. |
| **Kinase** | A protein enzyme that phosphorylates (adds phosphate), and thereby activating a target protein. |
| **Koch's postulates** | Criteria proposed by Koch for proving the pathogenicity of an organism; (1) the suspected causal organism must be constantly associated with the disease; (2) it must be isolated and grown in pure culture; (3) when inoculated into a healthy plant it must reproduce the original disease. |
| **Latent infection** | The state in which a host is infected with a pathogen but does not show any symptoms. |
| **Latent period** | The period after a vector has acquired a virus before it can transmit it. Often observed in the case of persistent virus transmission. |
| **Latent virus** | A virus that does not induce symptom development in its host. |
| **Leaf spot** | A self-limiting lesion on a leaf. |
| **Lectins** | A group of plant proteins that bind to specific carbohydrates. |
| **Lenticel** | An opening in the stem of woody plants that has spongy cells at its base and allows for the exchange of gases between the plant and the atmosphere. |
| **Leucine-rich repeats (LRR)** | Repetitious segments of amino acids containing multiple copies of leucine on a protein. |

| Term | Terminology |
| --- | --- |
| Life-cycle | The stage or successive stages in the growth and development of an organism that occur between the appearance and reappearance of the same stage (e.g., spore) of the organism. |
| Lipids | Substances whose molecules consist of glycerin and fatty acids and sometimes certain additional types of compounds. |
| Local lesion | A localized spot produced on a leaf upon mechanical inoculation with a virus. |
| Longevity end point | The storage time after which a virus in a crude sap preparation loses its infectivity. Usually determined at 0°C or 20°C. |
| LRR proteins | Proteins containing leucine-rich repeats Macroscopic Visible to the naked eye without the aid of a magnifying lens or a microscope. |
| Lyophilization | A technique by which water is removed under vacuum while the preparation or tissue is frozen. Used to preserve viruses or antisera. |
| Malignant | Use of a cell or tissue that divides and enlarges autonomously, i.e., its growth can no longer be controlled by the organism on which it is growing. |
| Masked symptoms | Symptoms of a virus-infected plant that are absent under certain environmental conditions but appear when the host is exposed to certain conditions of light and temperature. |
| Mechanical inoculation | Inoculation of a plant with a virus through transfer of sap from a virus-infected plant to a healthy plant. |
| Meiospore | A spore produced through meiosis, a sexual spore. |
| Melanin | A dark brown to black compound found in the cell walls of some fungi and needed by them for pathogenicity. |
| Meristem-tip | The meristem dome of cell and one or two pairs of primordial leaves (0.5 to 1 mm in diameter), which comprises the explants removed from a bud and grown in tissue culture to produce a virus free plant. |

| Term | Terminology |
| --- | --- |
| **Messenger RNA (mRNA)** | A chain of ribonucleotides that codes for a specific protein. |
| **Metabolism** | The process by which cells or organisms utilize nutritive material to build living matter and structural components or to breakdown cellular material into simple substances to perform special functions. |
| | Organism utilizes nutrient material to build up living matter and structural components. |
| **Metal shadowing** | A technique used to prepare viruses for electron microcopy, in which the virus particles are exposed to the vapour of a heavy metal such as gold or platinum. Now replaced by the negative contrast staining method. |
| **Microarray analysis** | A molecular method employing large-scale hybridization of fluorescently labeled nucleic acids from biological samples to single- stranded cDNA sequences and used to study the degree of expression of thousands of genes in parallel during a certain treatment. |
| **Microbial genetics** | The study of the genetics heredity of microorganism. |
| **Microbiology** | The scientific study of micro-organisms including their taxonomy, morphology, physiology, biochemistry and genetics. |
| **Micrometer (μm)** | A unit of length equal to 1/1000 of a millimeter. |
| **Microscopic** | Very small; can be seen only with the aid of a microscope. |
| **Middle lamella** | The cementing layer between adjacent cell walls; it generally consists of pectinaceous materials, except in woody tissues, where pectin is replaced by lignin. |
| **Migratory** | Migrating from plant to plant. |
| **Mildew** | A fungal disease of plants in which the mycelium and spores of the fungus are seen as a whitish growth on the host surface. |
| **Millimeter (mm)** | A unit of length equal to 1/10 of a centimeter (cm) or 0.03937 of an inch. |

| Term | Terminology |
|------|-------------|
| **Mitosporic fungi** | Producing spores only through mitosis (imperfect fungi or deuteromycetes). |
| **Modal length** | The length that occurs most frequently in a population of virus particles |
| **Mold** | Any profuse or woolly fungus growth on damp or decaying matter or on surfaces of plant tissue. |
| **Molecular marker** | A molecular characteristic (a land mark) on a piece of DNA that can be used to compare that DNA for degrees of similarity with those of other microorganisms. |
| **Molt** | The shedding or casting off of the cuticle in a nematode or insect. |
| **Monoclonal antibodies** | Identical antibodies produced by a single clone of lymphocytes and reacting only with one of the antigenic determinants of a pathogen or protein. |
| **Monocyclic** | Having one cycle per season. |
| **Monoecious aphid** | An aphid that spends its complete life cycle on a single plant specie. |
| **Monophagous** | When and insect such as an aphid feeds on a specific type of host plant. |
| **Mosaic** | Symptom of certain viral diseases of plants characterized by intermingled patches of normal and light green or yellowish colour. |
| **Mottle** | An irregular pattern of indistinct light and dark areas. |
| **Movement protein** | One or more proteins of a virus that facilitate the movement of the virus through the plant and/or by the vector. |
| **Multicomponent virus** | A virus whose genome is divided into two or more parts, each par being separately encapsulated. Hence two or more components are needed to initiate an infection. |
| **Mutalism** | An interaction in which both the organisms benefit. |
| **Mutant** | An individual possessing a new, heritable characteristic as a result of a mutation. |

| Term | Terminology |
| --- | --- |
| **Mutation** | An abrupt appearance of a new character is tic in an individual as the result of an accidently change in a gene or chromosome. |
| **Mycelium** | The hypha or mass of hyphae that make lip the body of a fungus. |
| **Mycoplasma** | Pleomorphic prokaryotic microorganisms that lack a cell wall. |
| **Mycoplasma-like organisms** | Microorganisms found in the phloem and phloem parenchyma of diseased plants and assumed to be the cause of the disease; they resemble mycoplasmas in all respects except that they cannot yet be grown on artificial nutrient media. Now called phytoplasmas or spiroplasmas. |
| **Mycoplasmas** | Pleomorphic prokaryotic microorganisms that lack a cell wall. |
| **Mycorrhiza** | A symbiotic association of a fungus with the roots of a plant |
| **Mycotoxicoses** | Diseases of animals and humans caused by consumption of feed and foods invaded by fungi that produce mycotoxins. |
| **Mycotoxins** | Toxic substances produced by several fungi in infected seeds, feeds, or foods; and capable of causing illnesses of varying severity and death to animals and humans that consume such substances. |
| **Mycovirus** | A virus that infect fungi. |
| **Nanometer (nm)** | A unit of length equal to 1/1000 of a micrometer. |
| **Necrotic** | Dead and discolored. |
| **Nectarthode** | An opening at the base of a flower from which nectar exudes. |
| **Nectrotroph** | A microorganism feeding only on dead organic tissues. |
| **Negative contrast staining** | A staining procedure used to prepare virus particles for examination in an electron microscope. |
| **Nematicide** | A chemical compound or physical agent that kills or inhibits nematodes. |

| Term | Terminology |
|------|-------------|
| **Nematode** | Generally microscopic, worm like animals that live saprophytically in water or soil or as parasites of plants and animals. |
| **Neutralism** | The two microorganisms behave entirely independently |
| **Nitrogen fixation** | The conversion of atmospheric N into ammonia, nitrate and other N containing compounds by N fixing bacteria. |
| **Noisiness** | Errors that may occur during the replication of a virus's genome. |
| **Non-persistent transmission** | A type of insect transmission in which the virus is acquired by thevector after very short acquisition feeding times' and which is transmitted during very short inoculation feeding period. The vector remains viruliferous for only a short period unless it feeds again on infected plant. |
| **Non-host resistance** | Inability of a pathogen to infect a plant because the plant is not a host of the pathogen due to lack of something in the plant that the pathogen needs or to the presence of substances incompatible with the pathogen. |
| **Non-infectious disease** | A disease that is caused by an abiotic agent, i.e., by an environmental factor, not by a pathogen. |
| **Nuclear-binding site (NBSI)** | A protein whose config-uration of surface virus is acquired by xprotein to bind to and activate a protein in its nucleus. |
| **Nucleic acid** | An acidic substance containing pentose, phosphorus, and pyrimidine and purine bases. Nucleic acids determine the genetic properties of organisms. |
| **Nucleoprotein** | Referring to viruses: consisting of nucleic acid and protein. |
| **Nucleoside** | The combination of a sugar and a base molecule in a nucleic acid. |
| **Nucleotide** | The phosphoric ester of a nucleoside consisting of a base (purine or pyrimidine), a sugar, and phosphate. Nucleotides are the building blocks of DNA and RNA. |

| Term | Terminology |
|---|---|
| **Obligate aerobes** | Organisms which require oxygen for their normal growth and reproduction. |
| **Obligate anaerobes** | Organisms which do not require oxygen for their normal growth and reproduction; presence of oxygen inhibits, sometimes kills the growth of such characters. |
| **Obligate parasite** | A parasite that in nature can grow and multiply only on or in living organisms. |
| **Obligate photoautotrophs** | Is strictly dependent on light for its energy source and on $CO_2$ for its principle carbon source. |
| **Odontostyle** | The feeding stylet of a nematode. |
| **Oligogenic** | A character controlled by a few genes. |
| **Ontogenic resistance** | When the degree of resistance of a plant to a pathogen varies with age and the developmental stage of the plant. |
| **Oogonium** | The female gametangium of oomycetes containing one or more gametes. |
| **Oomycete** | A fungus-like chromistan that produces oospores; a water mold. |
| **Oospore** | A sexual spore produced by the union of two morphologically different gametangia (oogonium and antheridium). |
| **Operon** | A cluster of functionally related genes regulated and transcribed as a unit. |
| **Osmosis** | The diffusion of a solvent through a differentially permeable membrane from its higher concentration to its lower concentration. |
| **Ostiole** | A pore-like opening in perithecia and pycnidia through which the spores escape from the fruiting body. |
| **Ovary** | The female reproductive structure that produces or contains the egg. |
| **Oxidative phosphorylation** | The utilization of energy released by the oxidative reactions of respiration to form high-energy ATP bonds. |

| Term | Terminology |
|---|---|
| **Ozone (O$_3$)** | A highly reactive form of oxygen that may injure plants in relatively high concentrations |
| **Papilla** | A nipple-like protuberance of the cell wall on the inside of a cell being attacked by a fungus, apparently serving as a defense mechanism against infection |
| **Paraphysis** | A sterile hypha present in some fruiting bodies of fungi. |
| **Parasexualism** | A mechanism whereby recombination of hereditary properties occurs within fungal heterokaryons. |
| **Parasite** | An organism living on or in another living organism (host) and obtaining its food from the latter. |
| **Parenchyma** | A tissue composed of thin-walled cells that usually leave intercellular spaces between them. |
| **Pasteurization** | The process of heating, at a controlled temperature, liquid food or beverage to enhance the keeping quality as well as to destroy harmful micro-organisms. |
| **Pathogen** | An entity, usually a microorganism that can incite disease. |
| | An entity that can incite disease. |
| **Pathogenesis** | It is the sequence of progress in disease development from the initial contact between a pathogen and its host to the completion of the syndrome. |
| **Pathogenicity** | The capability of a pathogen to cause disease. |
| **Pathogenicity factors** | These factors are produced by pathogenicity genes, are essential, and are involved in all crucial steps in disease induction and development. |
| **Pathogenicity genes** | Genes those are essential for a pathogen to be able to cause disease. |
| **Pathotype** | This is a subdivision of a species distinguished by the common characters of pathogenicity, particularly in relation to the range of hosts. |
| **Pathovar** | In bacteria, a subspecies or group of strains that can infect only plants within a certain genus or species. |

| Term | Terminology |
|------|-------------|
| **Pectin** | A methylated polymer of galacturonic acid found in the middle lamella and the primary cell wall of plants. |
| **Pectinase** | An enzyme that breaks down pectin. |
| **Penetration** | The initial invasion of a host by a pathogen. |
| **Perfect stage** | The sexual stage in the life cycle of a fungus; the teleomorph. |
| **Periplasm** | The area between the plasma membrane and the cell wall. |
| **Perithecium** | The globular or flask-shaped ascocarp of the Pyrenomycetes, having an opening or pore (ostiole). |
| **Persistent transmission** | A type of insect transmission in which the virus is acquired by the vector only after a long acquisition feeding period, and in which there may be a latent period, following the acquisition feed, before the vector can transmit the virus. The vector remains viruliferous for a long period, often throughout its life span. The virus sometimes multiplies within the vector. |
| **Phage** | A virus that attacks bacteria; also called bacteriophage. |
| **Phenolic** | Applied to a compound that contains one or more phenolic rings. |
| **Phenotype** | The external visible appearance of an organism. |
| **Phloem** | Food-conducting tissues, consisting of sieve tubes, companion cells, phloem parenchyma, and fibers. |
| **Photoautotrophs** | Using light as an energy source and $CO_2$ as the principle carbon source. Example, photosynthetic organisms, algae. |
| **Photoheterotrophs** | Using light as the energy source and organic compound as principle carbon source. *e.g.* purple and green bacteria. |
| **Phototrophs** | The organisms that are able to use sunlight as an energy source. |
| **Phyllody** | Excessive production of leaves in place of shoots and blossoms. |

| Term | Terminology |
|---|---|
| **Physiotype** | It is a population of pathogen in which an individual have a particular character of physiology (but not pathogenicity) in common. |
| **Phytoalexin** | A substance that inhibits the development of a fungus on hypersensitive tissue formed when host plant cells come in contact with the parasite. |
| **Phytoanticipins** | Inhibitory antimicrobial compounds present in plant cells before infection. |
| **Phytopathogenic** | Term applicable to a microorganism that can incite disease in plants. |
| **Phytoplasmas** | Mollicutes .that infect plants and cannot yet be grown in culture, as contrasted to spiroplalsmas, which can be cultured. |
| **Phytotoxic** | Toxic to plants. |
| **Plant pathogenesis related proteins (PR)** | Groups of proteins with different chemical hproperties produced in a cell within minutes or hours following inoculation, but all being more or less toxic to pathogens. |
| **Plantibodies** | Antibodies produced in transgenic plants expressing the antibody-producing gene(s) of a mouse that had been injected previously with a pathogen (usually a virus) that infects the plant. |
| **Plasmalemma** | The cytoplasmic membrane found on the outside of the protoplast adjacent to the cell wall. |
| **Plasmid** | A self-replicating, extra chromosomal, hereditary circular DNA found in certain bacteria and fungi, generally not required for survival of the organism. |
| **Plasmodesma (plural plasmodesmata)** | A fine protoplasmic thread connecting two protoplasts and passing through the wall that separates the two protoplasts. |
| **Plasmodium** | A naked, slimy mass of protoplasm containing numerous nuclei. |
| **Plasmolysis** | The shrinking and separation of the cytoplasm from the cell wall due to exosmosis of water from the protoplast. |
| **Plerome** | The plant tissues inside the cortex. |

| Term | Terminology |
|------|-------------|
| **Polyclonal antibodies** | The usual mix of antibodies present in the serum of the blood of an animal that has been injected with a pathogen or protein that generally has many antigenic determinants. |
| **Polycyclic** | Completes many (life or disease) cycles ill one year. |
| **Polyetic** | Requires many years to complete one life 1' disease cycle. |
| **Polygenic** | A character controlled by many genes. |
| **Polyhedron** | A spheroidal particle or crystal with many plane faces. |
| **Polymerase** | An enzyme that joins single small mole-cules into chains of such molecules (e.g., DNA, RNA). |
| **Polymerase chain reaction** | A technique that allows an almost infinite amplification (multiplication) of a segment of DNA for which a primer (short piece of that DNA) is available. |
| **Polymorphism** | An organism having several spore forms (more than two) in the life cycle is called polymorphic and the phenomenon of several spore forms formed by an organism is called polymorphism. |
| **Polysaccharide** | A large organic molecule consisting of many units of a simple sugar. |
| **Polysome (or polyribosome)** | A cluster of ribosome's associated with messenger RNA. |
| **Population genetics** | Population genetics is the study of allele frequency distribution and change under the influence of the four main evolutionary processes, natural selection, genetic drift, mutation and gene flow. |
| **Precipitin** | The reaction in which an antibody causes visible precipitation of antigens. |
| **Primary host** | The plant on which the sexual forms of an aphid mate and lay eggs to overwinter. |
| **Primary infection** | The first infection of a plant by till' overwintering or over summering pathogen. |
| **Primary inoculum** | The overwintering or over summering pathogen or its spores that cause primary infection. |

| Term | Terminology |
| --- | --- |
| **Primary symptoms** | The symptoms that develop at the site of virus entry are called as primary symptoms. |
| **Prisons** | Prisons are small, proteinaceous infections particles that resist inactivation by procedures which affect nucleic acids. To date, no detectable nucleic acids of any kind and no virus like particles have been associated with prisons. Prisons cause 5 crappie and other spongiform encephalopathies of animals and humans. |
| **Probe** | A radioactive nucleic acid used to detect the presence of a complementary strand by hybridization. |
| **Programmed cell death** | Death of specific cells of an organism, the initiation and execution of which is controlled by the organism. |
| **Prokaryote** | A microorganism whose genetic material is not organized into a membrane bound nucleus, for example bacteria or millicutes. |
| **Promoter** | A region on DNA or RNA recognized by the RNA polymerase in order to initiate transcription. |
| **Promycelium** | The short hypha produced by the teliospore; the basidium. |
| **Propagative virus** | A virus that multiplies in its insect vector. |
| **Propagule** | The part of an organism, such as a spore or a bacterium, that may be disseminated and reproduce the organism. |
| **Proteasome** | An extremely large protein complex that carries out most protein degradation in the nucleus and the cytoplasm. |
| **Protectant** | A substance that protects an organism against infection by a pathogen. |
| **Protein kinases** | Proteins that act as signal transducers and amplifiers by responding to the size of the input signal through a proportional increase in activity and corresponding cellular response. |
| **Protein subunit** | A small protein molecule that is the structural and chemical unit of the protein coat of a virus. |

| Term | Terminology |
|------|-------------|
| **Proteome** | The total of proteins produced by an organism, or produced under certain developmental or environmental conditions. |
| **Proteomics** | The study of the identity and function of the proteins produced by an organism. |
| **Protocooperation** | An association of mutual benefit to the two species but without the cooperation being obligatory for their existence or for their performance of some reaction. |
| **Protoplast** | A plant cell from which the cell, wall has been removed. The organized living unit of a single cell; the cytoplasmic membrane and the cytoplasm, nucleus, and other organelles inside it. |
| **Protoplast fusion** | A tissue culture procedure for somatic hybridization that is used in cell manipulation studies. |
| **Protozoa** | Individual organisms of the kingdom Protozoa or of the phylum Protozoa of the kingdom Protista. Among the plant pathogens, it includes Myxomycetes, Plasmodiophoromycetes, and Flagellate protozoa. |
| **Pseudo-recombinants** | New strains of a virus that result from the reassortment of genome nucleic acids during the replication of viruses with divided genomes in mixed infections. |
| **Pseudofungi** | A name formerly used for Myxomycetes, Plasmodiophoromycetes, and Oomycetes, all of which were thought to be fungi until about 1990, but now the first two are considered protozoa (protista) and the Oomycetes are considered chromista. All three, however, continue to be studied along with the true fungi (Chytridiomycetes, Zygomycetes, Ascomycetes, and Basidiomycetes). |
| **Pseudothecium** | The ascocarp of the Loculoascomycetes (ascostromatic ascomycetes) in which asci is formed directly in cavities within a stroma (matrix) of mycelium; Pseudothecium also called an ascostroma. |
| **Pure culture** | A culture containing only one species of organism. |

| Term | Terminology |
| --- | --- |
| **Purification** | The isolation and concentration of virus particles in a pure form, free from cell components. |
| **Pustule** | Small blister-like elevation of epidermis created as spores form underneath and push outward. |
| **Pycnidium** | An asexual spherical or flask-shaped fruiting body lined inside with conidiophores and producing conidia. |
| **Pycniospore** | Also called a spermatium. A spore produced in a pycnium (sperm gonium). |
| **Pycnium** | Also called a sperm gonium. In some basidiomycetes, it contains spermatia and receptive hyphae. |
| **Quarantine** | Control of import and export of plants to prevent spread of diseases and pests. |
| **Quorum sensing** | Dependence of bacterial or spore behaviour and pathogenicity on their cells reaching a certain density by sensing the concentration of certain signal molecules in their environment. |
| **Race** | A genetically and often geographically distinct mating group within a species; also a group of pathogens that infects a given set of plant varieties. |
| **Radial diffusion** | An immunodiffusion serology test in which liquid antigen (or antibody) is placed in a well cut in gel containing the other reactant, and allowed to diffuse out into the gel. |
| **Reactive oxygen radicals** | Oxygen species much more reactive than molecular oxygen ($O_2$), which, upon contact of a resistant cell with a pathogen, react with and quickly oxidize various cellular components into compounds toxic to the pathogen. |
| **Recessive gene** | A gene that is not expressed in the phenotype of the heterozygote. |
| **Recognition factors** | Specific receptor molecule or structures on the host (or pathogen) that can be recognized by the pathogen (or host). |
| **Recombinant** | A new strain of a virus that occurs as a result of the breakage and renewal of covalent links in a nucleic acid chain, so that the nucleic acid is rearranged in the chain. |

| Term | Terminology |
| --- | --- |
| **Reflective mulch** | A polythene or straw layer placed on the soil surface around the crop plant to control air borne insect vectors. |
| **Resistance** | The ability of an organism to exclude or overcome, completely or in some degree, the effect of a pathogen or damaging factor. |
| **Resistant** | Possessing qualities that hinder the development of a given pathogen; infected lime or not at all. |
| **Resting spore** | A sexual or other thick-walled spore of a fungus that is resistant to extremes in temperature and moisture and which often germinates only after a period of time from its formation. |
| **Restriction enzymes** | A group of enzymes from bacteria that break internal bonds of DNA at highly specific points. |
| **Reverse transcription** | Copying of RNA into DNA. |
| **Rhizoid** | A short, thin hypha growing in a root-like fashion toward the substrate. |
| **Rhizosphere** | The soil near a living root. |
| **Ribonuclease (RNase)** | An enzyme that breaks down RNA. |
| **Ribonucleic acid (RNA)** | A nucleic acid involved in protein synthesis; also the most common nucleic acid (genetic material) of plant viruses. |
| **Ribosome** | A sub cellular particle involved in protein synthesis. |
| **Rickettsiae** | Microorganisms similar to bacteria in most respects but generally capable of multiplying only inside living host cells; parasitic or symbiotic. |
| **Ring spot** | A circular area of chlorosis with a green center; a symptom of many virus diseases. |
| **Rosette** | Short, bunchy habit of plant growth. |
| **Rot** | The softening, discoloration, and often disintegration of a succulent plant tissue as a result of fungal or bacterial infection. |
| **Russet** | Brownish roughened areas on skin of fruit as a result of cork formation. |

| Term | Terminology |
| --- | --- |
| **Rust** | A disease giving a "rusty" appearance to a plant and caused by one of the Uredinales (rust fungi). |
| **Sanitation** | The removal and burning of infected plant parts, decontamination of tools, equipment, hands, etc. |
| **Saprophyte** | An organism that uses dead organic material for food. |
| **Satellite nucleic acids** | They are nucleic acids which have no (or very little) sequence similarity with the viral genome, yet depend on the virus for replication and encapsulation. They are mainly associated with plant viruses but are dispensable as they are not part of the genome. They interfere with virus replication and may or may not the symptom expression. |
| **Satellite viruses** | These are also viruses (having own cont protein) which depend for their replication on helper viruses. |
| **Scab** | A roughened, crust-like diseased area on the surface of a plant organ; a disease in which such areas form. |
| **Scion** | A piece of twig or shoot inserted on another by grafting. |
| **Sclerotium** | A compact mass of hyphae with or without host tissue, usually with a darkened rind, and capable of surviving under unfavourable environmental conditions. |
| **Scorch** | "Burning" of leaf margins as a result of infection or unfavorable environmental conditions. |
| **Screening test** | A test to observe the response of a range of plant cultivars or types to virus infection. |
| **Secondary infection** | Any infection caused by inoculums produced as a result of a primary or a subsequent infection; an infection caused by secondary inoculums. |
| **Secondary infection** | Any infection caused by inoculums produced as a result of a primary or a subsequent infection; an infection caused by secondary inoculums. |

| Term | Terminology |
|------|-------------|
| **Secondary inoculums** | Inoculums produced by infections that took place during the same growing season. |
| **Secretome** | The total of proteins secreted by an organism or sets of proteins secreted under certain conditions. |
| **Sedentary** | Staying in one place; stationary. |
| **Sedimentation coefficient** | The rate of sedimentation of a virus per unit centrifugal field measured in Svedberg units. |
| **Selection** | The process by which populations of the fittest variants in a particular environment increase in frequency while those of less fit variants decrease. |
| **Semi-persistent transmission** | Virus transmission by an insect vector that is intermediate between non-persistent and persistent transmission. |
| **Septate** | Having cross walls. |
| **Septum** | A cross wall (in a hypha or spore). |
| **Serology** | A method of identifying microorganisms and viruses, their chemical components and relations to one another. |
| **Serology** | A method using the specificity of the antigen-antibody reaction for the detection and identification of antigenic substances and the organisms that carry them. |
| **Serotype** | A serotype is a population of a pathogen (usually a I bacterium or virus) in which all individuals possess a given character of serology in common. On the basis of serological tests, differences between apparently similar organisms or viruses may be found and the subdivisions thus formed are called serotypes or strains. |
| **Serum** | The clear, watery portion of the blood remaining after coagulation. |
| **Sexual** | Participating in or produced as a result of a union of nuclei in which meiosis takes place. |
| **Shock symptoms** | The severe, often necrotic symptoms produced on the first new growth following infection with some viruses; also called acute symptoms. |

| Term | Terminology |
| --- | --- |
| **Shot hole** | A symptom in which small diseased fragments of leaves fall off and leave small holes in their place. |
| **Sieve plate** | Perforated wall area between two phloem sieve cells through which they are connected. |
| **Sieve tube** | A series of phloem cells forming a long cellular tube through which food materials arc transported. |
| **Sign** | The pathogen or its parts or products seen on a host plant. |
| **Signal molecules** | Host molecules that react to infection by a pathogen and transmit the signal to and activate proteins and genes in other parts of the cell and of the plant so they will produce the defense reaction. |
| **Signal transduction** | The means by which cells construct and deliver responses to a signal, generally involving intracellular Ca and protein kinases. |
| **Signaling genes** | Genes that respond to changes in the environment and set off signaling cascades that alter the expression of the genes of the organism. |
| **Signaling pathways** | The series of compounds involved in the transmission of cellular signals, often involving several protein kinases functioning in series. |
| **Slime molds** | Formerly fungi, now protozoa of the class Myxomycetes; also superficial diseases caused by these pseudofungi on lowlying plants. |
| **Smut** | A disease caused by smut fungi (Ustilaginales) characterized by masses of dark, powdery and some-times odorous spores. |
| **Soft rot** | A rot of a fleshy fruit, vegetable, or ornamental in which the tissue becomes macerated by the enzymes of the pathogen. |
| **Soil inhabitants** | Microorganisms able to survive in the soil indefinitely as saprophytes. |
| **Soil solarization** | Attempt to reduce or eliminate pathogen populations in the soil by covering the soil with clear plastic so that sun rays will raise the soil temperature to levels that kill the pathogen. |
| **Soil transients** | Parasitic microorganisms that can live in the soil for short periods. |

| Term | Terminology |
|------|-------------|
| **Somaclonal variation** | Variability in clones generated from a single mother plant, leaf, etc., by tissue culture. |
| **Somatic hybridization** | Production of hybrid cells by fusion of two protoplasts with different genetic makeup. |
| **Sooty mold** | A sooty coating on foliage and fruit formed by dark hyphae of fungi that live in the honeydew secreted by insects such as aphids, mealy bugs, scales, and whiteflies. |
| **Sorus** | A compact mass of spores or fruiting structure found especially in rusts and smuts. |
| **Spermagonium (formerly pycnium)** | A fruiting body of rust fungi in which gametes or gametangia are produced. |
| **Spermatium (formerly pycniospore)** | The male gamete or gametangium of rust fungi. |
| **Spiroplasmas** | Pleomorphic, wall-less microorganisms present in the phloem of diseased plants; often helical in culture and thought to be a kind of mycoplasma. |
| **Sporagiophore** | A specialized hypha bearing one or more sporangia. |
| | Non-motile, asexual spore borne in a sporangium. |
| **Sporangium** | A container or case of asexual spores. In some cases it functions as a single spore. |
| **Spore** | The reproductive unit of fungi consisting of one or more cells; in function, it is analogous to the seed of green plants. |
| **Sporidium** | The basidiospore of smut fungi. |
| **Sporodochium** | A fruiting structure consisting of a cluster of conidiophores woven together on a mass of hyphae. |
| **Sporophore** | A hypha or fruiting structure bearing spores. |
| **Sporulate** | To produce spores. |
| **Spur precipitation line** | An antibody-antigen precipitation line formed when two antigenic ally distinct strains of a virus are placed in adjacent wells in a gel double-diffusion test. |

| Term | Terminology |
|------|-------------|
| **Stem pitting** | A symptom of some viral diseases characterized by depressions on the stem of the plant. |
| **Sterigma** | A slender protuberance on a basidium that supports the basidiospore. |
| **Sterile fungi** | A group of fungi that are not known to produce any kind of spores. |
| **Sterilization** | The process of making things sterile through killing or excluding of microorganisms or their spores with heat, filters, chemicals or other sterilants. |
| | The elimination of pathogens and other living organisms from soil, containers, etc., by means of heat or chemicals. |
| **Strain** | The descendants of a single isolation in pure culture; an isolate. Also a group of similar isolates; a race. In plant viruses, a group of virus isolates having most of their antigens in common. |
| **Stroma** | A compact mycelial structure on or in which fructifications are usually formed. |
| **Stylet** | A long, slender, hollow feeding structure of nematodes and some insects. |
| **Stylet borne** | A virus borne on the stylet of its vector; a non-circulative virus. |
| **Substrate** | The material or substance on which a microorganism feeds and develops; also a substance acted upon by an enzyme. |
| **Suppressive soils** | Soils in which certain diseases are suppressed because of the presence in the soil of microorganisms antagonistic to the pathogen. |
| **Suspect** | Any plant that can be attacked by a given pathogen; a host plant. |
| **Susceptibility** | It is the inability of a plant to resist the effect of a pathogen or any other damaging factor. Susceptibility of a given individual may be increased or decreased by environmental factors; It may also vary between individuals, varieties or species because of differences in the inherited characteristics which affect susceptibility. |

| Term | Terminology |
|------|-------------|
| **Susceptibility** | The inability of a plant to resist the effect of a pathogen or other damaging factor. |
| **Susceptible** | Lacking the inherent ability to resist disease or attack by a given pathogen; non-immune. |
| **Svedberg units(S)** | The units used to measure the rate of sedimentation of a virus to determine its sedimentation coefficient. |
| **Symbiosis** | The two symbionts behave relying on one another and both benefiting by the relationship. |
| | A mutually beneficial association of two or more different kinds of organisms. |
| **Symptom** | The external and internal reactions or alterations of a plant as a result of a disease. |
| **Symptomless carrier** | A plant that, although infected with a pathogen (usually a virus), produces no obvious symptoms. |
| **Symptoms** | The external and internal reactions or alternations of a plant as a result of a disease. |
| **Syncytium** | A multinucleate mass of protoplasm surrounded by a common cell wall. |
| **Syndrome** | The overall development and expression of a disease in a plant. |
| **Synergism** | The concurrent parasitism of a host by two pathogens in which the symptoms or other effects produced are of greater magnitude than the sum of the effects of each pathogen acting alone. |
| **Systemic** | Spreading internally throughout the plant body; said of a pathogen or a chemical. |
| **Systemic acquired resistance** | Systemically activated resistance after primary infection with a necrotizing pathogen accompanied by increased levels of salicylic acid and pathogenesis-related proteins. |
| **Teleomorph** | The sexual or so-called perfect growth stage or phase in fungi. |
| **Teliospore** | The sexual, thick-walled resting spore of rust and smut fungi. |
| **Telium** | The fruiting structure in which rust teliospores are produced. |

| Term | Terminology |
|------|-------------|
| **Thermal death point** | Temperature at which the organism is killed at 10 minute exposure. |
| **Tissue** | A group of cells of similar structure that perform a special function. |
| **Tolerance** | The ability of a plant to sustain the effects of a disease without dying or suffering serious injury or crop loss; also the amount of toxic residue allowable in or on edible plant parts under the law. |
| **Toxicity** | The capacity of a compound to produce injury. |
| **Toxin** | A compound produced by a microorganism; being toxic to a plant or animal. |
| **Transcapsidation** | The encapsulation of the nucleic acid of one virus strain with the protein of another, during simultaneous infection and replication of two strains. |
| **Transcription** | Copying of a gene into RNA; also copying of a viral RNA into a complementary RNA. |
| **Transduction** | The transfer of genetic material from one bacterium to another by means of a bacteriophage. |
| **Transfer RNA** | The RNA that moves amino acids to the ribosome to be placed in the order prescribed by the mRNA. |
| **Transformation** | The change of a cell through uptake and expression of additional genetic material. |
| **Transgenic (or trans- -formed) plants** | Plants into which genes from other plants or other organisms have been' introduced through genetic engineering techniques and are expressed, i.e., produce the expected compound or function. |
| **Translation** | Copying of mRNA into protein. |
| **Translocation** | Transfer of nutrients or virus through the plant. |
| **Transmission** | The transfer or spread of a virus or other pathogen from one plant to another. |
| **Transovarial transmission** | When virus is transmitted through the egg of the infected vector to its progeny. |
| **Transpiration** | The loss of water vapor from the surface or leaves and other above ground parts of plants. |

| Term | Terminology |
| --- | --- |
| **Transposable element** | A segment of chromosomal DNA that can move around (transpose) in the genome and integrate at different sites on the chromosomes. |
| **Transstadial** | When virus is retained through the moult of its insect vector. |
| **Tumor** | An uncontrolled overgrowth of tissue or tissues. |
| **Tylosis** | An overgrowth of the protoplast of a parenchyma cell into an adjacent xylem vessel or tracheid. |
| **Ubiquitin** | A small protein found in plants involved in the degradation of proteins. |
| **Ubiquitination** | The attachment of one or more ubiquity molecules to proteins destined for degradation and delivery to the proteasome where they are degraded. |
| **Uredium** | The fruiting structure of rust fungi in which; uredospores are produced. |
| **Variability** | The property or ability of an organism to change its characteristics from one generation to the other. |
| **Vascular** | Term applied to a plant tissue or region consisting of conductive tissue; also a pathogen 111.11 grows primarily in the conductive tissues of a plant. |
| **Vector** | An animal able to transmit a pathogen. III genetic engineering, vector (or cloning vehicle), a sell replicating DNA molecule, such as a plasmid or virus, used to introduce a fragment of foreign DNA into 1 host cell |
| **Vegetative** | Asexual; somatic. |
| **Vegetative incompatibility** | Failure, of the hyphae of strains of the same species of a fungus to fuse and form anatomizes. |
| **Vertical resistance** | Complete resistance to some races of a pathogen but not to others. |
| **Vesicle** | A bubble-like structure produced by a zoos sporangium in which zoospores are released or are differentiated. |
| **Vessel** | A xylem element or series of such elements whose function is to conduct water and mineral nutrients. |
| **Viable count** | A determination of number of cells in a population which are capable of growth and reproduction. |

| Term | Terminology |
|------|-------------|
| **Viperously** | A method of a sexual reproduction occurring in aphids by which the young are borne alive and active. |
| **Virescent** | A normally white or colored tissue that develops chloroplasts and becomes green. |
| **Virion** | A single complete viral particle. |
| **Viroids** | Small, low molecular weight ribonucleic acids (RNA) that can infect plant cells, replicate themselves, and cause disease. |
| **Virulence** | The degree of pathogenicity of a given pathogen. |
| **Virulence factors** | Coded for by virulence genes those are helpful but not essential for induction and development of disease. |
| **Virulence genes** | Enable a pathogen to express increased virulence on only one or a few related hosts. |
| **Virulent** | Capable of causing a severe disease; strongly pathogenic. |
| **Viruliferous** | Said of a vector containing a virus and capable of transmitting it. |
| **Virus** | A submicroscopic obligate parasite consisting of nucleic acid and protein. |
| **Virus cryptogram** | A descriptive code summarizing the main properties of a virus. |
| **Virus inactivator** | A chemical that inactivates a virus causing loss of infectivity. |
| **Virus inhibitor** | A chemical that prevents virus transmission without inactivating the virus. Inhibition of transmission may be temporary and can sometimes be avoided by dilution. Inhibitors frequently act on the recipient host rather than on the virus itself. |
| **Virusoid** | The extra-small circular RNA component of some isometric RNA viruses. |
| **Volunteer** | A plant from a previous season's crop that regenerates in a subsequent crop. |
| **Wilt** | Loss of rigidity and drooping of plant parts, generally caused by insufficient water in the plant. |

| Term | Terminology |
| --- | --- |
| **Witches' broom** | Broom-like growth or massed proliferation caused by the dense clustering of branches of woody plants. |
| **Xylem** | A plant tissue consisting of tracheids, vessels, parenchyma cells, and fibers; wood. |
| **Yellows** | A plant disease characterized by yellowing and stunting of the host plant. |
| **Zoosporangium** | A sporangium which containing or producing zoospores. |
| **Zoospore** | A spore bearing flagella and capable of moving in water. |
| **Zygospore** | The sexual or resting spore of zygomycetes produced by the fusion of two morphologically similar gametangia. |
| **Zygote** | A diploid cell resulting from the union of two gametes. |

# 3

# ABBREVIATIONS

| Abbreviation | Full form |
| --- | --- |
| % | per cent |
| μ | microlitre |
| AMI | Association of Microbiologist of India |
| ATCC | American Type Culture Collection |
| BIS | Bureau of Indian Standards |
| BNF | Biological Nitrogen Fixation |
| BOD | Biochemical oxygen demand |
| BSI | Botanical survey of India |
| cm | Centimeter |
| CMV | Cauliflower inosaic virus |
| COD | Chemical oxygen demand |
| DBT | Department of Biotechnology |
| DEP | Dilution end point |
| dia | Diameter |
| DIP Act | Destructive Insect and Pest Act |
| DMC | Direct Microscopic Count |
| DNA | Deoxyribonucleic acid |
| DT | Doubling Time |
| e.g. | For example |
| EDTA | Ethylene Diamine Tetra Acetic acid |
| ELISA | Enzyme linked Immunosorbent Assay |
| EM | Electron microscopy |
| EST | Expressed Sequence Tag |
| et al | And others (et al) |
| etc | Excreta |
| FRI | Forest Research Institutes |
| hr | Hour(s) |

| Abbreviation | Full form |
|---|---|
| HR | Horizontal resistance |
| i.e. | That is |
| ISEM | Immunosorbent electron microscopy |
| ITCC | Indian Type Culture Collection, IARI |
| LIV | Longevity in vitro |
| LRR | Lucien rich repeats |
| MPN | Most Probable Number |
| MRNA | Messenger RNA |
| NBDC | National Biofertilizer Development Centre |
| NBPGR | National Bureau of Plant genetic Resources |
| NBSI | Nuclear binding site |
| Nm | Nanometer |
| No | Number |
| PCR | Polymerase Chain Reaction |
| PGPR | Plant Growth Promoting Rhizobacteria |
| PPB | Plant Pathogenic Bacteria |
| PPLO | Pleuropneumonia like organisms |
| PSB | Phosphorus solubilizing Bacteria |
| PSM | Phosphate Solubilizing Microorganisms |
| PVX | Potato virus X (Pot X) |
| PVY | Potato virus Y (Pot Y) |
| RBDC | Regional Biofertilizer Development Centre |
| RFLP | Restriction fragment length polymorphism |
| RNA | Ribonucleic acid |
| RNase | Ribonucleabe |
| S | Svedberg units |
| SPC | Standard Plate Count |
| TIP | Thermal inactivation point |
| TMV | Tobacco Mosaic virus |
| VAM | Vesicular Arbuscular Mycorrhizae |
| viz. | Videlicet (Namely) |
| VR | Vertical resistance |

# 4

# DISTINGUISH

## Difference between Bacteria and Viruses

| No. | Bacteria | Viruses |
|-----|----------|---------|
| 1. | Possess cellular organization | Do not possess cellular organization |
| 2. | Grow on inanimate media | Do not grow on inanimate media |
| 3. | Multiply by binary fission | Do not multiply by binary fission |
| 4. | Possess RNA and DNA | Possess RNA or DNA |
| 5. | Posses ribosome's | Do not possess ribosomes |
| 6. | Sensitivbe to antibiotic | Insensitive to antibiotic |
| 7. | Insensitive to interferon | Sensitive to interferon |

## Difference between Virus and Viroid

| Virus | Viroid |
|-------|--------|
| 1. Viruses are nuclear protein | Contains nucleic acids only (lacks protein coat) |
| 2. Nucleic acid may be RNA or DNA | RNA alone |
| 3. Molecular weight 1,000,000 to 10,000,000 (in self replicating viruses) Daltons | 75,000 to 120, 000 daltons |
| 4. Viruses are shown to encode a viral specific polymerase | Viroids lack mRNA activity |
| 5. Viruses are bigger than viroids | Viroids are as small as 246 nucleotides (CCCV) and less than 1 to $10^{th}$ size of the genome of the smallest plant virus viz., maize streak (2 single stranded DNA 2681 or 2687 nucleotides) |
| 6. In plan virus, high temperatures have been used as a method of freeing bulbs cuttings and seeds from viruses. | Increased viroid yields and symptoms are observed at higher or temperatures above 20°C and at least up to 35°C. |
| 7. In most viruses meristematic tissues have been used for production of healthy plants from infected plants. | Viroids are meristematic tissue pathogens. Their concentration and translocation are higher in growing parts of plants and they are usually present up to 0.2 mm of apex. |
| 8. Virus multiplication and retention in callus and cell suspension culture is less stable. | Viroid multiplication and retention in callus and cell suspension culture is much more stable. In some cases viroid concentration in cell suspension culture is more than that found in leaf tissue. |

## Difference between Viruses and PLOs (phytoplasma like Organisms)

| *Viruses* | *PLOs* |
|---|---|
| 1. Viruses are mesobiotic entities | PLOs are biotic agents |
| 2. Obligate parasite | Not obligate parasite |
| 3. Nucleic acid is either RNA or DNA and never both | Nucleic acid is DNA, and RNA |
| 4. Not visible under and optical microscope | Visible under an optical microscope |
| 5. The size is variable; spheres 10-80 nm; rods 190-1250 nm | Size ranges from 0.1-1.0 $\mu$m |
| 6. Rigid in shape; rod or spherical cannot be cultured in cell free medium | Pleiomorphic |
| 7. Cannot be cultured in cell free medium | Can be cultured in cell free medium. The colony shows fried egg appearance |
| 8. Protein is not synthesized by their own enzymes | Protein is synthesized by own enzymes |
| 9. Viruses depend on host nucleic acid for multiplication | PLOs do not depend on host nucleic acid for multiplication |
| 10.Inclusion bodies are produced within the cells of infected plants | No inclusion bodies are produced |
| 11. Light green and dark green spotting appear on leaves as symptoms | Entire leaf becomes pale or yellow |
| 12. Typical local lesions are produced on several herbaceous plants | No local lesions are produced |
| 13. Viruses rarely breaks dormancy | PLOs break dormancy of all auxiliary buds of the plants. |
| 14. Virescent flower production not changed and are normal | Production of virescent flowers (Green flower parts) and eventual cessation of flowers. |
| 15. Angles of branches of plants are not changed and are normal | PLOs change the angle of growth of the branches in diseased plants. As a result the branches become stiff and become upright |
| 16. Transmission by aphids, leafhopper, thrips, beetles, mites and nematodes | Transmission by leafhoppers |
| 17. Transmitted mechanically also | None of them are transmitted mechanically |
| 18. They are not sensitive to tetracyclines | PLOs are sensitive to tetracyclines and choramphenical. |

### Difference between Mycoplasma and Spiroplasma

| *Mycoplasma* | *Spiroplasma* |
|---|---|
| 1. Lacks cell wall | Lack true cell wall |
| 2. Shape-spheroidal to ovoid | Spiral or Helical |
| 3. Completely resistant to penicillin, sensitive to tetracycline and chloramphenicol | Resistant to penicillin but sensitive to tetracycline. |

### Difference between Non-Persistent, Semi-persistent, persistent viruses

| *Test* | *Non-ciculative* | | *Circulative* |
|---|---|---|---|
| | *Non-Persistent* | *Semi- Persistent* | *Persistent* |
| 1. Pre-acquisition fasting | Enhances chances of acquisition | No such effect | No such effect |
| 2. Tissue of acquisition | Epidermis | Epidermis, Mesophyll | Mesophyll, phloem |
| 3. Duration acquisition access | Longer the access, lesser the chance of acquisition | Longer the access, better the chance of acquisition | Longer the access, better the chances of acquisition |
| 4. Latent period | None | None | Always have a latent period |
| 5. Effect of moulting | Stop transmission | Stops transmission affected | Transmission not |
| 6. Injection of purified virus into the haemolymph | The insect not rendered infective | Not rendered infective | The insect becomes infective |
| 7. Retention of infectivity | Minutes to hours, rarely longer | Hours to days | Days to life long |
| 8. Vector specificity | Generally low | Medium | Medium to high |

### Difference between Ployclonal and Monoclonal antibodies

| *Polyclonal Antibodies* | *Monoclonal Antibodies* |
|---|---|
| 1. Every time we have to inject the animal in several doses | A small quantity of antigen is the antigen into enough to start with. |
| 2. Antibodies are produced in response to several antigens (Polyspecific). | Produced for a single antigen (unspecific) can be obtained even if the preparation is impure |
| 3. Continuous supply of antibody not possible | It is possible as the hybridomas can be preserved under liquid nitrogen |
| 4. Failure of recognition of antigens by the polyclonal antisera may occur at times | Highly specific antibodies are produced. The antigens not recognized by polyclonal antisera can be accurately recognized. |
| 5.Antibody production is completely in vivo | Antibody production completely in vitro |

*Contd...*

| Polyclonal Antibodies | Monoclonal Antibodies |
|---|---|
| 6. The quality and quantity of anti-varies from animal to animal & even from one bleeding to the next in a given animal. | Such situation never arises as bodies obtained the antibody production is specific to a single antigen |
| 7. May have cross fraction with other antigens | Cross reactivities to other antigens never occurs |
| 8. As a result of the polyclonal nature of conventional antiserum, no antiserum is precisely reproducible, and the antiserum developed in one laboratory is often different from the antiserum generated against the same pathogen in another laboratory. This commonly leads to confliction results. | Variability in the nature of antibodies never occurs even when prepared by different laboratories |
| **9. Advangate of pcAb** | **Disadvantage of mcAb** |
| Capacity to form large insoluble complexes with antigens or to agglutinate cells readily so that reactions can be seen and measured visually or photometrically. | Due to their monospecificity immune mcAb's reactions are not readily visible. However, this problem can be rectified. |

### Difference between Cross protection and Induced Resistance

| Cross protection | Induced resistance |
|---|---|
| 1. First organism (mild strain) in the plant acts directly (as an antagonist) against the second organism (Virulent strain of the pathogen) | The biotic or abiotic stress initiates the defense process whereby the host plant itself inhibits the challenger |
| 2. Mild strain acts by any one or a combination of antibiosis, competition of sites, nutrients, hyphal interference or parasitism of the second organism by the first within the host tissue. | It may be any combination of the forms of active self-defense of plants pathogens |

### Difference between SAR and ISR

| SAR | ISR |
|---|---|
| 1. Plastic in nature | Elastic in nature |
| 2. Induced by mild strains of pathogens and chemicals | Mediated by saprophytes and plant growth promoting rhizobacteria |
| 3. Salicylic acid pathway is involved | SA, JA and ethylene are involved |
| 4. Related to resistance against pathogens | Related to resistance against insects and pathogens |

### Difference between Horizontal and Vertical Resistance

| Horizontal Resistance | Vertical Resistance |
|---|---|
| 1. Controlled by many genes (Polygenic or multigene resistance) | Controlled by one or a few genes (monogenic or oligogenic) |

*Contd...*

| Horizontal Resistance | Vertical Resistance |
| --- | --- |
| 2. Each gene plays a minor role (minor gene resistance) | The gene plays a major role in expression of resistance (major gene resistance) |
| 3. Non-differential type of resistance (non-specific) | Differentiates between races of a pathogen (host variety resistant to one race but not to another race) |
| 4. Affected by environmental conditions | Less affected by environmental conditions |
| 5. Does not provide complete protection but slows down the disease spread | Provides complete resistance |
| 6. Quantitative in nature | Qualitative in nature |
| 7. Not completely broken down by a new race of the pathogen | Can be completely broken down by a new race of the pathogen. |

### Difference between R genes and Defense related genes

| R genes | Defense related genes |
| --- | --- |
| 1. Regulatory genes (metabolic regulation) | Non regulatory (Pathogenesis related) |
| 2. Non in all plants – in some plants only or cultivated only or wild plants. | Found in all plants |
| 3. Activated only during incompatible interaction | They are inducible by any agents (biotic or abiotic agents). |

### Difference between hrp gene and avr gene

| hrp gene | avr gene |
| --- | --- |
| 1. hrp gene functions on any plant | Function on plants carrying R genes |
| 2. Induces hypersensitive response and involved in pathogenesis | Induces hypersensitive response but are not involved in pathogenesis |
| 3. Regulated the transcription of avr gene | Regulated by hrp gene |

### Difference between Enzymes and Phytotoxins

| Enzymes | Phytotoxins |
| --- | --- |
| 1. All enzymes are proteinaceous in nature | All are secondary metabolites, polysaccharide proteins and organic acids |
| 2. Single/few genes involved | Large number of genes are involved in pathway process |
| 3. High molecular weight | Low molecular weight |
| 4. Sensitive to high temperature | Tolerant/resistant to high temperature |

## Difference between Host specific and host non-specific toxin

| Host specific toxin | Host non-specific toxin |
|---|---|
| 1. Narrow host range e.g. Pyricularin Rice blast) pyricularia grisea | Wide host range and toxic to several host plants, e.g. Phaseolotoxin, fusaric acid |
| 2. Few genes are involved | Several genes are involved |
| 3. Act as pathogenicity factor | Act as virulence factor |
| 4. Causes susceptibility by interrupting the biosynthetic pathway | Causes susceptibility |

## Difference between General suppressive soil and Specific suppressive soil

| General suppressive soil | Specific suppressive soil |
|---|---|
| **Quantitative** | **Quantitative** |
| 1. Total amount of microbiological activity at a time, critical to propagate germination of pathogen | Specific effects of indi vidual (or) selective group of microbial antagonistic to pathogen during any stages of pathogen |
| 2. More than, one microorganism is responsible for general suppression | Only one group of microorganism is responsible |

## Difference between Direct and Indirect – ELISA (Enzyme-Linked Immunosorbent Assay)

| Direct ELISA | Indirect – ELISA |
|---|---|
| 1. Ab of virus +virus + Enzyme conjugated | Virus+Ab of virus + Enzyme conjugated Ab |
| 2. Antibody, or antigen is immobilized on a solid surface | Antigen is bound to bottom of a microtitier plate. |
| 3. Amount of bound antigen, or by antibody, respectively is detected by colour changes in the case of enzyme linked detection | Specific antibody (primary antibody) is added that binds to the antigen. |
| | An enzyme linked secondary antibody is added |
| | Intensity of colour produced is proportional to the amount of bound primary antibody. |

## Difference between PCR, Southern, Northern and Western blot.

| Particulars | Polymerase chain Reaction (PCR) | Southern Blotting | Northern Blotting | Western Blotting |
|---|---|---|---|---|
| 1. Discovery | Kary Mullis, (1983) | Southern, (1975) | Alwine et. al. (1979) | Towbin et. al. (1979) |
| 2. Purpose | Amplification and specific sequencing of DNA and RNA (by RT-PCR) | Confirmation of specific DNA fragments | Confirmation of specific RNA fragments | Confirmation of gene expression at protein level |

*Contd...*

| 3. Steps involved | 1. Denaturation<br>2. Annealing<br>3. Extension | 1. Depurination<br>2. Denaturation<br>3. Neutralization<br>4. Hybridization | -<br>-<br>-<br>- | 1. Protein extraction<br>2. PAGE<br>3. Membrane transfer<br>4. Serology |
| --- | --- | --- | --- | --- |
| 4. Enzymes used | Taq polymerase | - | - | - |
| 5. Hybridization | No hybridization | Hybridization with known DNA probe | Hybridization with known RNA probe | Hybridization with specific antibodies |
| 6. Radioactive detection | No | Yes | Yes | No |

# 5

# PRINCIPLES AND THEORIES

## (1) PRINCIPLES OF NOMENCLATURE

1. The essential points in the nomenclature are: (a) to aim at fixing the names (b) to avoid or to reject use of names which may cause error or ambiguity or throw scientists into confusion.

2. In the absence of a relevant proof or where consequences of a rule are doubtful, established custom must be followed. In case of doubt a resume in which all pertinent facts are outlined should be submitted to the Judicial Commission for an opinion.

3. Nomenclature of bacteria is independent of botanical or zoological nomenclature.

4. Scientific names of all taxa are usually taken from Latin or Greek. When taken from any language other than Latin, they are treated as they were Latin. The classification rules for Latinization of Greek and other words of non-Latin origin should be followed.

5. Nomenclature deals with : (a) the terms which denote the categories of taxa such as species, genus, family and other relative ranks of these categories, (b) the names which are applied to individual taxa like *Bacillus subtilis* (a species), *Salmonella* (a genus), Spiralaceae (a family), Pseudomonales (a order), etc.

6. The rules and the recommendations of the bacteriological nomenclature are applied to all bacteria.

7. The term which denote the rank of taxa is defined as the principle category in ascending sequences are species, genus, family, order, class and division. In some families rank like tribe is applied.

8. The primary purpose of giving a name to a taxon is not to indicate the character or the history of the name of the taxon but to supply a meance of referring to it.

## (2) PRINCIPLES OF PURE CULTURE/TECHNIQUES

There are 5 steps/ principles in identification of micro-organisms.

1. Inoculation     2.    Incubation
3. Isolation       4.    Inspection
5. Identification

1. **Inoculation:** A suspension of cells or inoculum is introduced into artificially prepared nutrient material. In the case of Plant Pathogenic Bacteria, the PPB can be isolated readily using young lesions. The inoculum is mixed in the molten but cool medium and poured into the plates or alternatively the inoculum can be streaked using a inoculating needle over the surface of the solidifying medium. The medium is normally poured into sterile petriplates or glass tubes.

2. **Incubation**: The inoculated plates orb tubes are kept for incubation, i.e., at appropriate temperature and oxygen concentration for the development of culture. The controlled chambers used for the purpose are called incubators.

3. **Isolation:** Sub-culturing refers to the transfer of inoculum from the colony under aseptic conditions to fresh medium to get a pure culture. The single colonies are touched with an inoculating needle and transferred to agar slants in a tube. The isolation and purification can also be done by serial dilution plating. Pure cultures have only one type of colony with original culture characteristics.

4. **Inspection:** The cultures are observed for morphological and biochemical characters etc.

5. **Identification:** The identification of bacteria is done by stains, biochemical tests, DNA/RNA analysis.

## (3) TISSUE CULTURE TECHNIQUES USED FOR VIRUS ERADICATION

Various types of tissues have been cultured to produce virus-free plants from infected plant material, including callus protoplast various reproductive tissues meristem tips.

**Callus culture:** Several workers have been shown that healthy plantlets can be regenerated from TMV-infected tobacco callus. These virus free plantlets probably result from 'sectoring' of the cultured callus tissues into healthy and infected cells. It has been suggested that the virus free areas of tissue arise because virus replication is slower than cell proliferation.

Various studies suggest that the use of callus culture to produce virus free plants should be avoided if clonally uniformity is required in the generated plants. Plants generated from callus are frequently genetically different from their parent clone.

**Protoplast Culture:** Shepard (1975) has shown that virus free plants may be regenerated from protoplasts taken from PVX-infected tobacco leaves. He found that of 4140 plants regenerated, 7.5 per cent were virus free. The reason for this loss of virus appears to be the failure of the virus to infect every cell. Unfortunately, plants regenerated from protoplasts are also likely to be genetically variable.

**Culture of reproductive tissues:** A few workers have successfully cultured floral tissues to produce virus-free plants. This method has been particularly useful for citrus species, in which most of the viruses are not seed transmitted. The failure of citrus viruses to enter nucellar and ovular tissues has been used to produce healthy oranges.

The culture of floral meristem has been used to produce cauliflowers free of turnip and cauliflower mosaic viruses (Walkey et. al., 1974). In this species the primordial floral meristem (curd) reverts to vegetative growth in tissue culture enabling many plantlets to be regenerated from a single plant, which normally has only a single terminal bud and no auxiliary buds in the vegetative phase.

**Meristem-tip culture:** The most important and effective method of tissue culture for the production of virus free plants has been meristem tip culture. On the suitable medium, meristem-tips grow more quickly into plantlets than cultured tissues from other sources. The regenerated plantlets usually retain the genetic characteristics of the parent plant. The greater genetic stability of plants regenerated from meristem tip is probably due to the more uniformly diploid nature of the tips cells. Various terms have been used to describe the technique including bud tips, auxiliary bud, shoot-apex, meristem-tip, and meristem or simply tip culture.

**Contact Slide Technique:** It is one of the most interested methods to study the qualitative estimation of soil microflora. This method was proposed by Rossi in 1929 and subsequently by chlodny in 1930 and hence name is given as Rossi colonies barred slide technique.

## (4) THEORY OF SPONTANEOUS GENERATION

This theory held that the living organisms could arise spontaneously from inanimate or non-living matter. In 1748, Needham, an English clergyman and a naturalist, supported the theory of spontaneous generation. To prove his point, he boiled meat in a flask which was corked. He demonstrated that the contents had spoiled due to micro-organisms. We now know that, while he killed the vegetative bacteria, the bacterial spores were not killed at the temperatures used. In 1775, Spallanzani, an Italian investigator, challenged Needham and proved that the decay of meat by bacteria and other lower forms of life could be prevented by heating the material in a flask which was sealed in such a way as to exclude contamination from the air. His experiments supported his contention. Others failed to get similar results consistently. Nevertheless, the views put forward by Spallanzani at that time were not accepted. Antonie Laurent Lavosier (1143-1794) discovered the presence of oxygen in the air and its indispensability to sustain macroscopic forks of life. In 1775, he argued that this essential gas is vital for life, and therefore, sealed flasks that did not permit the entrance of oxygen were incapable of allowing spontaneous generation to occur.

Schwann and Cagniard Latour discovered the yeast cell and maintained that it caused fermentation. On the other hand, in 1839, Berselius, a Swedish chemist, strongly, defended the mechanistic explanation of fermentation and he was supported by Liebig, a German agricultural chemist. It was in 1860 that Pasteur furnished irrefutable evidence that micro-organisms arise from pre-existing living entities and that fermentation is a biological phenomenon rather than a purely chemical one. It may be mentioned that the autogenetic theory of disease continued to be dominant throughout the first half of the 19th century. The most influential proponents were Unger, Meyen and Liebig. Unger was an Australian Physician and Professor of Botany. He thought that fungi, associated with plant diseases, arose from the diseased plant because of abnormalities in plant juices and were thought to be the products rather than the causes of diseases. Pasteur was born in 1822 of humble parents in the village of Dole in France. He began his life as a chemist. He worked as a teacher at Strassburg and then at the Ecole des Beaux Arts in Paris. He never studied medicine, though his discoveries about, the protozoan nature of the silkworm and the confirmation of the bacterial origin of anthrax buried for ever the controversy over the theory of spontaneous generation.

## Koch's Postulate

The germ theory in relation to diseases of man and animals was established in 1876 by Robert Koch. Koch was a Germen physician and a co-worker of Pasteur. The first final proof of the causal relation of the anthrax bacillus to anthrax disease was published by Robert Koch in Germany in 1876. Koch (1843-1910) was trained as a physician; however he turned his attention to the new science of bacteriology. He began his work by establishing laboratory procedures and techniques that would yield reproducible results. Koch established the principles of pure culture technique, although it was Lister, the English surgeon, who pioneered work in the field of aseptic surgery. He was impressed by Pasteur's work and kept it as a basis of his own research. He was the first man to culture a bacterium capable of souring milk. In the medical field, Robert Koch discovered the bacterium that causes tuberculosis. He enunciated certain rules or criteria that should be satisfied before the identity of the disease producing organism in a particular disease can be established.

These rules are known as Koch's postulates and can be briefly summarized as follows:

1. A specific organism must always be associated with a disease.

2. The organism has to be isolated in pure culture.

3. The organism must be identified.

4. The organism must produce disease in a healthy susceptible host.

5. The organism must be isolated from the experimental host, again in pure culture and its identity established.

# 6

# SHORT EXPLANATIONS

**True Resistance:** Disease resistance that is genetically controlled by the presence of one or many genes for resistance in the plant in known as true resistance. There are two kinds of true resistance, i.e., horizontal and vertical.

**Horizontal resistance:** All the plants have certain, but not always the same, level of possibly unspecific resistance that is effective against each of their pathogens. Such resistance is sometimes called nonspecific, general, quantitative, adult plant, field, or durable resistance, but it is most commonly referred to as horizontal resistance. Horizontal resistance is controlled by many genes thereby the name polygenic or multigene resistance.

**Vertical resistance:** Many plant varieties are quite resistant to some races of a pathogen while they are susceptible to other races of the same pathogen. In other words, depending on the race of the pathogen used to infect a variety, the variety may appear strongly resistant to one pathogen race and susceptible to another race under a variety of environmental conditions. Such resistance differentiates clearly between races of a pathogen since it is effective against others, such resistance is sometimes called strong, major, specific, qualitative or differential resistance, but it is more commonly referred to as vertical resistance. Vertical resistance is always controlled by one or a few genes thereby the name is monogenic or oligogenic resistance.

## PLANT QUARANTINE

Plant quarantine is the legal restrictions on the movement of plant material/ agricultural commodities from one country to another country or from one region/ state to another region/state within the country for exclusion, prevention and minimizing the spread of plant disease in uninfected areas. The need for adoption of quarantine regulation act has risen with the fact that extensive and sudden damages have been caused by exotic pathogens, which have been introduced along with plants, plant parts or seeds. In 1914, Government of India passed quarantine law under the name DIP Act (Destructive Insect and Pest Act). In addition to the quarantine station at air ports, seaport is and land borders, the research materials are examined by National Bureau of Plant Genetic Resources (NBPGR), New Delhi for agricultural horticultural crops, Forest Research Institutes (FRI), Dehradoon for forest plants, and Botanical Survey of India, Calcutta for all other plants of economic importance.

# HYPERSENSITIVE REACTION

Most bacterial plant pathogens can induce a hypersensitive response when injected into the tissues if a non-host plant. Non-pathogenic bacteria and some plant pathogens, particularly those that are opportunistic or gall forming, don't incite this hypersensitive response. The reaction can provide an extremely useful presumptive diagnostic technique.

A variety of plants may be used, but for many bacteria is the preferred plant, since it is easy to cultivate and maintain. Its large cavities beside leaf veins make it relatively easy to infiltrate inoculum and its reaction to many pathogens is well known. But for *Xanthomonas* tomato is preferred. The procedure for tobacco hypersensitive tests is as follows. Dilute culture suspension of the test bacterium is inoculated by injection infiltration method in mesophyll of leaf blades to tobacco leaves with a hypodermic syringe with a needle of 0.4 mm diameter. Different culture suspension may be inoculated in different interveinal sections of the same tobacco leaf. The cultures which are plant pathogenic cause quick necrosis within 24-48 hrs with rapid collapse and water soaking of inoculated tissues. Results are visible as light localized necrosis within three days as positive. Only yellowing or browning with out collapse is negative result.

**Phytoalexins:** Phytoalexins are toxic antimicrobial substances produced in appreciable amounts in plants only after stimulation by various types of phytopathogenic microorganisms or by chemical and mechanical injury. Phytoalexins are produced by healthy cells adjacent to localized damaged and necrotic cells in response to material diffusing from the damaged cells. Phytoalexins are not produced during compatible biotrophic infections. Phytoalexins accumulate around both resistant and susceptible necrotic tissues. Resistance occurs when one or more phytoalexins reach a concentration sufficient to restrict pathogen development. Most known phytoalexins are toxic to and inhibit the growth of fungi pathogenic to plants, but some are also toxic to bacteria, nematodes and other organisms. Some of the better studied phytoalexins include phaseollin in bean, pisatin in pea, glyceollin in soybean, alfalfa, and clover, rishitin in potato, gossypol in cotton and capsidiol in pepper.

**Trichoderma:** Bio-control agent in the management of soil borne plant diseases:

1. *Trichoderma* is ubiquitous in nature.

2. *Trichoderma* grows rapidly and inhibit the mycelial growth of other fungi.

3. *Trichoderma* spp. can effectively control the soil borne fungi e.g. *Sclerotium rolfsii, Rhizoctonia solani, Pythium* spp. *Fusarium* spp. *Aspergillus niger*

4. Easy to isolate and non pathogenic to higher plants.

5. Acts as mycoparasite competes well for food and site and produces antibiosis, mycoparasitism and competition.

6. Not producing any adverse effect during handling.

7. Primarily reported as antifungal as well as antinematode.

## TYPES OF MICROSCOPY

### Stereoscopic Microscopy

Direct illumination is usually used but transmitted light from a substage mirror is most useful for examining thin material such as leaves for the presence of embedded fruit bodies or sclerotia. Whole plant organs can be examined and these microscopes may be mounted on extendible stands to cater for large objects such as logs or big fruits. These microscopes are available with fitted cameras for photography.

## HIGH POWER COMPOUND MICROSCOPY

This is essential for the detailed examination of tissues at the cellular level for the presence fungal hyphae, bacterial pathogens and other manifestations of disease, and for the identification of fungal and bacterial pathogens. At high power oil is used as a medium between the objective and the glass to reduce unwanted diffraction by air interfaces. The resolution of light microscope is limited by the wavelength of light so that objects less than about 0.2 ìm cannot be clearly seen and magnifications above about 1200X do not produce clear images.

### (a) Bright Field Illumination

Most compound microscopes used for plant pathology are set up for bright field illumination. For nearly all specimens ex4mined in a bright field and in white light, the microscope should be at constant image luminosity. Although direct light may be used, illumination is almost always transmitted through the specimen by focusing a beam of light onto it by means of a substage condenser; the double diaphragm method is the most frequent illumination method. Focusing of the light from the lamp through tt1e condenser, specimen and objective is critical for optimum performance.

### (b) Dark Field Illumination

The aim of dark field illumination is to use transmitted light from below to illuminate specimens so that they glow with a brilliant luminosity against a dark background. Various methods have been developed to highlight specimens by manipulating the light diffracted around their edges or through them, as observed on a dark field. This technique is particularly useful for examining pathogens that are translucent, branched, hairy or spiny. In a simple version of dark field illumination, a specimen may even be examined by low power microscopy by placing it on a sheet of dark cloth such as velvet across a hole in a table above a source of illumination. Among the more sophisticated techniques, incident or reflected light with an oblique or ring light source is often used with stereoscopic microscopes or binocular magnifiers. The dark field is not noticeable with this system, when the object fills the entire field of observation. With central dark field, a central or annular screen in the objective: stops direct light, but with annular dark field, annular light is transmitted from a numerical aperture greater than that of the lens. Similar effects can be achieved by placing a birefringent object between two crossed polarizers or examining the specimen at right angles to the direction of the beam of light in a flat field.

## (c) Phase Contrast

Most objects seen under a microscope are either coloured or dark, absorbing sufficient incident light so that their density and colour are illuminated as amplitude objects. Although, specimens who are thin and colourless appear transparent, their contours and details can be clearly seen in a dark field. However, even these specimens may be contrasted without staining if an annular phase plate is placed within the internal focal plane of the objective. This causes the phase shift to revert to that given by an 'amplitude object' as the phase contrast is converted to amplitude contrast. When using a phase contrast microscope, thin unstained specimens without thick membranes can be examined providing they cover only part of the field, the optical surfaces of both the microscope and the preparation are absolutely clean and he light source is more intense than that used for bright field microscopy. The phase contrast microscope is also known as the one-wave interference microscope.

## Fluorescence Microscopy

Fluorescence occurs when a specimen or part of a specimen absorbs light of one frequency and re-emits a different one usually displaced to the red (lower) end of the spectrum. Primary microfluoresence is caused when the specimens are fluorescent themselves, induced or secondary microfluoresence results when a stain selectively attached to a specimen fluoresces. Fluorochromes are attached to labeled antibodies in immunofluorescence. Frequently, direct examination by phase contrast is combined with fluorescence microscopy in order to accurately locate the fluorescent sources in the specimen in microscopy or host tissues. Similarly, fluorescence enables the use of ultraviolet (UV) light. Because of the sorter wavelength of UV and appropriate lenses are used in combination with UV fluorescent staining of the specimen to make it visible.

## Electron Microscopy

Although a compound light microscope is usually adequate for examining fungal pathogens, the detailed microscopic examination of tissue for viruses and sometimes bacteria requires the use of a transmission electron microscope. Electron microscopes rely on a beam of electrons produced from a hot filament (e.g. tungsten), rather than light as the source of illumination and their power of resolution is consequently very much greater as it is not limited by the wavelength of light. After the electrons are accelerated down the microscope by a potential difference (accelerating voltage) between the cathode (the filament) and the anode, electromagnetic lenses (comparable to a condenser in a light microscope) focus them into a narrow beam. However, since gas molecules in the air deflect and disperse the electron beam, the interior of electron microscopes must be maintained under vacuum. Electron, microscopes are of two types, transmission electron microscopes (TEMs) or scanning electron microscopes (SEMs). While a TEM is analogous to a compound light microscope. An SEM is more similar to a binocular stereoscopic microscope. Unfortunately, both are rather costly for ordinary use and unless the pathogen is abundant, the task of finding it in a section or on a coated grid can be very tedious and time-consuming, so immunological or other specific staining techniques may be used. Computer processing of electron microscope images is also a potentially valuable technique.

### *Transmission Electron Microscopy (TEM)*

When using a TEM, a beam of focused electrons is passed through a thin section of the specimen. Electromagnetic lenses magnify the image formed in the beam (like an objective an, eyepiece in a light microscope before the final image is projected onto a fluorescent screen where, it can be seen and photographed.

### *Scanning Electron Microscopy (SEM)*

The beam of focused electrons in an SEM is scanned across the specimen by a process similar to forming a television picture, when the beam of electron strikes the specimen emitting, secondary electrons which are attracted towards a positively charged grill and detection system. These generate variable strength electronic signals which, after amplification and processing produce the final image on a cathode-ray tube screen that can be photographed. Specimen examined with the SEM is often coated with a thin film of an electric conductor such a carbon or gold applied with a sputter coater; both reduce charging and increase electron emission. Although most material for the SEM is not sectioned, frozen host tissue can be fractured so that internal hyphae and other structures are exposed.

## BACTERIOCINS

Bacteriocins are proteinaceous substances which are lethal to other strains of bacteriocoins. The bacteriocarcinogeny is attributed to the plasmids and a bacterium is immune to its own bacteriocin. Bacteriocin Agrocin K84 produced by *Agrobacterium radiobacter* strain K84 is a substituted nucleotide adenosine. Bacteriocins which are heat stable and inactivated by trypsin are S type while those resistant to trpsin and heat labile are R type. Bacteriocins are adsorbed on the surface of bacterial cell and they stop biosynthesis of nucleic acids and result in cell death. Bacteriocins are named according to the bacteria producing them.

**The Coliform Group :** The coliform group of bacteria includes all the aerobic and facultatively anaerobic, gram-negative, non-sporulating bacilli that produce acid and gas from the fermentation of lactose. The classical species of this group are *Escherichia coli* and *Enterobacter aerogenes.* The relationship of these organisms to others of the enteric group—*Salmonella, Shigella, Proteus, Pseudomonas,* and *Alcaligenes,* all of which are gram negative, nonsporulating bacilli

*E. coli,* as we have already pointed out, is a normal inhabitant of the intestinal tract of man and other animals. *Ent. aerogenes* is most - frequently found on grains and plants but may occur in the feces of man and other animals. These species bear a very close resemblance to each other in their morphological and cultural characteristics. Consequently, it is necessary to resort to biochemical tests for differentiation. Tests with the following four characteristics are especially important for this purpose:

1.  Ability to produce indole. *E. coli* does, and *Ent. aerogenes* does not.

2.  Amount of acidity Produced in a special glucose-broth medium and detected by the indicator methyl red. Both organisms produce acid from glucose. However, *E. coli* produces a lower pH, which turns the indicator red, whereas *Ent. aerogenes* cultures do not produce this color change.

3.  Abilty to produce the compound acetylmethylcarbinol in a glucose-peptone medium. This chemical is detected by the Voges-Proskauer test procedure. *E. coli* does not produce acetylmethylcarbinol, but *Ent. aerogenes* does.

4.  Utilization of sodium citrate. *Ent. aerogenes* is capable of utilizing sodium citrate as its sole source of carbon; i.e., it will grow in a chemically defined medium in which sodium citrate is the only carbon compound. *E. coli* does not grow under the same circumstances.

For convenience, these tests collectively are designated as the *IMViC reactions* (I = indole, M = methyl red, Vi = Voges-Proskauer reaction and   C = citrate). The reactions for a typical strain of each species are shown in table. The reactions for all coli-aerogenes organisms are unfortunately not as clear cut as those described here. Some cultures give other combinations of reactions to this scheme of testing and are usually referred to as intermediate types. Furthermore, there are species of bacteria in the genera *Klebsiella* and *Citrobacter* which resemble the coliforms and for which more detailed biochemical and serological data may be needed for differentiation.

| Organism | TEST | | | |
| | Indole | Methyl Red | Voges-Proskauer | Citrate |
| --- | --- | --- | --- | --- |
| *Escherichia coli* | + | + | − | − |
| *Enterobacter aerogenes* | − | − | + | + |

## PRIMARY INOCULUM

The overwintering or over summering stage may be in or on the seed of the host, in perennial wild host, in debris of infected plant of the previous season, in dormant spores, sclerotia or other organs of the soil, or in the form of an inhabitant of the soil flora. Whatever the source, it causes primary infection. It may serve as an infective agent and thus becomes the primary inoculum, or it may multiply as a saprophyte (or as a parasite on perennial host) to produce spores which become the primary inoculum. The rust fungus over summers in India in the uredial stage on summer wheat in the hills from where the primary inoculum is carried by air currents to the next crop grown the foot of the bills and then to the plains. When the host becomes diseased, the organism commonly produces another crop of spores or infective bodies which are known as the secondary inoculum and serve to cause secondary infection. This last cycle may be repeated several times. The uredial pustules formed in the wheat crop as a result of primary infection become a source of secondary infection to other plants as the urediospores are blown down to these leaves and plants.

## COMMENSALISM

It is the way many polysaccharides are transformed to nutrients supporting non specialized produce from cellulose a number of organic acids that severe as carbon sources for non-cellulytic bacteria and fungi.

A second type of commensal association arises from the need of many microorganisms for growth factors. These compounds are synthesized by certain microorganism and their excretion permits the proliferation of nutritionally fastidious soil inhabitants.

The relationship may be commensalisms if only one of the associates benefits as when the secondary population synthesizes metabolites without which the primary heterotrophs will not multiply.

## PROTOCOOPERATION

It may be involved as when the secondary organism favours their associates by removing toxic wastes but simultaneously get carbon in the form of products made by their associates.

Nutritional protocooperation has been demonstrated frequently in culture for example, in a medium deficient acid and biotin, neither *Proteus vulgour* nor *Bacillus polymyxa* will multiply as former bacterium requires nicotinic acid and the latter biotin. In mixed culture in the same medium, however, both grow since the partner bacterium synthesizes the missing vitamin.

## SYMBIOSIS ASSOCIATION

There are evident in soil among several groups of organisms: Algae and fungi in lichens, bacteria residing within protozoan cells, bacteria and roots in the rhizobium- legume symbiosis, fungi and roots in mycorrhizae, protozoa in underground termites, and above rather than beneath the surface – fungi and ants. In lichens, the algae and fungi are in such as intimate physical and physiological relationship that the lichen they make up are classified as distinct organism. The Alga benefits in past because of the protection afforded to it by the hyphae that envelop and protect it from environmental stresses, while the fungus gains by making use of the $CO_2$ fixed by its photosynthetic partner where blue green algae are participants in the lichen associations, the heterotrophy benefits from the fixed nitrogen acquired from the possessor of nitrogenase. That the symbiosis is ecologically successful is clear from the ubiquity of lichens on rocks, acid and semiarid sites, and other locations where neither of the free-living organisms is detected.

The mutual benefits gained by legumes and rhizobia includes nitrogen delivered to the plant by the $N_2$ – metabolizing bacteria and organic carbon transferred to the rhizobia by the $CO_2$ – metabolizing host.

## MICROBIAL COMPETITION

The categories of deleterious interactions are summarized by the terms competition, ammensalism, Parasitism and Predation that is , (a) the rivalry for limiting nutrients or other common needs, (b) the release by one species of products toxic to its neighbors, and (c) the direct feeding of one organism on a second. Because the supply of nutrients in soil perennially inadequate competition for carbon, inorganic nutrients, or $O_2$ is quite common.

Alteration of the environment to the detriment of certain microbial groups may occur through the synthesis of metabolic products that inhibit or kill microbial cells by the utilization of $O_2$ which leads to the suppression of obligate aerobes or by the autotrophic formation of nitric acid and sulfuric acids, which affect the proliferation of acid sensitive microorganisms.

Predation and parasitism are observed in the feeding on bacteria by protozoa, the attack on nematodes by predacious fungi, the digestion of fungal hyphae by bacteria, and the lyses of bacteria and actinomycetes by bacteriophages.

Of considerable practical significance is the competition between strains of *Rhizobium* derived from soil and those applied with legume seeds at the time of sowing?

## AMENSALISM

An antibiotic is a substrate formed by one organism that, in low concentrations, inhibits the growth of another organism. The capacity of an individual colony on a dilution plate to produce an antibiotic is confirmed by streaking the culture on fresh agar, and after two to three days, crossing the line of growth with perpendicular streaks of one or more test species. Following a suitable period of incubation, Antibiosis is observed as suppression to the test organism. Antibiosis is especially common among *Streptomyces* isolates; *Nocardia* and *Micromonospora, Bacillus, Pseudomonas, Penicillium, Trichoderma, Aspergillus, Fusarium* and others also secrete antibiotic substances. The antimicrobial substance fungistasis, which was also present in the soil, the agent that inhibits but does not kill fungus. Identified microbial products generated in soil also are known to be harmful to the activities of native population, such as $CO_2$, $NH_3$, nitrite, ethylene ,and sulfur compound.

$CO_2$– inhibits conidial germination

Ammonia–inhibitor of Nitrobacter

Nitrite-deleterious to fungi

**Pasteurization:** It is the process by which the medium is heated with steam up to 60°C for 1 hr, on three successive days. In the Koch or Arnold sterilizer, steam is used to sterilize the material at atmospheric pressure, which would otherwise be destroyed by steam under pressure. Sometimes the material for sterilization is heated with circulating steam at 100°C for 20 minutes on three consecutive days. After the first heating the vegetative cells are killed. The spores which are left over, germinate and develop vegetative cells. These cells are killed by heating the next day. On the third day the germinating spores which escaped second heating are killed. Thus, intermittent heating helps in completely eliminating even the resistant microbes without seriously affecting the medium which is being sterilized. This method was first used by Tyndall and is now commonly known as tyndallization.

**Bacteriophage:** A virus that infects bacteria. Phage genomes consist of either RNA or DNA. The phage nucleic acid is packaged with phage-encoded proteins that determine the phage structure, and interact with specific receptors on the bacterial surface to initiate infection. Each phage requires the presence of a particular receptor, and bacteria lacking that receptor are immune to infection by that particular phage. Depending upon the species of bacteriophage, two types of infections can be seen: lytic infection and lysogenic infection. In a lytic infection, the virus produces progeny and kills the host cell. Some bacteriophages are only capable of causing lytic infections.

Some bacteriophages are capable of either lysogenic or lytic infection. During lysogenic infection, the virus integrates into the host chromosome or replicates in plasmid form. Copies are passed to daughter cells when the host cell goes through cell division. Lambda phage is a classic example of a lysogenic phage that integrates into its host chromosome.

## BACTRIOPHAGE

❑   A bacteriophage (phage) is a virus that infects bacteria.

❑   The phage T4 and the phage lambda, for example both infect *E. coli.*

❑   Phages, like all viruses, are obligatory intracellular parasites and must invade a host cell in order to reproduce.

❑   Viruses can multiply by two alternate mechanisms : the lytic or the lysogenic cycle.

❑   T4 multiplies by the lytic cycle which kills the host and lambda multiplies by the lysogenic cycle which does not cause death of the host cell.

❑   In lysogeny, the phage DNA remains latent in the host until it breaks out in a lytic cycle.

## BDELLOVIBRIOS

Bdellovibios are vibriod bacteria which parasitise other bacteria. Phylogenetically they fall into Purple bacteria. They exhibit lytic activity on gram negative bacteria of Pseudomonadaceae and Enterobacteriaceae. They attack *Erwinia, Agrobacterium, Pseudomonas* and *Rhizobium.* There are no bdellovibrios reported on gram positive bacteria. They are obligate aerobes. Taxonomically there are 3 species of bdellovibrios *viz., Bd starrii, Bd. Stolpii* and *Bd. Bacteriovorus.* They are common shaped, single polar sheathed flagellum, gram negative, strict aerobes, and chemoorganotrophs. The bdellovibrios con not use free sugars and therefore, they use host amino acids as major energy source.

## MEMBRANE FILTERS TECHNIQUE

The membrane-filter technique for the bacteriological examination of water consists of the following steps:

1.   A sterile filter disk is placed in a filtration unit.

2.   A volume of water is drawn through this filter disk, the bacteria being retained on the surface of the membrane.

3.   The filter disk is removed and placed upon an absorbent pad that; has previously been saturated with the appropriate medium. Special petri dishes of a size to accommodate the absorbent pad and filtration disk are employed for incubation.

4.  Upon incubation, colonies will develop upon the filter disk wherever bacteria were entrapped during the filtration process.

This technique has several desirable features, some of which are:

1.  A large volume of water sample can be; examined; theoretically almost any volume of water could be filtered through the disk, the organisms from any given volume being deposited on the disk.

2.  The membrane can be transferred from one medium to another for purposes of selection or differentiation of organisms.

3.  Results can be obtained more rapidly than by the conventional standard methods.

4.  Quantitative estimations of certain bacterial types, e.g., coli forms, can be accomplished when appropriate media are used.

This technique, with modifications, has been adopted for many microbiological procedures other than the examination of water.

## USES OF SPIRULINA

1.  As food supplement: Some of the *Spirulina* applications in food are: (i) in milk products, (ii) in jams and pickles (iii) as in the form of vitamin tablets serve as food supplement, health tonic or health drink. The World Health Organization has found *Spirulina* to be an excellent food for human consumption and *Spirulina* has been approved as the Food item by Food and Drugs Authority of the United States for being sold as a natural food.

2.  **Modern Medicine:** *Spirulina* has been easily incorporated in medical formulations owing to its excellent natural composition. What is more, it has no side efforts and is non-habit forming. Today *Spirulina* is used as therapeutic agent to treat and prevent diseases with good results. *Spirulina* has properties known to lower blood-sugar levels in diabetics, have curative effects on patients suffering from pancreatic, hepatitis and cirrhosis, in curing glaucoma and cataracts, gastric ulcers, night blindness, liver and circulation disorders and in controlling anemia. In Vietnam, *Spirulina* tablets are used to increase lactation in nursing mothers and as multivitamin tablets in the US and Japan.

3.  **As Specialized Animal Feed:** The use of *Spirulina* as feed for fish, poultry, and silkworms has shown to improve the yield from such breeding farms.

4.  **As Cosmetics:** *Spirulina* is used in cosmetics and these prepartions are totally natural in composition. It is fast replacing synthetic contents in the preparation of cosmetics. It is ideally suited to the manufacturer of

face creams as it has high protein content, a range of natural pigments and vitamins-all essential for maintaining a healthy skin.

**Forecasting:** Forecasting is advance warning of infection by a plant pathogen or the development of disease. Thereby, it aids in effective and efficient control of plant diseases. The nature and magnitude of disease problems vary from season to season and where routine use of expensive control measures is uneconomic. Low cost, need based control measures are economically justified through forecasting system. The wake of the catastrophic potato blight epidemics in western Europe during the middle of the nineteenth century led to the realization for the need of forecasting systems.

## IMPORTANCE OF FORECASTING

(1)  To plan advance preventive measures for possible losses.

(2)  A reliable forecasting system would develop confidence in farming community.

(3)  Insure timely spraying of chemicals, and

(4)  Reduce pollution by hazardous chemicals out of excessive use of chemicals.

## USAGE OF FORECASTING

(a)  For highly destructive diseases of major crops or cash crops, e.g., rusts, powdery and downy mildews.

(b)  For diseases which can be reliably and timely predicted early enough for effective and economical control measures to be employed.

(c)  For diseases against which economical control measures are available and disease management practices would be financially beneficial to farmers.

## PROPERTIES OF GOOD CARRIER FOR BIOFERTILIZER

1.  High water holding and retention capacity (e.g., Peat)

2.  Non toxic to rhizobia.

3.  Ease of sterilization.

4.  Ready availability in ample quantities at moderate cost.

5.  Adhesive properties for coating seeds.

6.  Good buffering capacity.

7.  Suitable pH.

8. Fine particle size for better adherence to seed (70 to 100% through 200 mesh screen and to facilitate through mixing with other components.

9. Should have no heat of wetting.

10. Should be free of lump forming materials.

11. Should be amenable to nutrient supplement.

12. Should have high surface area.

13. Should be uniform chemically as well as physically.

14. Should have organic matter content around 40%.

15. Should be non-toxic, biodegradable and none polluting.

16. Should be easily manageable in mixing, curing and packing.

17. Should rapidly release rhizobia in the soil.

18. Should support good rhizobia growth.

19. Should be suitable for all rhizobia.

### Cross Inoculation Groups of *Rhizobium*

| Cross inoculation group | Rhizobium, Species | Host involved |
|---|---|---|
| Alfalfa | *Rhizobium meliloti* | Medicago, Melilotus, Trigonella |
| Clover | *Rhizobium trifolii* | Trifolium |
| Pea | *Rhizobium leguminosarum* | Pisum, Vicia, Lens & Lathyrus spp. |
| Bean | *Rhizobium phaseoli* | Phaseolus |
| Lupine | *Rhizobium lupini* | Lupinus, Ornithopus |
| Soybean | *Rhizobium japonicum* | Glycine |
| Cowpea | *Rhizobium* spp., | Vigna, Crotaloria, *Arachis hypogea, Cajanus cajan, Dolichus etc.* |

## VERMICOMPOST

There are about more than 500 species of earth worms in India with various food and burrowing habits. *Eisenia fetida, Eudrilus eugeniae* and *Perionyx excavatius* are some of the species which are voracious feeders and prolific breeders having higher multiplication rate, shorter life cycle and easy to handle.

## BENEFITS OF VERMICOMPOST

❑ It is now well documented that the use of chemical fertilizers and pesticides will lead to adverse conditions. The alternative to avoid usage of chemical fertilizers, pesticides, growth regulators etc. is organic farming. For quick change from chemical farming to organic farming is, the use of vermicompost.

❑ The optimal carbon/nitrogen (C/N) ratio is available in vermicompost which determines the quality of compost.

❑ Vermicompost increases the soil texture, fertility and soil moisture and reduces the water requirement in long run.

❑ It improves the pH of the soil. Earthworm is having characteristics of changing acidic or alkaline soil to neutral soil. Waste land can also be converted to fertile land.

❑ Provide safe and clean environment.

## BIOFERTILIZERS

Biofertilizers are carrier-based preparations containing beneficial microorganisms in a viable state intended for seed or soil application and designed to improve soil fertility and help plant growth by increasing the number and biological activity of desired microorganisms in the root environment. According to Motsara et al. (1995) biofertilizer is known to make a number of positive contributes in agriculture. For example.

❑ They supplement fertilizer supplies for meeting the nutrient needs of crops.

❑ They can add 20 – 200 kg N/ha (by fixation) under optimum conditions and solubilize/mobilize, 30-50 kg $P_2O_5$/ha.

❑ They liberate growth promoting substances and vitamins and help to maintain soil fertility.

❑ They suppress the incidences of pathogens and control diseases.

❑ They increase crop yield by 10-50%, N-fixers reduce depletion of soil nutrients and provide sustainability to the farming system.

❑ They are cheaper, pollution free and based on renewable energy sources.

They improve soil physical properties, tilth, and soil health in general.

### Methods of Sterilization of Various Essential Tools and Glassware

Sterilization of apparatus and working area involves the inactivation or physical elimination of all living cells from the environment. It's achieved by exposing the material to lethal agent which may be chemical physical or ionic in nature. It does not include the destruction or elimination of constitutive the distraction metabolic byproduct or removal of dead cells. Selection of the method depends upon the desired efficiency, its applicability and toxicity, ease to use, availability and cost and effect on the properties of the object to the sterilizer.

The methods commonly used for sterilization are heat, gas, and in case of liquids ultra filtrations. Various methods used for sterilization are as follows.

## (i) Flame sterilization

Flame sterilization is done for metal objects such as needles, tips of forceps, lips of flasks and culture tubes, microscopic slides and cover slips. The needles or forceps tips are dipped in alcohol or rectified spirit and flamed. Similarly, lips of flasks culture tubes the slides and cover slips are passed through flames for sterilization.

## (ii) Dry heat sterilization

Glassware such as Petridishes, pipettes, flasks and other usable without graduations can be sterilized by dry heat in hot air oven. The method and precautions required are mentioned in section under General equipment, their uses and precautions in handling.

## (iii) Moist heat sterilization

Culture media are generally subjected to moist heat sterilization. Moist heat has more penetrating power and with the increase in pressure, the desired temperature is attained rather much more quickly, the process thus, takes less time.

## (iv) Filtration

Filtration physically separates microorganisms, cells and debris from liquids but not viruses and metabolic byproducts. Except for these limitations sterilization by filtration is superior *to* other methods since there is no change in the properties of aqueous solutions, oils and organic solutions. Filters used are sintered glass filter, asbestos filter, unglazed porcelain and cellulose ester membranes. Microorganisms and other large particles are retained on the filter in part by the small size pores and dry absorptions onto pore walls.

## (v) Gas sterilization

Some gases such as ethylene oxide, propylene oxide, formaldehyde, methyl bromide, â propiolacetone and ozone are generally used for sterilization of object such as container of rubber or plastic, plastic or metal drums.

The major disadvantage of the gas sterilization is that it takes longer time as compared to heat sterilization. Materials used are flammable, highly toxic and cost more than heat sterilization. This method should only be used when the other methods are not feasible or practically possible.

## (vi) Surface sterilization by chemicals

Objects that can not be sterilized with heat or filtration or through gas may be sterilized with chemicals such as alcohols, mercuric chloride and sodium hypochlorite. This method of sterilization is commonly used to disinfect the tissue surface and plant materials and disinfest the non-living surfaces as knife, blade, inoculation needle, etc.

Surface sterilization of plant material is effected by immersion in 1:1000 solutions of mercuric chloride/Sodium hypochlorite approximately for 5 minutes and then washing in several changes of sterile water or by immersion in sodium hypochlorite solution.

Chromium plated or stainless steel instruments such as scalpels, forceps, scissors etc. should be dipped in alcohol and flamed.

## ROLE OF MICRO-ORGANISMS IN ORGANIC FARMING

Addition of organic matter to soil or retaining the farmer waste in the field it self as through green manuring was naturally meeting the nutritional requirement of the crop only partially and slowly. Through this natural process it was found possible to enhance the process of degradation of organic waste through the addition of microorganisms. The research shows many organisms and their specific activities in decomposition of organic waste or producing quality or good FYM has been developed. The important microorganisms useful in producing FYM are *Trichoderma, Chaetomium* and *Aspergillus* spp. Besides these fungi, the bacteria present present in the soil also acts as scavengers, *Bacillus, Pseudomonas, Nitrosomonas, Nitrobacter, Thiobacillus* is some of the potential bacteria which help to degrade specific component of organic matter. As a result of the combine action of the microorganisms various important elements, i.e., nitrogen, phosphorus, potassium, sulphur etc. required for the growth of crop are released in the soil.

Atmospheric nitrogen the major nutrient for the plant growth could be utilized by number of microorganisms including fungi, bacteria, algae in soil and then this nitrogen become available to plants. Basically the organisms responsible for nitrogen fixation have been identified and their pure culture have been developed and employed for use as seed treatment for adoption in the soil.

There are different types of nitrogen fixing organisms some are symbiotic and other non-symbiotic. *Rhizobium* is symbiotic and *Azotobacter* is non-symbiotic.

Phosphorus is normally fixed in the soil it can be converted in to available form for plant roots to absorb by the action of phosphorus solubilizing bacteria the bacteria type.

Besides nitrogen fixation and release of P, Some of the most important fungi like *Trichoderma viride, Trichoderma harzianum, Paceiliomyces lilicinus,*

*Aspergillus* and bacteria like *Pseudomonas fluorescence, Bacillus subtilis* etc. have been cultured and used in the disease management. Commercial product of these bioagent is available in the market.

## 1. Orchid mycorrhizas

Some types of orchid are non-photosynthetic; others only produce chlorophyll when they have grown past the seedling stage. In all cases, the plant depends on sugars derived from a fungal partner for at least part of its life. The minute orchid seeds, with negligible nutrient reserves, will not germinate unless a fungus infects them, although the seeds can germinate aseptically if supplied with the 'fungal sugar' **trehalose**.

These mycorrhizas are unusual because, in effect, the plant parasitises the fungus that invades it. The fungi in these associations resemble the common plant pathogen *Rhizoctonia solani,* but recent taxonomic studies have assigned them to several related genera. They are mainly **saprotrophic** - they grow by degrading organic matter in soil - but they might obtain trace elements or some other factor from the plant.

An interesting variation on this theme is shown by some orchids and other non-photosynthetic plants (*Monotropa* species) that have wood rotting species of *Armillaria* as their mycorrhizal symbionts. In some of these cases the fungus can even be a pathogen of tree roots, so that the non-photosynthetic plant gains its nutrients by indirect parasitism of a living tree (see *Armillaria*).

## 2. Arbuscular mycorrhizas

Arbuscular mycorrhizas are found on the vast majority of wild and crop plants, with an important role in mineral nutrient uptake and sometimes in protecting against drought or pathogenic attack. Structures resembling those of the present-day AM fungi have been found in fossils of primitive pteridophytes of the Devonian period. It is thought that these fungi colonised the earliest land plants and that mycorrhizal associations could have been essential for development of the land flora.

The fungi involved are members of the zygomycota (related to *Mucor*). They are classified currently in six genera (*Acaulospora, Entrophospora, Gigaspora, Glomus, Sclerocystis* and *Scutellospora*) and they seem to be obligate symbionts: none of them can be grown in axenic culture, i.e. in the absence of their hosts.

## 3. Ectomycorrhizas

Ectomycorrhizas (sometimes termed ectotrophic mycorrhizas) are characteristic of many trees in the cooler parts of the world - for example pines, spruces, firs, oaks, birches in the Northern Hemisphere and eucalypts in Australia.

However, some trees, (e.g., willows) can have both ectomycorrhizas and arbuscular mycorrhizas, and most tropical trees have only arbuscular mycorrhizas.

The fungi involved are mainly Ascomycota and Basidiomycota, including many that produce the characteristic toadstools of the forest floor. Most of these fungi can be grown in laboratory culture but, unlike the wood-rotting fungi, they are poor degraders of cellulose and other plant wall materials. So they gain most of their sugars from the living plant roots in natural conditions.

## *PSEUDOMONAS* PLANT—MICROBE INTERACTIONS

### 1. Colonization

Motile and non-motile mutants of *Pseudomonas* spp. are being used to identify the significance of flagella in survival, spread and attachment of bacteria to surfaces in soil and on roots.

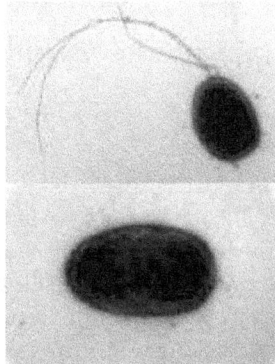

Motile *Pseudomonas putida* (top) and non-motile mutant (bottom)

### 2. Antibiotic production

The role of homoserine-lactone signalling (HSL) in the regulation of antibiotic production in *Pseudomonas* spp. on roots is under investigation. The development of bioassays for HSL, phenazine and the use of marker genes are key elements of this project.

### 3. Release of GMMs

The first UK field trial of a marked, free living bacterium was performed at HRI in the spring of 1993. *Pseudomonas fluorescens* SBW25 was genetically tagged with the reporter genes *lacZY* and *xylE* in addition to resistance to the antibiotic kanamycin. Following release on to wheat, the survival, spread and impact of this bacterium was monitored for 4 years.

## CHARACTERIZATION OF STRAINS

The bacterial flagellin protein has been widely used to serotype bacteria, and in this project, PCR amplification of the flagellin gene of *Pseudomonas* spp. is being used as a tool to differentiate strains within a species. Comparisons of this method to genomic RFLP studies, phage typing and serological tests indicate that it has great potential. This work has recently been expanded to include *Burkholderia* strains (formerly *Pseudomonas*).

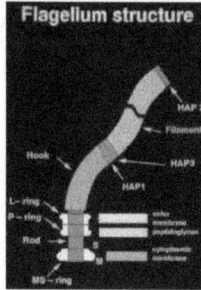

**PCR amplification of flagellin gene**

## MYCOPLASMAS

PPLO (pleuro pneumonia like organisms) were discovered to be the agents of bovine pleuropneumonia. They have now been renamed as mycoplasmas. Mycoplasma group are organisms that lack cell walls and contain a very small genome. Phylogenetically the mycoplasmas are closely related to clostridia and hence to gram positive bacteria. Mycoplasmas are able to resist osmotic lysis (unlike protoplasts). This is probably due to the presence of substances such as sterols in the membrane. The mycoplasma cannot synthesise sterols but require them to be available in the medium. Based on the sterol reqirement the mycoplasmas can be divided into those which require sterol (*Mycoplasma, Spiroplasma* and *Ureaplasma)* and those which do not require sterols *(Acholeplasma, Thermoplasma* and *Asreroleplasma).* Mycoplasma may contain lipoglycans which help to stabilise the membrane. Mycoplasma cells are small, pleomorphic and divide by budding. Colonies of mycoplasmas on agar exhibit a characteristic fried egg appearance because of the formation of dense central core surrounded by a lighter circular spreading area. Growth of mycoplasmas is not inhibited by penicillin, cyclosterine or other antibiotics that inhibit cell wall synthesis but are sensitive to antibiotics which act on targets other than the cell wall. *Spiroplasma* genus has corkscrew shaped cells associated with various plant disease conditions. They are motile and exhibit undulating or rotating movement. *S. cirri* has been isolated from the leaves of citrus plants where it causes citrus stubborn disease and from com plants where it causes com stunt disease.

**Cyanobacteria:** Cyanobacteria (earlier grouped as blue-green algae) are a large and heterogenous group of phototrophic bacteria. They are oxygenic and related to gram positive bacteria. Cyanobacteria are divided into five groups based on Bergeys manual. Group I consists of unicellular single cells or cell aggregates *(Gloeotheca)*. Group II reproduce by small spherical cells called baeocytes produced through multiple fission *(Pleurocapsa)*. Group III consists of filamentous cells that divide by binary fission in a single plane *(Oscillatoria)*. Group IV consist of filamentous cells that produce heterocysts *(Nostoc)* and Group V are the branching types *(Stigonema)*.

Cyanobacteria produce extensive mucilaginous envelopes or sheaths. They have chlorophyll a and phycobilins (phycocyanin or phycoerythrin). Some form heterocysts which is the centre of nitrogen fixation. They are low in phycobilin and lack photosystem II. Cyanobacteria contain cyanaphycin which is a simple polymer of aspartic acid. It is an energy (food) reserve. Many cyanobacteria exhibit gliding motility. In some species resistant spores or akinetes are formed.

## ACTINOMYCETES

The actinomycetes are gram positive bacteria which ramify to form a mycelium. Spore forming, 63-78% GC. *Streptomyces* is an important genus of acinomycetes. It has 500 species. *Streptomyces* occurs in alkaline and neutral soils and are higher in number in well drained soil. A number of species produce antibiotics such as macrolides, aminoglycosides, tetracyclines and polyenes. They also produce a number of extracellular enzymes that permit the use of proteins, polysaccharides and hydrocarbons. Several *Streptomyces* spp. have been found to contain large linear plasmids of over 500 kb in length. *Streptomyce* filaments are usually 0.5 im in diameter and are of indefinite length, and often lack cross walls in the vegetative stage. Growth occurs at the tip of the filament and is accompanied by branching. As the colony ages, characteristic aerial filaments called sporophores give rise to spores. The spores and sporophores are often pigmented and contribute a characteristic colour to the mature colony. Differences in the shape and arrangement of aerial filaments and spore bearing structures of various species are among the features used to separate *Streptomyces groups*.

## FLAGELLA

Flagella are organs of bacteria responsible for motility. They are delicate and fragile and the cultures should be handled carefully if these organs are to be stained. The width of these organs (10-12 nm) is smaller than the wavelength of light and therefore they cannot be seen by ordinary staining. In order to make the flagella visible under the microscope after staining, it is necessary to use mordents like tannic acid, potassium sulfate or mercuric chloride. Mordants precipitate on the flagella making the width of these organs more than the wavelength of light and render them visible.

There are two principal modes of insertion of flagella: peritrichous and polar.

(a) *Peritrichous:* Flagella are present on all sides of the cell, e.g., *Erwinia*

(b) *Polar:* Flagella are present at one or both the ends of the bacterium; polar flagella may be

- Monotrichous: single flagellum at one end only *(Xanthomonas)*

- Amphitrichous: several flagella at both the ends *(Pseudomonas)*

- Lophotrichous: several (tuft of) flagella at one or both end (poles) of a cell (*Spirillum)*

## FLAGELLAR STRUCTURE

The flagellar apparatus has three distinct regions.

1. **Filament:** The outermost region of flagellum is known as the filament. This is helical and composed of flagellin. Flagellin has molecular weight of 30-40000 and is synthesised in the cell which moves through the hollow core of the flagellum to the tip where it assembles. Flagellin is an incomplete protein with 14 amino acids and characterised by high content of acidic amino acids, low content of aromatic amino acids and the absence of cysteine in many cases. Flagellin in different species show systematic differences which are useful in serology. The filament is attached to a hook.

2. **Hook:** The hook is slightly wider than the filament. It is 45 nm wide and made up of different types of proteins. Hook of the gram-positive bacterium is longer than that of gram negative bacterium.

3. **Basal Body:** The basal body consists of a small central rod which is inserted into a system of rings. The gram positive and gram-negative, bacteria differ in the number of the rings. The inner pair of rings (S and P rings) are embedded in the cell membrane and are found in both gram positive and gram negative bacteria. However, the outer rings (L and P) are found only in the case of gram negative bacteria. The rings of the basal body are described below:

   ❏ **L Ring:** L ring is found in the lipopolysaccharide layer of gram negative bacteria.

   ❏ **P Ring:** P ring is embedded in peptidoglycan layer of the gram-negative bacteria. Land P rings are just bushings and are not necessary for flagellar movement.

   ❏ **S Ring:** S Ring (stator ring) is in the periplasmic space just outside the cytoplasmic membrane.

   ❏ **M Ring:** M (motor) ring is embedded in the cytoplasmic membrane.

The S and M rings are the ones which are important for the movement of the flagella.

**Functions of Flagella:** Flagella are the locomotor organs of the bacteria. They are capable of spinning in a clockwise or anticlockwise direction. Movement is imparted by a basal motor. The motor operates by causing the Sand M rings to rotate relative to each other. The energy source for this is probably a proton motive force. During motility, the flagellum is a rigid left handed helix rotating in counter clockwise direction; flagella are directed backward to the flow of bacterial movement. Bacterial cells show chemotaxis i.e. movement towards attractants and away from chemical repellents. Aerobic responses are also shown by bacteria when viewed under a slide; since aerobic bacteria tend to accumulate near the edges of the coverslip. The microaerophilic bacteria accumulate a little away from the edge while the anaerobic bacteria accumulate at the center of the coverslip where there is least aeration.

**Endospores:** Bacteria are capable of forming a range of sporulation structures but the main is endospore. Spores are resting bodies produced by some species of bacteria within the cell. Most of the phytopathogenic bacteria do not produce spores (resting bodies). However, the most characteristic spore structure known as endospore is formed only by gram positive genera. Endospores are characteristic because they are intracellular, extremely refractile, resist staining by aniline dyes, possess chemicals like dipicolinic acid, muramic acid lactam and typically one endospore per cell. They are formed under nutrition limitation and released by the lysis of the cells. The free endospores are practically metabolically inactive and may retain viability for several years. This total dormancy is known as cryptobiosis. Endospores are highly heat resistant and tolerant to UV and ionising radiations, toxic chemicals etc. Most spore formers are found in soils and may be classified based on their oxygen requirement into:

1. Aerobic: *Sporosarcina* and *Bacillus*

2. Anaerobic: *Clostridium* and *Desulfotomaculum*

**Endospore Formation:** Endospore formation is the formation of a new type of cell within the vegetative cell. The endospore is different from the vegetative cell in fine structure, chemical composition and physiological properties. Massive sporulation occurs at the end of log phase, and the various steps are:

1. **Nuclear division:** The nuclear body divides and the two nuclear bodies then rearrange to form an axial chromatin thread.

2. **Formation of septum:** Onset of sporulation is marked by the formation of a transverse septum near one cell pole, which separates the cytoplasm and DNA of the smaller cell from the rest of the cell.

3. **Production of forespore:** The membrane of the larger cell grows rapidly around the smaller cell which becomes completely engulfed within the cytoplasm of the larger cell to produce the forespore. Forespore is clear, non-refractile and free of granular inclusions. Forespore is protoplast

enclosed by two concentric sets of unit membranes, its own bounding membrane and the membrane of the mother cell which has grown around it. Once the forespore is formed the cell is committed to undergo sporulation.

4. **Spore maturation:** Cortex is deposited between the inner and outer membrane of the forespore. A spore coat is formed exterior to the outer unit membrane surrounding the cortex. Sporulation is complete in 7 hours in *Bacillus subtilis.*

## POLYMERASE CHAIN REACTION (PCR)

PCR is an advanced technique of amplifying the amount of specific DNA segment. The purpose of a PCR is to make a huge number of copies of a gene within a short period of time . PCR is an unique technique of *in vitro* DNA synthesis/multiplication. PCR is performed with the help of automated thermocyclers/PCR machines for which reaction mixture with some important components are needed.

## IMPORTANT COMPONENTS OF PCR REACTION MIXTURE

1.  Template DNA
2.  Heat stable DNA Taq polymerase
3.  A pair of primers (Forward and reverse)
4.  dBTPs (deoxynucleoside triphosphates)

## STEPS IN PCR

1.  **Denaturation :** In this step, reaction is mixture is heated to 94-95°C for a very short period of about 15-30 sec. leading to the separation of double strand DNA into single strands which can act as templates for DNA synthesis.

2.  **Annealing :** After denaturation, the reaction mixture is cooled rapidly to a defined temperature (ranges from 45-58°C ) for 45-60 seconds allowing the primers to bind to the sequence on each of the two strands flanking the target DNA.

3.  **Elongation/Extension:** This is the last step where in the temperature of reaction mixture is again raised to 72°C for 1-2 minutes to allow Taq DNA polymerase to elongate each primer by copying the single stranded templates.

These three steps together constitute one cycle where in two strands are denatured primers are annealed and extended and these steps are repeated. In second cycle four strands denatured, primers are annealed and extended.

Likewise, in each cycles original DNA will be amplified to a billion fold. The cycling process will be completed by 2-3 hours.

## APPLICATIONS OF PCR

1.  DNA sequencing can be simplified by using PCR.
2.  PCR generally facilitates the analysis of gene expression and function by creating mutations with the help of suitable primers.
3.  Its use in the field of medicine is immense especially in screening human genetic diseases.
4.  PCR has great use in cloning.

Its application can also be extended to forensic studies (DNA finger printing).

## SEROLOGICAL TECHNIQUES

**Microbial degradation of pesticides**

**RAPD**

**RFLP**

**Microflora and Microfauna (Agro 50-51)**

5.  Formulas:

**Formula for calculating dilutions:**

$$\frac{\text{Volume of sample}}{\text{Volume of diluent + volume of sample}}$$

## COMPUTATION OF DISEASE INDEX

1.  When class rating is expressed in per cent :

$$\text{Disease Index} = \frac{\text{Frequency} \times \text{mean of rating category}}{\text{Total number of observations}}$$

2.  When class rating is expressed in arbitary numbers :

$$\text{Disease Index} = \frac{\text{Sum of all disease rating}}{\text{Total number of observations (sample)} \times \text{Maximal disease rating grade}} \times 100$$

3. Severity of estimates for larger area :

$$\text{Disease Index} = \frac{\text{Field class rating} \times \text{Number of hectare in a class}}{\text{Total number of hectare}}$$

$$\text{Resolving power} = \frac{\text{Wavelength of light in nanometers}}{2 \times \text{numerical aperture of the objective lens}}$$

Where, numerical aperture is a mathematical constant that describes the relative efficiency of a lens in bending light. It is 0.1 for the low power lens and 1.25 for ...... oil immersion lens.

## 6. EQUATIONS / REACTIONS

Reactions in Nitrogen Cycle

1. Proteolysis

$$\text{Proteins} \xrightarrow{\text{proteinase}} \text{Peptides} \xrightarrow{\text{proteinase}} \text{amino acids}$$

2. Ammonification

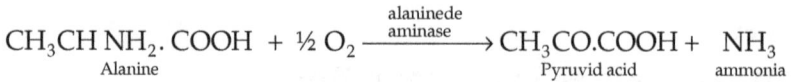

$$\underset{\text{Alanine}}{CH_3CH\,NH_2.\,COOH} + \frac{1}{2}\,O_2 \xrightarrow{\underset{\text{aminase}}{\text{alaninede}}} \underset{\text{Pyruvid acid}}{CH_3CO.COOH} + \underset{\text{ammonia}}{NH_3}$$

3. Nitrification

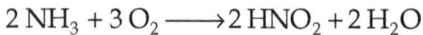

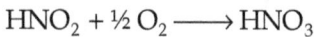

$$2\,NH_3 + 3\,O_2 \longrightarrow 2\,HNO_2 + 2\,H_2O$$

$$HNO_2 + \frac{1}{2}\,O_2 \longrightarrow HNO_3$$

(i) Oxidation of ammonia to nitrite by *Nitrosomonas*

(ii) Oxidation of nitrite to nitrate by *Nitrobacter*

4. Denitrification

$$NO_3 \longrightarrow NO_2 \longrightarrow N_2O \longrightarrow N_2$$

5. Nitrogen fixation

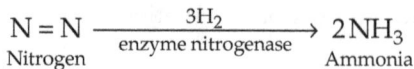

$$\underset{\text{Nitrogen}}{N = N} \xrightarrow[\text{enzyme nitrogenase}]{3H_2} \underset{\text{Ammonia}}{2\,NH_3}$$

## ENZYMES

The various oxido-reduction processes involved in the metabolism of microorganisms are carried out by a network of highly specialized chemical substances, known as enzymes. These enzymes play a pivotal role in microbial life.

If the substrate protein is acted upon, then the enzyme responsible is known as proteinase, and if lactose is acted upon, the enzyme is lactase.

Chemically, all enzymes are proteins or protein containing substances. The active enzyme complex is known as holoenzyme, which is usually composed of a protein part referred to as apoenzyme.

The following are some of the examples of important chemical reactions accomplished by the enzyme systems :

1. *Oxidation* : Removal of hydrogen or the addition of oxygen by *oxidases*.

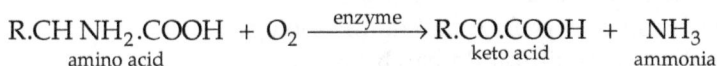

$$R.CH\ NH_2.COOH + O_2 \xrightarrow{enzyme} R.CO.COOH + NH_3$$
$$\text{amino acid} \qquad\qquad\qquad \text{keto acid} \qquad \text{ammonia}$$

2. *Reduction* : Addition of hydrogen or removal of oxygen by *hydrogenases*.

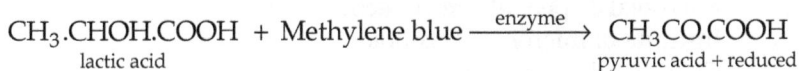

$$CH_3.CHOH.COOH + \text{Methylene blue} \xrightarrow{enzyme} CH_3CO.COOH$$
$$\text{lactic acid} \qquad\qquad\qquad\qquad \text{pyruvic acid + reduced}$$

3. *Dehydration* : Removal of $H_2O$ by *dehydrogenases*.

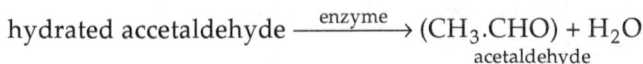

$$\text{hydrated accetaldehyde} \xrightarrow{enzyme} (CH_3.CHO) + H_2O$$
$$\text{acetaldehyde}$$

4. *Hydrolysis* : Addition of H2O by *hydrolytic* enzymes.

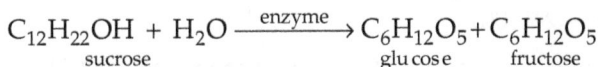

$$C_{12}H_{22}OH + H_2O \xrightarrow{enzyme} C_6H_{12}O_5 + C_6H_{12}O_5$$
$$\text{sucrose} \qquad\qquad\qquad \text{glucose} \qquad \text{fructose}$$

5. *Deamination* : Removal of amino group in the molecule by *deaminases*.

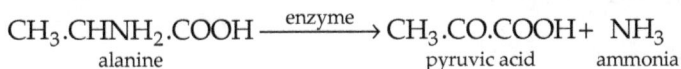

$$CH_3.CHNH_2.COOH \xrightarrow{enzyme} CH_3.CO.COOH + NH_3$$
$$\text{alanine} \qquad\qquad\qquad \text{pyruvic acid} \qquad \text{ammonia}$$

6. *Decarboxylation* : Removal of $CO_2$ from carboxyle group by *decarboxylases*.

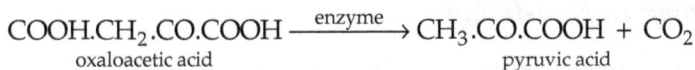

$$COOH.CH_2.CO.COOH \xrightarrow{enzyme} CH_3.CO.COOH + CO_2$$
$$\text{oxaloacetic acid} \qquad\qquad\qquad \text{pyruvic acid}$$

7. *Phosphorylation* : Addition of phosphate group to an organic molecule by *phosphatases*.

$$\text{glycogen} + PO_4 \xrightarrow{enzyme} C_6H_{12}O_6 + \text{glucose phosphate}$$

8. *Dephosphorylation* : Removal of phosphate group from organic molecules by *dephosphatases*.

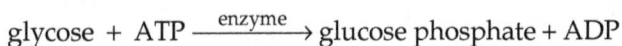

$$\text{glycose} + ATP \xrightarrow{enzyme} \text{glucose phosphate} + ADP$$

9. *Group-transference :* Transfer of a chemical group from one molecule to another molecule by *transaaminases*.

glutamic acid + pyruvic acid $\xrightarrow{\text{enzyme}}$ α-keto glutaric acid + alanine

10. *Isomerization* : Conversion of one compound into another of the same chemical composition but of different structure by *isomerases*.

dihydroxyacetone phosphate $\xrightarrow{\text{enzyme}}$ phosphoglyceraldehyde

## CULTURE COLLECTION CENTERS IN INDIA

Indian Type Culture collection (ITCC)
Division of Mycology & Plant Pathology
Indian Agricultural Research Institute,
New Delhi – 110 012

Microbial Type Culture collection (MTCC)
Institute of microbial Technology,
Chandigarh – 160 031

Defence Research laboratory (DRL)
Research and Development Organization,
Kanpur – 208 004

Forest Pathological Branch (FRI)
Forest Research Institute & Colleges,
Dehra Dun – 248 001

National Collection of Industrial Micro-organisms (NCIM)
National Chemical Laboratory,
Pune – 411 008

## VIROLOGY

### *Viruses: Living or Non-living*

"A living organism is the unit element of a continuous lineage with an individual evolutionary history." Viruses become quite respectable as living organisms because:

They most definitely replicate, so they can be fit in a continuous lineage.

Their evolution can (within limits) be traced quite effectively, and

They are independent in terms of not being limited to a single organism as host, or even necessarily to a single species, genus or phylum of host.

## VIRUS PROPERTIES

Infectious – must be transmissible horizontally

Intracellular – require living cells

RNA or DNA genome, not both

Most all have protein coat

May of may not have lipid envelope

May have broad or narrow host range

Replication involves eclipse (breaking apart of virus particles) and reassembly

Use host factors for to complete replication cycle

## VIRUS PARTICLE PROPERTIES

Shape of the virus particle: Rigid Rod, Flexuous rod, isometric, bacilliform, geminate or pleomorphic

Size of the virus particle: Isometric particle sizes vary from ~17 nm to ~400 nm diameter, Rod shaped sizes vary from ~300 nm to ~2000 nm One or many particle (multipartite) encapsidate one or many segments of genome (multisegmented). May have one or many proteins in particles. Proteins can arrange in a helical fashion or in icosahedrons.

## GENOME PROPERTIES

Nature of the genome: circular (as in all known plant DNA viruses) or linear.

Genome sizes: 0.3 - 1200 kb; average genome sizes vary with host organism types.

**Nature of genome:** May have single-stranded (ss) or double-stranded (ds) RNA or DNA genome.

**Sense of genome:** If ssRNA, may be + or – sense Number of genome components: This varies from a single component (e.g., in the genera *Potyvirus* and *Tobamovirus*) to 11 (in some members of the genus *Nanovirus*). Individual components vary in size from about 1 kb (*Nanovirus* components) to about 20 kb (in the genus *Closterovirus*).

ÀÛ™Ü Number of genes in the genome: These vary considerably. Most plant viruses have at least 3 genes: 1 (or more) concerned with replication of the nucleic acid, 1 (or more) concerned with cell-to-cell movement of the virus and 1 (or more) encoding a "capsid" protein. There may also be additional genes that have a regulatory function or which are required for transmission between plants (association with a vector).

# PLANT VIRUS INDUCED SYMPTOMS

## (A) Symptoms in leaves: Changes in leaf morphology

(1) **Leaf rolling and curling:** Leaf rolling is used to describe folding of leaves along their mid axes resulting in a more or less tube-like structure. When the folding is more irregular or does not result in a tube-like structure it is usually referred to as leaf curling.

(2) **Leaf distortion:** Leaf distortion means that the growth and development of the leaf has been disturbed resulting in changes ranging from small deviations from normal leaf shape to severe deformations.

(3) **Rugose:** Rugose means: "Rough leaves". This term covers both crinkling (leaves looking edged or wrinkled) and leaf puckering (blister-like irregularities formed on the leaf).

(4) **Enation:** Abnormal outgrowth of vascular tissue in leaves or on the stem.

## (B) Symptoms in leaves: Changes in leaf colour

(1) **Mottling:** If leaves display a pattern of abnormal colouration this is referred to as a mottle. Mottling can be both light and dark. Mottling patches have a distinct boundary.

(2) **Mosaic:** Dark and light colouration intermingled to each other (no distinct boundary). Mosaic may be green mosaic, pale yellow mosaic, yellow mosaic or golden mosaic. In monocots mosaics usually consist of elongated light areas referred to as "streaks" or "stripes".

(3) **Chlorosis:** The term, chlorosis, is used to describe reduced amount of chlorophyll resulting in light color. If the leaf is light green it is referred to as "mild chlorosis" whereas "severe chlorosis" or "yellowing" is used to describe a total depletion of chlorophyll giving a yellow colour.

(4) **Vein clearing:** In some viral diseases veins become translucent and more distinct.

(5) **Vein yellowing:** In some viral diseases veins become yellow and a netting appearance of yellow vein symptoms observed.

(6) **Vein banding:** Veins sometimes resist the yellowing process resulting in veins having a dark green appearance.

(7) **Leaf spots:** Some infections are associated with leaf spots which can be either chlorotic or necrotic spot.

Often some of the spots are ring formed - these are simply referred to as ring spots. After some time nercrotic spots can appear as holes in the leaves. If the inoculated plant initiates a local defense reaction this may be observed as a spot or a ring spot.

**(C)  Symptoms in Stem:** Stem pitting by *Citrus tristeza virus*, Bark scaling caused by *Citrus psorosis virus*

**(D)  Symptoms in Flower:** Colour breaking of tulip by *Potyvirus*.

**(E)  Symptoms in Fruit:** Fruit distortion on eggplant fruit caused by *Tomato bushy stunt virus*, Ring spot in cucurbit by *Potyvirus*, Ring spot in tomato by *Tospovirus*.

**(F)  Symptoms in Seed/Tuber:** Tuber cracking and ring spot by *Potato mop top virus*, Seed mottling due to *Potyvirus*.

**(G)  Symptoms in Plant growth:** Stunting and Dwarfing. Stunting means normal growth is stopped resulted in a incompletely grown plant. Dwarfing means shortening (roots, leaves and stem) and particular in dicots shortening of internodes resulted in a small bushy plant (here growth is complete but plant is weaker and smaller than a full grown plant).

# PLANT VIRUS CLASSIFICATION

*What is the purpose of classification?*

- To make order
- To be able to communicate with each other
- To assemble like members with each other
- Various virus classification schemes have been used
- Host & symptoms have been important considerations
- Particle morphology was important after EM developed
- Physico-chemical properties became important later
- Sedimentation coefficient and density of particles
- Protein compositions and sizes
- Nucleic acid types, numbers, and sizes

Sequence analysis confirmed most relationships that were inferred otherwise, and revealed new ones.

- Molecular phylogeny now a primary tool for classifying viruses.
- Complete genomes can be analyzed relatively quickly.

   Naming Virus Diseases

   First host + prominent symptom

   Tobacco mosaic

   Alfalfa latent

Cowpea chlorotic mottle

Tomato spotted wilt

Taxonomy of Viruses

Tobamovirus: Tobacco mosaic virus

Tospovirus: Tomato spotted wilt virus

Cucumovirus: Cucumber mosaic virus

## HISTORY OF CLASSIFICATION

- Latin binomials were proposed first by Holmes in 1939
- Various other schemes proposed between 1940 and 1966
- 1966 the International Committee for the Nomenclature of Viruses formed; met in 1970
- Changed to the International Committee for the Taxonomy of Viruses in 1973
- 7th Report of the ICTV was published in 2000
- 56 families, 9 sub-families, 233 genera, and 1550 virus species
- Includes retrotransposons, satellites, viroids, prions

Plant virus classification through cryptogram analysis

- Proposed by A.J. Gibbs, B.D. Harrison, D.H. Watson & P. Wildy, (Nature, UK 209: 450-454, 1966)
- It is a descriptive code summarizing some of the main properties of a virus.
- Cryptograms aid identification and add precision to the vernacular names of viruses.
- They also show how much is known about a given virus and they help to overcome the difficulties of synonymy, homonymy, and translation into other languages.
- Each cryptogram consists of four pairs of symbols (for example, R/1:2/ 5:E/E:S/* for tobacco mosaic virus) with the following meanings: 1st pair. Type of nucleic acid/strandedness of nucleic acid. Symbols for type of nucleic acid: R = RNA; D = DNA. Symbols for strandedness: 1 = single-stranded; 2 = double stranded 2nd pair: Molecular weight of nucleic acid (in millions)/percentage of nucleic acid in infective particles 3rd pair. Outline of particle/outline of 'nucleocapsid' (the nucleic acid plus the protein most closely in contact with it). Symbols for both properties: S = essentially spherical; E = elongated with parallel sides,

ends not rounded; U = elongated with parallel sides, end(s) rounded; X = complex or none of the above 4th pair. Kinds of host infected/kinds of vector. Symbols for kinds of host: A = actinomycete; B =bacterium; F = fungus; I = invertebrate; S = seed plant; V = vertebrate. Symbols for kinds of vector: Ac = mite and tick (Acarina, Arachnida); Al = white-fly (Aleyrodidae, Hemiptera, Insecta); Ap = aphid (Aphididae, Hemiptera, Insecta); Au = leaf-, plant or tree-hopper (Auchenotthyncha, Hemiptera); Cc = mealy-bug (Coccidae, Hemiptera); Cl = beetle (Coleoptera, Insecta); Di = fly and mosquito (Diptera, Insecta); Fu = fungus (Chytridiales and Plasmodiophorales, Fungi); Gy = mirid, piesmid, or tingid bug (Gymnocerata, Hemiptera); Ne = nematode (Nematoda); Ps = psyllid (Psyllidae, Hemiptera); Si = flea (Siphonaptera, Insecta); Th= thrips (Thysanoptera, Insecta); Ve = vectors known but none of above. Symbols for all pairs: * = property of the virus is not known; () enclosed information is doubtful or unconfirmed.

- Present criteria for classification:

The order of presentation of the virus families and genera of the VIIth ICTV Report follows four criteria:

(1) the nature of the viral genome,

(2) standedness of the viral genome,

(3) the fact that some viruses are reverse transcribed, and

(4) the polarity of the virus genome.

As there are no known ssDNA, nor dsRNA reverse transcribed viruses, and there are negative sense viruses only for ssRNA viruses, these four criteria give rise to seven clusters comprising the 56 families and 233 genera of viruses. In addition, subviral agents, namely the satellites, viroids and agents of spongiform encephalopathies (prions) are included, in most cases official taxonomic status.

- Modified binomial is used

- Taxonomy depends on particle properties, nucleic acid properties and especially sequence

- Family is the highest taxonomic level that is commonly used; ends in viridae, e.g., Tobamoviridae

- Genus ends in suffix virus, e.g., Tobamovirus

- Species is usually the commonly used virus name; it is italicized in formal usage, e.g., Tobacco mosaic virus

- Small genome sizes, gene shuffling make broad taxonomic schemes difficult

# MAJOR GROUPS OF PLANT VIRUSES

(1)   Positive sense ssRNA Viruses

(2)   Negative sense ssRNA Viruses

(3)   Ds RNA Viruses

(4)   ss DNAviruses

(5)   Ds DNA Viruses

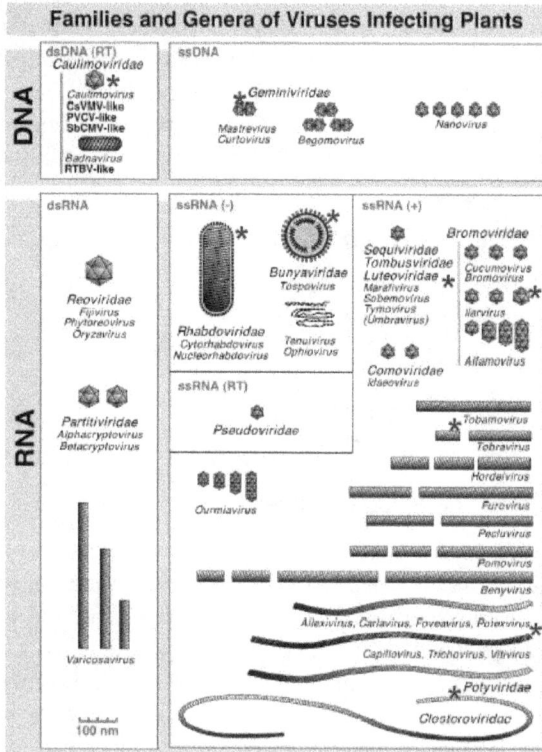

**Families and Genera of Viruses Infecting Plants**

Certain virus families/groupings cross "kingdom" or phylum boundaries For example, virus families infecting two kingdoms of organisms are:

Bunyaviridae (animals and plants)

Partitiviridae (plants and fungi)

Reoviridae (animals and plants)

Rhabdoviridae (animals and plants)

Phycodnaviridae (protozoa and plants)

(Picornaviridae) (plants and animals – tentative)

(Totiviridae) (Protozoa/fungi and insects tentative)

# PLANT VIRUS ENTRY, MOVEMENT, TRANSMISSION OTHER THAN INSECT VECTOR

## Plant virus entry

No cellular receptor

Do not appear to specifically interact with host cell membranes or cell walls, as do bacterial and animal viruses

The mechanisms employed to enter cells rather appear to be passive carriage through breaches in the cell wall in the first instance, followed by later cell-to-cell spread in a plant by means of specifically-evolved "movement" functions, and perhaps spread via conductive tissue as whole virions.

## CELL-TO-CELL MOVEMENT

Plant viruses move cell-to-cell slowly through plasmodesmata

The viral protein that facilitates movement is called the "movement protein" (MP). Most plant viruses move cell-to-cell as complexes of non-structural protein and genomic RNA.

MPs act as host range determinants.

MP alone causes expansion of normally constricted plasmodesmata pores; MPs then traffic through rapidly.

## MECHANISM OF PLANT VIRUS MOVEMENT

MP complexed with viral RNA moves along microtubules from ER-associated sites of viral replication; actin microfilaments deliver MP–RNA complexes to putative cell wall adhesion sites and plasmodesmata. These viruses do not require CP for cell-to-cell movement. Pectin methyl esterase-cell wall-associated protein that specifically binds the viral MP.

NSP is a nuclear shuttle protein that moves newly replicated viral ssDNA genomes from the nucleus to the cytoplasm. A movement protein, MPB, associated with ER- derived tubules, traps the NSP–ssDNA complexes in the cytoplasm and guides these along the tubules and through the cell wall into adjacent cells. These viruses also do not require CP for cell-to-cell movement

Some plant viruses move as whole particles through highly modified plasmodesmata (tubules).

These viruses require CP in addition to MP for cell-to-cell movement.

# PLANT VIRUS TRANSMISSION

Most plant viruses are absolutely dependent on a vector for plant-to-plant spread. A number of different types of organisms work as vectors for different plant viruses. Plants, as sessile organisms, cannot transmit viruses except for some instances of mechanical transmission, graft transmission, transmission through propagating materials, seed or pollen transmission and the movement of plants resulting from human intervention. Thus, the great majority of plant viruses are dependent for their spread upon efficient transmission from plant to plant by specific vectors. Vector transmission is a specific event in the virus life cycle. Virus-encoded determinants specifically interact with the vector, thereby facilitating virus transmission, and various plant viruses utilize different, but specific, vectors to facilitate their spread. Different organisms such as insects, fungi, nematodes, animals and arthropods are recognized as vectors for various plant viruses. Following are the different means of transmission of plant viruses.

# INSECTS

They are the most important means of plant virus transmission and next chapter will exclusively deal with insect transmission.

## Mechanical Transmission

The mechanical transmission of plant viruses by direct transfer of sap from plant to plant by contact in nature is unimportant. One exception to this seems to be potato virus X (PVX) in potato. Transmission of PVX through sap inoculation is readily accomplished by the contact of leaves in the field due to wind or machinery. Tomato mosaic virus is also readily sap transmissible in the field when sap is accidentally transferred from infected tomato plants to other tomato plants on tools, hands, clothes or machinery. The ability of these viruses to be spread by sap in the field is due to their extreme stability. Sap or mechanical transmission is most important in the study of plant viruses *in vitro*, since all investigations outside of the natural host require the ability to demonstrate and measure infectivity of the agent. When a homogenate of infected tissue or purified virus is gently rubbed onto the leaf surface of an appropriate herbaceous host, many viruses will produce symptoms on the inoculated leaves in 1 to 3 weeks.

# VEGETATIVE PROPAGATION

Viruses present in a mother plant from which tubers, corms, bulbs, rhizomes, cuttings, bud wood or other tissue explants are taken will almost always be transmitted to their progeny. A meristem may be a source of virus-free plants through meristem culture. Potatoes are propagated by tubers, and many important viruses are transmitted via potato seed pieces. Virus symptoms may be seen on the tubers as in the case of tobacco rattle virus. More frequently, symptoms of virus infection are not evident until the plant produces shoots and leaves. Leaf-roll symptoms are evident in plants established with potato leaf roll virus-infected tubers.

## GRAFT TRANSMISSION

Most fruit trees are propagated by budding or grafting. Virus transmission through the use of virus-infected bud wood or rootstocks is important. In apple, tomato ring spot virus may become established in an orchard through the use of infected rootstock. A line of dead tissue can be seen at the union between rootstock and scion of an infected, debarked apple tree. In orchard and forest environments, tree roots spread widely and often form natural root grafts. These natural grafts may transmit viruses as in the case of Tulare apple mosaic virus in apple.

### Pollen Transmission

When a virus is transmitted by pollen, it may infect the seed and the seedling that will grow from that seed, or it may infect the plant through the fertilized flower. The plant-to plant transmission of virus by pollen is known to occur in fruit trees such as sour cherry. Ilar viruses are commonly transmitted by pollen.

### Seed Transmission

A number of important virus diseases are known to be seed-transmitted. Seed transmission results in the earliest possible infection of the young seedling. This often results in increased severity of the virus infection. *Pea seed borne mosaic virus* has been disseminated worldwide in infected seed.

## DODDER TRANSMISSION

Dodder (Cuscuta spp.) is a yellowish, vine-like parasitic plant. A number of species of dodder transmit viruses. The parasite forms haustoria which penetrate the host. The dodder acts as a continuous cytoplasmic strand through which virus may move to other susceptible plants. Virus transmission by dodder in nature is not of economic importance. For experimental purposes dodder transmission is very much important for phytoplasmas because all the phytoplasmas are transmitted by dodder and produces symptoms in Periwinkle (Cathranthus roseus) plant which helps in detection of phytoplasma infection.

**Nematode Transmission:** Approximately a dozen widespread and important viruses have been shown to be transmitted via soil by nematodes. Three genera of nematodes of the order Dorylaimida are known to transmit plant viruses, *Xiphinema, Logidorus,* and *Trichodorus.* Dagger nematode, *Xiphinema,* feeding at a root tip with stylet. *Xiphinema* spp. are long and slender and are vectors of polyhedral-shaped NEPO viruses such as grape fan leaf virus, tobacco ringspot virus, tomato ringspot virus, cherry leaf roll and other viruses. The virus is acquired within 15 to 60 minutes and can be retained in the nematode's gut for up to a year. *Xiphinema* spp. prefers to feed on woody species such as grape and peach. *Longidorus* spp. is also long and slender and vectors of polyhedral-shaped NEPOviruses such as raspberry ringspot virus and tomato black ring virus. *Longidorus* spp. prefers to feed on herbaceous perennial hosts. Nematode vectors feed on cells at root tips with their stylet, acquiring virus. The virus is retained within the esophagus and/or gut and transmitted when the nematode feeds again. *Trichodorus* spp. are short and plump nematodes and vectors of short, tubular shaped TOBRA viruses such as tobacco rattle virus and pea early browning virus. *Trichodorus* may remain infective for as long as 2 years after acquiring virus from infected plants.

## Fungal Transmission

Thirty soil borne viruses or virus-like agents are transmitted by five species of fungal vectors. Ten polyhedral viruses, of which nine are in the family *Tombusviridae,* are acquired in the in vitro manner and do not occur within the resting spores of their vectors, *Olpidium brassicae* and *O. bornovanus.* Fungal vectors for other viruses in the family should be sought even though tombusviruses are reputed to be soil transmitted without a vector. Eighteen rod-shaped viruses belonging to the furo- and bymovirus groups and to an unclassified group is acquired in the in vivo manner and survives within the resting spores of their vector, *O. brassicae, Polymyxa graminis, P. betae,* and *Spongospora subterranea. Polymyxa graminis* is the vector of several economically important viruses of wheat such as *Wheat soil borne mosaic virus* and *Wheat spindle streak mosaic virus.* Another economically important crop potato is getting infected by *Potato mop top virus* which is transmitted by the fungal vector *Spongospora subterranean.* The viral coat protein has an essential role in vitro transmission. With in vivo transmission a site in the coat protein-read through protein (CP-RT) of beet necrotic yellow vein fur virus determines vector transmissibility as does a site in a similar 98- kDa polyprotein of *Barley mild mosaic bymovirus.* The mechanisms by which virions move (or are moved) into and out of the protoplasm of zoospores or of thalli remain unclear.

## MITES

Mites belong to the subclass Acarina and the class Arachnida. Among the different families of mite, those belong to family *Eriophyidae* are important vectors of different plant viruses. Viruses in the genera *Rymovirus* and *Tritimovirus* under the family *Potyviridae* are transmitted by different eriophyid mites. *Aceria tulipae* transmits *Wheat streak mosaic virus* and *Wheat spot mosaic virus*. *Aceria ficus* transmit *Fig mosaic virus*. Abacarus hystrix transmits *Ryegrass mosaic virus*. Transmission by mites is generally occurring in a semi-persistent manner. Recently, it was demonstrated that helper component-proteinase (HC-Pro) protein of the mite transmitted viruses play key determinant role for their vector specificity.

## INSECT TRANSMISSION OF PLANT VIRUSES

The ability of plant viruses to disperse from host to host is an essential part of the infection cycle. Plant viruses have evolved an array of specialization means ensuring their transmission and consequently improved fitness and survival. Among these are pollen, seed and mechanical transmission, but by far the most common mode of plant virus transmission involves a mobile biological vector that feeds and completes its life cycle on a living plant that is a host of the virus. Among the insects that transmit plant viruses, those that have evolved the ability to feed in the vascular tissues of higher plants are the most common and widespread vectors of plant viruses in both temperate and tropical ecosystems. Most plant viruses rely on an arthropod or nematode vector that feeds in the vascular, mesophyll, or epidermal tissues for transmission between hosts, with 70-80% having evolved an intimate relationship with their hemipteran insect vector. Aphids, thrips, leafhoppers, plant hoppers and whiteflies are the important insect vectors of plant viruses. Although many species of leafhoppers have been established as vectors of plant viruses, aphids continue to outrank all other groups in the number transmitted.

## PERSISTENT VS. NON-PERSISTENT TRANSMISSION

There are two distinct types of transmission by insects. In one type insects fail to infect the first series of exposed plants, but in later series begin to infect and may continue throughout their lifetime; in the other type the insects infect the first series of plants to which they are exposed, but few or none in later series. Watson and Roberts (1939) proposed the terms "persistent" and "no persistent" to designate these two types of transmission. The concept of persistent and no persistent modes of plant-virus transmission by vectors has proved to be a key contribution, first, as a guide in the technique of handling suspected vectors, second as a means of separating individual viruses from complexes, and finally, to help solve the basic problem of how vectors transmit viruses.

The persistent virus is characterized by an acquisition feeding period of several hours to several weeks followed by a latent period of similar length before the insect can transmit it. The vector retains the virus for long periods, usually for life. This virus is usually not mechanically transmissible, and there is a high degree of vector specificity. The delay in development of infectivity is characteristic of most leafhopper-borne viruses, all viruses transmitted by thrips and whiteflies, and a few of those transmitted by aphids. The basic features of non-persistent transmission include the ability of the virus to be acquired in seconds to minutes and to be transmitted in a similar time and with no latent period. The ability to transmit is lost within a matter of minutes to hours. Most no persistent viruses have several vectors, and they are usually mechanically transmitted. The term 'non-circulative' applies to viruses that do not enter the cells, but are retained in the alimentary tract. Viruses transmitted in the no persistent manner have a non-circulative relationship with their vectors

## PROPAGATIVE VS. CIRCULATIVE TRANSMISSION

Propagative viruses are those that have been proven to multiply in their vectors. More direct evidence of virus multiplication in the vector is obtained by quantities serology using ELISA that shows increase in virus titer in the vector after a relatively short AAP on a virus source. Propagative viruses usually require longer latent periods than circulative (non-propagative) viruses. For non-propagative viruses, the latent period is presumed to be the time necessary for the virus to circulate in the vector, i.e., from ingestion of virus from a virus source to inoculation of virus. However, with propagative viruses, the longer latent period may also be necessary for virus multiplication in various tissues of the vector before moving into the salivary secretions of the vector. Another important difference between propagative and non-propagative viruses is that transovarial transmission has been reported and confirmed only with propagative viruses.

Most tenuiviruses are transovarially transmitted to a large proportion of the progeny in their plant hopper vectors. For example, *Rice stripe virus* (RStV) was reported to pass through the eggs of a single infective female of *Laodelphax striatellus* for 40 generations, with 90% of the progeny insects of the 40th generation being inoculative with Rstv.

With plant viruses the term 'acquisition threshold' usually refers to the minimum time required for exposure or feeding on diseased plants, after which an insect vector can become viruliferous (virus carrying or virus infected). This 'acquisition threshold' possibly involves both the time necessary for the insect stylets to reach the plant tissue from which the virus can be acquired, e.g., mesophyll or phloem, as well as a 'threshold titer' of virus that must be ingested before infection of the vector can occur. This is indicated by several studies that show a positive correlation between longer AAP on diseased plants and greater efficiency of transmission for many propagative plant viruses by their vectors. The terms 'vector specificity', 'vector efficiency' and 'vector competence' are often used to describe the comparative ability of certain species, biotypes or lines of vector insects to transmit a certain virus or a virus strain. Before an insect can transmit any circulative/propagative virus they must have: (1) ingested a virus; (2) the virus must have entered the cells of the insect midgut; (3) the virus must then be released from these cells into the haemocoel; (4) virus must enter the salivary glands; (5) virus must be released into the saliva; and (6) the insect must feed on a susceptible host. Similarly four groups of barriers to transmission of propagative viruses in their vectors have been identified (1) midgut infection barrier; (2) dissemination (including midgut-escape and salivary gland infection) barriers; (3) salivary gland escape barrier; and (4) transovarial transmission barriers.

## IMPORTANT INSECT VECTORS

### *Leafhoppers and plant hoppers*

In 1883, a Japanese rice farmer suspected that a dwarfing disease of rice was associated with leaf hoppers and a year later he demonstrated the relationship between the hoppers and the occurrence of the disease. In subsequent studies, the disease agent was identified as the *Rice dwarf virus* and the vectors as *Resilia dorsalis* and *Nephotettix cincticeps.* Of the 60 plant viruses reported to be transmitted by the hoppers (as of 2002), less than 10% infect dicotyledonous plants. Many of the hopper vectors are delphacids and most of them belong to the family cicadellidae. As of 2002, three plant viruses are known to be transmitted by leafhoppers in a semipersistant manner. The *Rice tungro spherical virus* (RTSV) and *Rice tungro bacilliform virus* (RTBV) are the typical examples, which are transmitted by the leafhopper *Nephotettix virescens* in a semipersistant manner. Both sexes of the adult leafhoppers as well as nymphs can transmit these viruses, but inoculativity is lost upon moulting. The RTSV can be transmitted by *N. virescens* only from the plants co-infected with RTSV or from the plants infected with RTBV if vector had previous access to RTSV-infected plants. Eleven geminiviruses are persistently transmitted in a circulative (non-propagative) manner by cicadellid leafhoppers. The *Maize streak virus* (MSV) that is transmitted by the leafhopper *Cicadulina mbila* is the typical example for this kind of transmission. ELISA-positive insects occurred from 17 day post-AAP. The MSV titer was 0.36ng per leafhopper 3 days after AAP, whereas 14 days later there was only 0.20 ng of virus/insect. These results suggested that MSV does not multiply in *C. mbila.* MSV transmission is trans-stadial, i.e., nymphs do not lose inoculativity upon moulting, but apparently is not trans-ovarial or vertical, i.e., inoculativity is not passed from adult females to their progeny. The vector specificity of geminiviruses is governed by the viral coat protein. At least 41 plant viruses are transmitted either by leafhoppers or plant hoppers in a persistent propagative manner.

## APHIDS

More plant viruses are transmitted by aphids than by any other group of insects. *Myzus persicae,* one of the most efficient vectors (no persistent transmission), is reported to transmit more than 50 plant viruses. Most aphid-borne viruses are mosaics. A single aphid is capable of transmitting both persistent and no persistent viruses from the same host. Much of the understanding of determinants of non-persistent transmission has come from studies of potyviruses. The N-terminal portion of potyvirus CPs is exposed on the surface of the virion, and a comparison of the amino acid sequences near the N-terminus of aphid transmissible potyvirus isolates revealed a conserved *asp-ala-gly* (DAG) motif. In potyvirus isolates that had lost aphid transmissibility, mutation of one of these amino acids, usually *gly* (G) to *glu* (E) was found. Hence, it was suggested that changes in the DAG motif were responsible for loss of aphid transmissibility. In addition to virions with a transmission-competent N-terminal CP sequence, potyvirus transmission requires the acquisition by aphids of a virally encoded protein helper component. There are a number of lines of evidence which indicate that aphids acquire transmissible virus from the epidermis. The observable characteristics of non-persistent transmission are probably the natural result of the host-selection behaviour of aphids. In fact, it may be that aphids, and not other insects, act as the vectors of these viruses because of their behaviour and their ability to penetrate plant cells without causing serious damage. Aphids characteristically make one or more brief probes into epidermal cells (<10) micrometers) before making the deep penetration required to reach their feeding site, which is usually the phloem. These brief probes (usually less than 30 s) are presumably made to test a plant as a potential food source, but probes of this duration are also optimal for acquisition of non-persistent viruses. Presumably, non-persistent viruses evolved to become adapted to retention in the stylets of probing aphids at a site where they could both survive and subsequently be inoculated.

## THRIPS

In the realm of plant-infecting viruses, tospoviruses are seemingly among the most aggressive viruses. Since the creation of the genus *Tospovirus* in 1991, a dozen new viruses were added on the basis of molecular characterization and nucleocapsid genes. The genus *Tospovirus* is the only genus in the family Bunyaviridae containing plant-infecting viruses, which are transmitted by thrips. The finding that TSWV (*Tomato spotted wilt virus*) multiplies in *Frankliniella occidentalis* was a landmark contribution in terms of virus classification. With regard to virus classification, all the bunya viruses replicate in their arthropod vectors, thus, the earlier finding provided support to place the tospoviruses in the family Bunyaviridae. At least ten species of thrips transmit tospoviruses, all of which belong to the family thripidae of thysanoptera. *F. occidentalis,* the western flower thrips, is thought to be the most important vector species because of its global distribution and its capacity to transmit most of the tospoviruses. For tospoviruses to be transmitted by the thrips, they must be acquired by the larvae. Thus immature thrips that acquire tospoviruses or adults arising from such immature are important for the transmission of the virus. The transmission efficiency of the first instars, second instars and adults were estimated as 47.3%, 12.4% and 0%, respectively. This concept is extremely important in managing the tospoviruses because only the plants that serve as hosts for both the virus and thrips are important in epidemics. Thus plants serve as the hosts for thrips alone do not contribute infective thrips and vice-versa. The changing relationship between tospovirus and thrips vectors is a peculiar phenomenon. The most striking example is *Thrips tabaci*. This species transmitted all known isolates of TSWV five decades ago, but today *T. tabaci* is associated with the emergence of a new tospovirus, *Iris yellow spot virus* (IYSV), yet is apparently a non-vector of several modern TSWV isolates. Interestingly, *F. Occidentalis,* a species that transmits all the other tospoviruses does not transmit IYSV. The basis for the changes in the virus-vector relationship is still unknown. It is postulated that the GP2 (glycol protein) of the virion facilitates the receptor-mediated cell entry and subsequent replication in the thrips vector. The GP2 of the tospoviruses has highly conserved amino acid sequence RGD (Arg-Gly-Asp). This sequence is known as an important determinant in binding with cells for a number of mammalian pathogens, e.g., foot and mouth virus. The lack of suitable control measures leaves a major gap in the management plan for virtually all tospovirus epidemics. Control of thrips-transmitted tospoviruses can rarely be achieved with insecticides for several reasons. Firstly, relatively small numbers of thrips often results in high rates of tospovirus spread. Secondly, many thrips are resistant to insecticides. Thirdly, tospovirus inoculation occurs quickly; hence, the thrips seldom die from the insecticide treatment before transmitting the virus.

## WHITEFLIES

Whitefly vectors of plant viruses belong to the family *Aleyrodidae* of the order Hemiptera, and are specialized for feeding in the phloem. Historically, emphasis has been on aphids and cicadellids; attention to whiteflies as vectors is more recent. Presently, three whitefly species, *Bemisia tabaci* (Genn.), *Trialeurodes vaporariorum* (West and *T. abutilonea* (Hald.), are recognized as vectors of plant viruses; *B. tabaci* is the most important, having been associated with more than 100 plant viral diseases mainly in the tropics and subtropics. Forty years ago, nine whitefly species were considered to be vectors of virus-like diseases. This was probably due to the confounded taxonomy of whiteflies, which may have resulted in misidentification. Begomoviruses, Carlaviruses, Criniviruses and Potyviruses are the important plant viruses transmitted by whiteflies. Begomovimses are transmitted exclusively by *B. tabaci* in a persistent and circulative manner. Virus acquisition requires several hours or longer, and there is a latent period of up to 24 h for different virusvect or combinations. Begomoviruses generally do not replicate in their whitefly vector, with one possible exception. *Tomato yellow leaf curl virus* from Israel (TYLCV-IS) is passed transovarially in a colony of the B biotype of *B. tabaci.* Once ingested, begomoviruses are not immediately transmissible. Virions must first translocate from the digestive tract to the haemolymph, and from there, to the salivary glands from which virus can be secreted with saliva during feeding. The time it takes for a begomovirus to complete this path, referred to as the latent period, is reflected in the minimum time that elapses from the beginning of feeding on infected plants to transmission to test plants. The average minimum latent period for begomoviruses is approximately 17 h. *T. vaporariorum,* which also feeds in the phloem and has a host range similar to *B. tabaci,* is capable of ingesting begomoviruses, yet it is unable to transmit these viruses. Whitefly-transmitted geminiviruses infect dicotyledonous plants and are assigned to the genus *Begomovirus* within the family *Geminiviridae.* Begomoviruses originating in the New World have a bipartite genome organization, whereas those from the Old World have either bipartite or monopartite genomes. DNA-A and DNA-B of bipartite begomoviruses are each about 2.6 kb in size and share a common region (CR) of approximately 200 nucleotides that is highly conserved among cognate components of a single virus species. The CR contains modular c/s-acting elements of the origin of replication *(ori).* Five open reading frames (ORFs) capable of encoding proteins >10 kDa in size are conserved among the DNA-A component. The CP, encoded by AVI ORF, is the most highly conserved protein among begomoviruses, and is required for encapsulation, whitefly-mediated transmission and certain movement functions. The Rep protein encoded by the AC1 ORF initiates viral DNA replication, and replication specificity is mediated through interactions of Rep with cisacting elements of the *ori.* The begomovirus DNA-B component encodes BV1 and BCI proteins that are essential for systemic movement and can influence host range.

## BEETLE

Plant viruses belonging to the genera *Comomovirus, Bromovirus, Sobemovirus,* and *Tymovirus* are known to be transmitted by beetles. All these viruses have isometric particle morphology and contain positive sense RNA genome. These viruses are usually quite stable and present in high concentrations within infected tissues. Beetle vectors acquire virus quickly by chewing on infected leaf tissue. The virus can be transmitted with the first bite. Transmission occurs when the beetles regurgitate sap and virus onto feeding wounds.

## MEALY BUG

Mealy bugs are much less mobile on plants than other groups of vectors, making them relatively inefficient virus vectors. Mealy bugs feed only on the phloem; the virus is carried on or near the stylets. Nymphs move more readily than adults and crawl from plant to plant over their surfaces. Badnaviruses are transmitted by mealy bug.

# PLANT VIRUS GENOME STRATEGY AND REPLICATION

## *Genes that are coded by plant viruses*

One gene for replication: STNV

- **Highest:** Reoviruses (12 proteins)

Most ssRNA positive sense RNA viruses have 4-7 proteins. Essential genes for proteins: capsid protein, replication protein and movement proteins. Besides these genes many other genes for proteins like protease, helicase, proteins that involve in insect transmission etc. may be coded by viral genome. In addition to the coding region, genomic nucleic acids of viruses also contain nucleotide sequences for recognition and control functions that are important for replication. These sequences frequently present at the 5′ or 3′ terminus.

How the small genome of plant viruses can economically utilizes its genome:

Lack of introns

**Overlapping genes:** Geminiviruses

**Ambisense genome:** Geminiviruses

**Multifunctionality of the proteins:** AMV CP functions as protective coat as well as initiation of infection. Interaction of viral proteins with host proteins. Untranslated region have multifunction like ribosome recognition (viral sense), replicate recognition (in complementary sense) etc.

## Genome strategy

All the plant viruses utilizes host protein translational machinery (ribosome, amino acids tRNA, all the enzymes involved in translation etc.) for translating their proteins inside the hosts. The characterization feature of eukaryotic ribosome is it can translate the ORF just downstream of the 5′ end of the mRNA and finishes translation when termination codon is recognized. How all the ORFs of plant viruses are translated when required? To overcome such constrains. Plant viruses evolved five strategies by which they ensure that all their genes are accessible to the eukaryotic protein synthesizing system. These strategies are called genome strategies.

1. **Sub-genomic RNA:** The synthesis of one or more subgenomic RNAs enables the 5′ ORF on each such RNA to be translated, e.g., Tobamoviridae.

2. **Polyproteins:** Here the whole genome is translated as a single polyprotein. Then it is cleaved at specific sites by viral encoded proteases to give final gene product for each ORF. For example, Potyviridae.

3. **Multipartite genome:** The 5′ gene of each RNA segment can be translated. For example, comoviridae.

4. **Read-through proteins:** The termination codon of the 5′ gene may be leaky and allow ribosome to continue translation up to next stop codon and give rise a longer functional polypeptide. For example, Tobamoviridae.

5. **Transframe proteins or Frameshift:** Ribosomal frameshifting is a strategy frequently employed by various organisms to produce more than one protein from overlapping reading frames. The frame shift generally occurs near the stop codon where nucleotide repeat is present. For example, Luteoviruses.

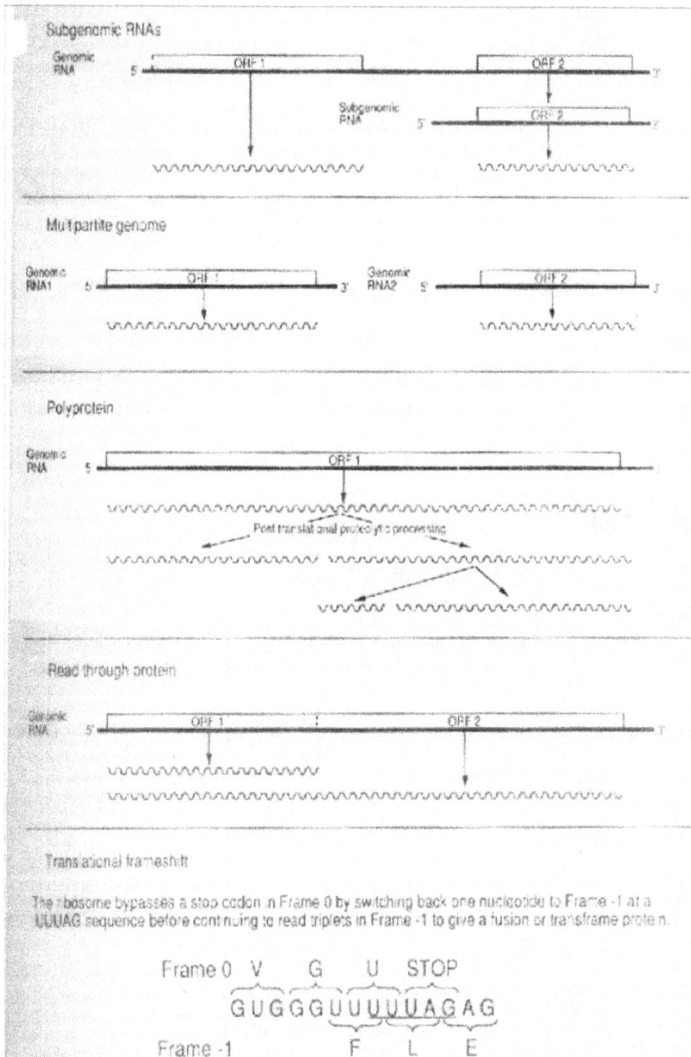

Genome organization, replication, transcription and translation of (+) ve sense ssRNA virus

Most common example is Tobacco mosaic virus.

Important disease of tobacco, tomato & other plants.

Easily mechanically transmitted (only means of transmission).

- Very high concentration in plants.
- First plant virus disease characterized (1898).
- First virus strains demonstrated.
- First cross protection shown.

First virus crystallized (1946 Stanley was awarded the Nobel Prize).

First demonstration of infectious RNA (1950s).

First virus to be shown to consist of RNA and protein

First virus characterized by X-ray crystallography.

- First plant virus genome to be completely sequenced.
- First virus used for coat protein mediated protection.
- First virus to have a resistance gene characterized.

## GNOME ORGANIZATION OF TMV

*(1)* *Tobacco mosaic virus* is typical, well-studied example of monopartite, rod shaped ssRNA + sense genome. Each particle (300 × 18 nm) contains only a single molecule of RNA (6395 nucleotide residues) and 2130 copies of the coat protein subunit (158 amino acid residues; 17.6 kilodaltons). 3 nt/subunit, 16.33 subunits/turn, 49 subunits/3 turns.

(2) At 5′ end there is an m7 Gppp cap followed by a 69 nt untranslated leader sequence.

(3) The first ORF produces a MW=126 K protein which has a mt/helicase activity

(4) The stop codon of this ORF (UAG) is leaky and produces a MW= 183 K protein which act as an RdRp (RNA dependent RNA polymerase).

(5) The terminal 5 codons of this read-through ORF overlaps with a third ORF coding for a MW=30 K   protein involve in cell-to-cell movement.

(6) A fourth ORF begins two nt after the termination codon of 3rd ORF and produce a MW= 17.6 K protein (coat protein).

(7) At 3′ end there is a tRNA like structure that can accept Histidine.

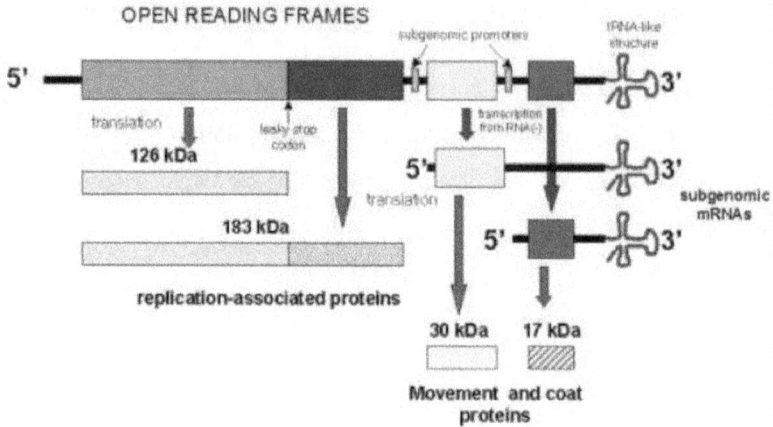

## TMV REPLICATION CYCLE

1. Virus enters injured cells.

2. Ribosomes strip off coat and begin to translate the RdRp. This is called co-translational disassembly.

3. RdRp binds to the tRNA-like 3′ end and initiates synthesis of minus strands to form RI RNA.

4. Plus-strand genomic RNAs and the movement protein (MP) & coat protein (CP) sgRNAs are synthesized.

5. MP and CP are translated.

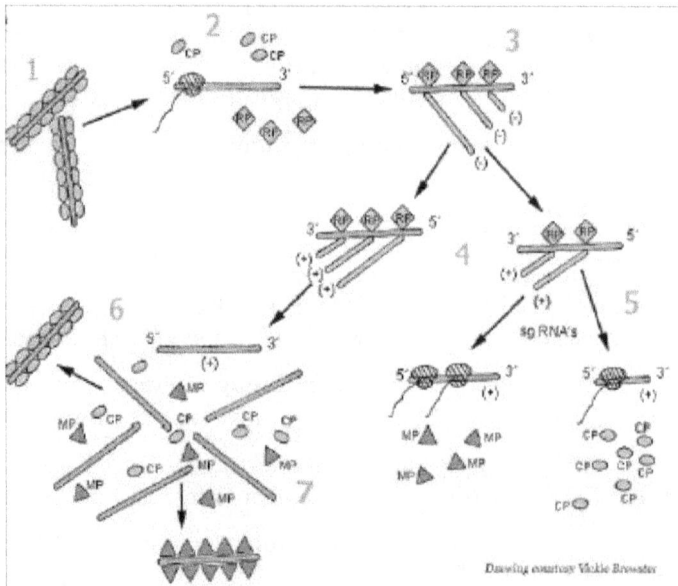

6. MP forms complexes with newly synthesized viral RNA and membranes.

7. The complex moves through plasmodesmata to adjacent cells.

8. CP & genomic RNAs accumulate to very high levels and virus crystals accumulate in the cytoplasm. Double-strand replicative form (RF) RNA accumulates in the cell and elicits host gene silencing responses.

Genome organization and Replication of ssDNA viruses of Plants

As an example genome organization and replication of geminiviruses are described.

**Genome Properties:** One or two ss ambisense circular DNA of 2.7 – 5.4 kb total genome size.

**Capsid Properties:** Twin icosahedral of size 20-30 nm × 30 nm.

**Host Range:** Both monocot and dicot.

**Vector:** Leaf hopper or whitefly.

## Classification:

Three genuses:

(1) **Mastrevirus:** They are monopartite, monocot-infecting viruses transmitted by leaf hopper. Type member: Maize streak virus

(2) **Curtovirus:** They are monopartite, dicot-infecting viruses transmitted by leaf hopper. Type member: Beet curly top virus

(3) **Begomovirus:** They are mono or bipartite, dicot-infecting viruses transmitted by whitefly. Type member: Bean golden mosaic virus

As a case study genome organization and replication of bipartite Begomovirus is described:

Genome has two ssDNA molecule designated as DNA A and DNA B which are separately encapsulated. Both the genomic fragments are ~2.7 kb long & differ from one another completely in nucleotide sequence, except for a shared 200 net non-coding sequence involved in DNA replication.

Because establishment of a productive infection requires both parts of the genome it is necessary for a minimum of two virus particles bearing one copy of each of the genome segments to infect a new host cell.

Although geminiviruses do not multiply in the tissues of their insect vectors (non-propagative transmission), a sufficiently large amount of virus is ingested and subsequently deposited onto a new host plant to favour such super infections.

## ORFS PRESENT IN DNA A: IN THE VIRAL SENSE TWO ORFS ARE PRESENT

AV 1: Coat Protein (CP). Involve in encapsulation. N-terminal region seems to play a major role in insect transmission. AV1 protein product also interacts with BVI protein product and thus has a role in movement of the virus. AV2: Pre-Coat protein. Exact role is not known. In the complementary sense four ORFs are present: AC 1: 39 kD Replication Initiation Protein (Rep). Mutational study reveals it involves in replication but do not have any polymerase activity. It has domain for DNA binding, site specific end nuclease and ligation, ATPase activity, ATP binding, Ren interaction and oligomerization. Rep is prerequisite for the replication of both DNA-A and DNA-B.Sequence specific recognition of the replication origin by the AC1 encoded protein, localizing the origin to a ~ 90 nucleotide segment, proximal side of the IR that includes the conserved geminiviral stem-loop structure and ~ 60 nucleotide of the 5′ upstream sequence. The Tyr residue in its end nuclease domain cleaves at TAATATT AC in the stem loop structure. AC 1 negatively regulates its own transcription. AC 2: Transcription Activator Protein (TrAP). Mutational studies showed that it does not hamper replication but hamper transcription of CP and BV1. It binds with CP and BV1 promoter and signals for transcription of CP and BV1. AC 3: Replication Enhancer Protein (Ren). Mutational study showed that in absence of Ren, viral DNA accumulation lowers down. It interacts with Rep protein and enhances AC 4: Function not known.

# 7

# DIAGRAMMATIC REPRESENTATION

Bio-fertilizers broadly classified as follows:

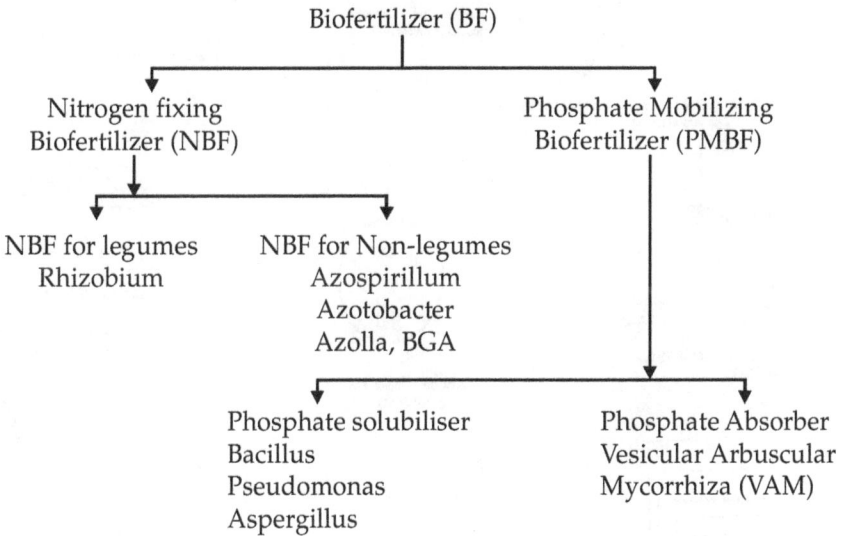

Biofertilizer (BF)

├── Nitrogen fixing Biofertilizer (NBF)

│   ├── NBF for legumes
│   │   Rhizobium

│   └── NBF for Non-legumes
│       Azospirillum
│       Azotobacter
│       Azolla, BGA
│
│       Phosphate solubiliser
│       Bacillus
│       Pseudomonas
│       Aspergillus

└── Phosphate Mobilizing Biofertilizer (PMBF)

    Phosphate Absorber
    Vesicular Arbuscular
    Mycorrhiza (VAM)

## Serial dil. Technique:

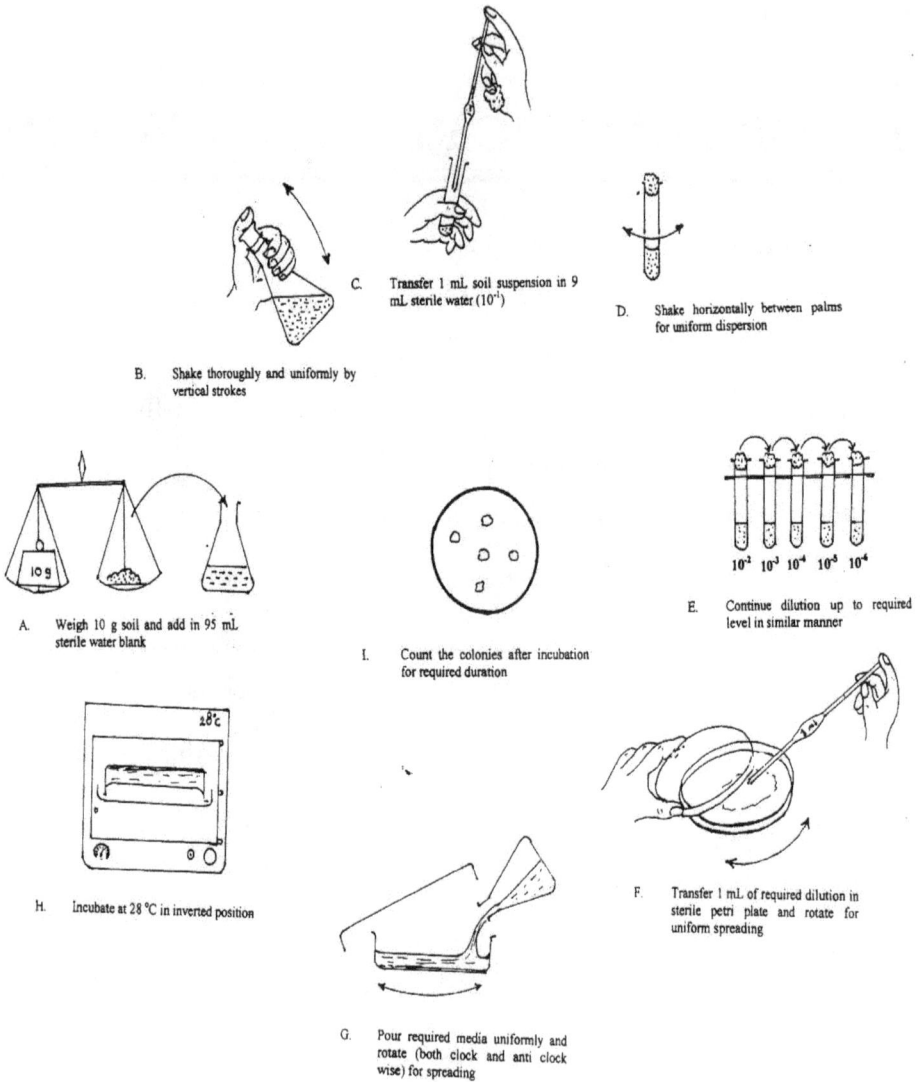

C.   Transfer 1 mL soil suspension in 9 mL sterile water ($10^{-1}$)

D.   Shake horizontally between palms for uniform dispersion

B.   Shake thoroughly and uniformly by vertical strokes

A.   Weigh 10 g soil and add in 95 mL sterile water blank

I.   Count the colonies after incubation for required duration

$10^{-2}$   $10^{-3}$   $10^{-4}$   $10^{-5}$   $10^{-6}$

E.   Continue dilution up to required level in similar manner

H.   Incubate at 28 °C in inverted position

F.   Transfer 1 mL of required dilution in sterile petri plate and rotate for uniform spreading

G.   Pour required media uniformly and rotate (both clock and anti clock wise) for spreading

**Fig. 1 :** Serial Dilution Technique for counting of microorganisms in soil.

# 8

# REASONING

## Constrains/Limitations of BF

The major limiting factors include:

- Narrow genetic base of mother cultures and lack of efficient and virulent strains suitable to various agro-environments.
- Unsatisfactory carrier material in respect of uniformity and good quality against imported peat material.
- High contamination in broth mixing and packing stages, not using completely closed system of production.
- Unsatisfactory packing material, which reduces shelf life.
- Unsatisfactory storing conditions, particularly during the distribution period. Exposure to high temperatures and sunlight destroy the microbial culture. They should be preferably kept in cold storage conditions.
- Not employing properly trained microbiologist.
- Lack of quality controls and certification procedures.
- Lack of awareness among farmers for its proper application.

## WHICH TYPE OF FACILITIES SHOULD BE NECESSARY TO PROVIDE QUALITY BIOFERTILIZERS TO THE FARMERS?

Though there are BIS standards for four species viz. Rhizobium (IS: 8268-2001) and Azotobacter (IS: 9138-2002), Azospirillum (IS: 14806:2000), PSB (IS: 14807:2000), there is no systematic quality certification system and monitoring mechanism. It is entirely an internal arrangement and voluntary system as of now. As the products being living microorganisms, the quality check up, and certification batch-wise even if it is internal is highly essential. Each unit should have lab infrastructure and plans/arrangements for the same. Each unit therefore should have the following facilities:

- Adequate microbiological lab and qualified microbiologist.
- Sampling and testing at various stages of production, including the quality of raw materials.

❑ Specify on the packets all contents and cell counts. The source of mother culture and the strain name should also be mentioned.

❑ The unit should fix their quality certificate and batch number, pack the products in proper packing material.

❑ Store the products in cooler places till they are sold to farmers.

❑ Ensure to have aseptic conditions, cleanliness and contamination free production lines and housing.

❑ Preferably use automatic and closed systems.

As per BIS specifications, certain tests are required to be conducted, like number of cells, colony character, reaction etc. Cell number at the time of manufacture should not be less than $10^7$ per gram of carrier material, for all biofertilisers. Similarly, the number of cell count and permissible contamination at expiry dates are also specified.

As certification arrangements are not in place at present, legislation for quality monitoring and accredited labs for testing may be needed in future to ensure proper quality and promote this products. Even there is a need of BIS specification for liquid culture, which is available in the market.

## VA MYCORRHIZAL (VAM) FUNGI

VA mycorrhizae [alias: vesicular arbuscular mycorrhizal fungi (VAM) and arbuscular mycorrhizal fungi (AMF)] are beneficial fungi that penetrate and colonize the root of the plant, then send out filaments (hyphae) into the surrounding soil. The term mycorrhizae literally mean fungus-root and VAM is considered endomycorrhizal since it colonizes the interior of the root.

## HOW DO VAM FUNGI WORK?

VAM fungi are associated with the plant in a mutually beneficial relationship. The VAM fungi, nestled inside the root, sends out long filaments or hyphae to explore up to 200 times the soil area available to the root alone. The hyphae literally form a bridge that connects the plant root with large areas of soil and serves as a pipeline to funnel nutrients back to the plant. In return, the plant must supply the VAM fungi with carbon for the fungal growth and energy needs.

This plant-fungal relationship is an elegant association and its development is evidently regulated by several factors. The promotion of root colonization by the VAM fungi is enhanced by different soil bacteria known as "mycorrhizal helper bacteria". The presence of these bacteria, many of which are aerobic heterotrophic bacteria such as *Bacillus* sp., probably affect VAM colonization and function by their production of vitamins, hormones, and other compounds. Plants produce discharges (exudates) sent out through their roots that contain specific compounds acting to stimulate the VAM hyphae growth.

## WHY DO YOUR PLANTS NEED VAM FUNGI?

Fortunately, most crops can be colonized by VAM fungi. The VAM fungi are best known for their ability to improve plant growth in low phosphate soils by exploiting large areas of soil and actively transporting the phosphate back to the plant. Other benefits to the plant supplied by the VAM colonization include increased absorption of nitrogen, potassium, magnesium, copper, zinc, sulfur, boron, molybdenum and other elements that are transported back to the plant. Due in part to the association of particular bacteria with the VAM fungi, many of these non-mobile elements are made water soluble by bacteria, absorbed by the VAM fungi, and translocated to the plant through the VAM "pipeline".

VAM fungi can increase the disease resistance of plants against root pathogens, especially when the VAM fungi can adequately colonize the root before the pathogen attacks. VAM fungi are important in forming stable soil aggregates by binding soil particles in the filamentous mass as well as producing sticky substances that hold the particles together. Due mainly to improved plant nutrition, VAM colonization can also improve plant drought resistance.

## VIRUSES

Viruses are very small (submicroscopic) infectious particles (virions) composed of a protein coat and a nucleic acid core. They carry genetic information encoded in their nucleic acid, which typically specifies two or more proteins. Translation of the genome (to produce proteins) or transcription and replication (to produce more nucleic acid) takes place within the host cell and uses some of the host's biochemical "machinery". Viruses do not capture or store free energy and are not functionally active outside their host. They are therefore parasites (and usually pathogens) but are not usually regarded as genuine microorganisms.

Most viruses are restricted to a particular type of host. Some infect bacteria, and are known as bacteriophages, whereas others are known that infect algae, protozoa, fungi (mycoviruses), invertebrates, vertebrates or vascular plants. However, some viruses that are transmitted between vertebrate or plant hosts by feeding insects (vectors) can replicate within both their host and their vector. This web site is mostly concerned with those viruses that infect plants but we also provide some taxonomic and genome information about viruses of fungi, protozoa, vertebrates and invertebrates where these are related to plant viruses.

We also provide information about viroids, which are infectious RNA molecules that cause diseases in various plants. Their genomes are much smaller than those of viruses (up to 400 nucleotides of circular single-stranded RNA) and do not code for any proteins.

## IMPORTANCE OF VIRUSES

Viruses cause many diseases of international importance. Amongst the human viruses, smallpox, polio, influenza, hepatitis, human immunodeficiency virus (HIV-AIDS), measles and the SARS coronavirus are particularly well known. While antibiotics can be very effective against diseases caused by bacteria, these treatments are ineffective against viruses and most control measures rely on vaccines (antibodies raised against some component of the virus) or relief of the symptoms to encourage the body's own defense system.

Viruses also cause many important plant diseases and are responsible for huge losses in crop production and quality in all parts of the world. Infected plants may show a range of symptoms depending on the disease but often there is leaf yellowing (either of the whole leaf or in a pattern of stripes or blotches), leaf distortion (e.g. curling) and/or other growth distortions (e.g., stunting of the whole plant, abnormalities in flower or fruit formation).

## TRANSMISSION OF VIRUS

Some important animal and human viruses can be spread through aerosols. The viruses have the "machinery" to enter the animal cells directly by fusing with the cell membrane (e.g. in the nasal lining or gut).

By contrast, plant cells have a robust cell wall and viruses cannot penetrate them unaided. Most plant viruses are therefore transmitted by a vector organism that feeds on the plant or (in some diseases) are introduced through wounds made, for example, during cultural operations (e.g., pruning). A small number of viruses can be transmitted through pollen to the seed (e.g., *Barley stripe mosaic virus*, genus *Hordeivirus*) while many that cause systemic infections accumulate in vegetative-propagated crops.

## MAJOR VECTORS OF PLANT VIRUSES

❏ **Insects:** This forms the largest and most significant vector group and particularly includes:

❏ **Aphids:** Transmit viruses from many different genera, including *Potyvirus*, *Cucumovirus* and *Luteovirus*. The picture shows the green peach aphid *Myzus persicae*, the vector of many plant viruses, including *Potato virus Y*.

❏ **Whiteflies:** Transmit viruses from several genera but particularly those in the genus *Begomovirus*. The picture shows *Bemisia tabaci,* the vector of many viruses including *Tomato yellow leaf curl virus* and *Lettuce infectious* *yellows virus.*

❑ **Hoppers:** Transmit viruses from several genera, including those in the families *Rhabdoviridae* and *Reoviridae.* The picture shows *Micrutalis malleifera,* the treehopper vector of *Tomato pseudo-curly top virus.*

❑ **Thrips:** Transmit viruses in the genus *Tospovirus.* The picture shows *Frankinella occidentalis,* the western flower thrips that is a major vector of *Tomato spotted wilt virus.*

❑ **Beetles:** Transmit viruses from several genera, including *Comovirus* and *Sobemovirus*

The virus-vector relationships are of several types:

❑ At one extreme, the association occurs within the feeding apparatus of the insect, where the virus can be rapidly adsorbed and then released into a different plant cell. The feeding insect looses the virus rapidly when feeding on a non-infected plant. Such a relationship is termed "non-persistent". The best studied examples are of potyvirus transmission by aphids.

❑ At the other extreme, the virus is taken up into the vector, circulates within the vector body and is released through the salivary glands. The vector needs to feed on an infected plant for much longer and there is an interval (perhaps several hours) before it can transmit. Once it becomes viruliferous, the vector will remain so for many days and such a relationship is therefore termed "persistent" or "circulative". The best studied examples are of luteovirus transmission by aphids. In some examples of this type (e.g., some hoppers and thrips), the virus multiplies within the vector and this is termed "propagative".

❑ **Nematodes:** These are root-feeding parasites, some of which transmit viruses in the genera *Nepovirus* and *Tobravirus.* The picture shows an adult female of *Paratrichodorus pachydermus,* the vector of *Tobacco rattle virus.* (Figure from

❑ **Plasmodiophorids:** These are root-infecting obligate parasites traditionally regarded as fungi but now known to be more closely related to protists. They transmit viruses in the genera *Benyvirus, Bymovirus, Furovirus, Pecluvirus* and *Pomovirus.* The picture shows *Polymyxa graminis,* the vector of several cereal viruses including *Barley yellow mosaic virus,* growing within a barley root cell.

❑ **Mites:** These transmit viruses in the genera *Rymovirus* and *Tritimovirus.* The picture shows *Aceria tosichella,* the vector of *Wheat streak mosaic virus.* (Figure from Description 393; bar represents 10 µm).

# CLASSIFICATION OF VIRUS

The highest level of virus classification recognizes six major groups, based on the nature of the genome:

- **Double-stranded DNA (dsDNA):** There are no plant viruses in this group, which is defined to include only those viruses that replicate without an RNA intermediate (see Reverse-transcribing viruses, below). It includes those viruses with the largest known genomes (up to about 400,000 base pairs) and there is only one genome component, which may be linear or circular. Well-known viruses in this group include the herpes and pox viruses.

- **Single-stranded DNA (ssDNA):** There are two families of plant viruses in this group and both of these have small circular genome components, often with two or more segments.

- **Reverse-transcribing viruses:** These have dsDNA or ssRNA genomes and their replication includes the synthesis of DNA from RNA by the enzyme reverse transcriptase; many integrate into their host genomes. The group includes the retroviruses, of which Human immunodeficiency virus (HIV), the cause of AIDS, is a member. There is a single family of plant viruses in this group and this is characterized by a single component of circular dsDNA, the replication of which is *via* an RNA intermediate.

- **Double-stranded RNA (dsRNA):** Some plant viruses and many of the mycoviruses are included in this group.

- **Negative sense single-stranded RNA (ssRNA-):** In this group, some or all of the genes are translated into protein from an RNA strand complementary to that of the genome (as packaged in the virus particle). There are some plant viruses in this group and it also includes the viruses that cause measles, influenza and rabies.

- **Positive sense single-stranded RNA (ssRNA+):** The majority of plant viruses are included in this group. It also includes the SARS coronavirus and many other viruses that cause respiratory diseases (including the "common cold"), and the causal agents of polio and foot-and-mouth disease.

Within each of these groups, many different characteristics are used to classify the viruses into families, genera and species. Typically, a combination of characters is used and some of the most important are:

- **Particle morphology:** The shape and size of particles as seen under the electron microscope.

- **Genome properties:** This includes the number of genome components and the translation strategy. Where genome sequences have been determined, the relatedness of different sequences is often an important factor in discriminating between species.

- ❑ **Biological properties:** This may include the type of host and also the mode of transmission.

- ❑ **Serological properties:** The relatedness (or otherwise) of the virion protein(s).

**Particle morphology**: Amongst plant viruses, the most frequently encountered shapes are:

**Isometric:** Apparently spherical and (depending on the species) from about 18nm in diameter upwards. The example here shows *Tobacco necrosis virus*, genus *Necrovirus* with particles 26 nm in diameter.

**Rod-shaped:** About 20-25 nm in diameter and from about 100 to 300 nm long. These appear rigid and often have a clear central canal (depending on the staining method used). Some viruses have two or more different lengths of particle and these contain different genome components. The example here shows *Tobacco mosaic virus*, genus *Tobamovirus* with particles 300 nm long.

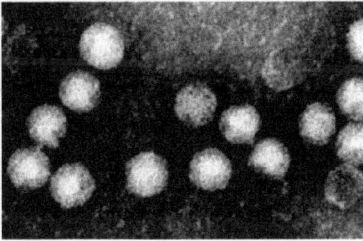

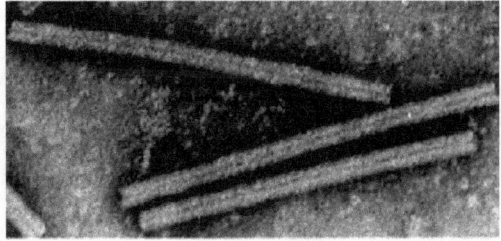

**Filamentous:** Usually about 12 nm in diameter and more flexuous than the rod-shaped particles. They can be up to 1000 nm long, or even longer in some instances. Some viruses have two or more different lengths of particle and these contain different genome components. The example here shows *Potato virus Y*, genus *Potyvirus* with particles 740 nm long.

**Geminate:** Twinned isometric particles about 30 x 18 nm. These particles are diagnostic for viruses in the family *Geminiviridae* which are widespread in many crops especially in tropical regions. The example here shows *Maize streak virus*, genus *Mastrevirus*.

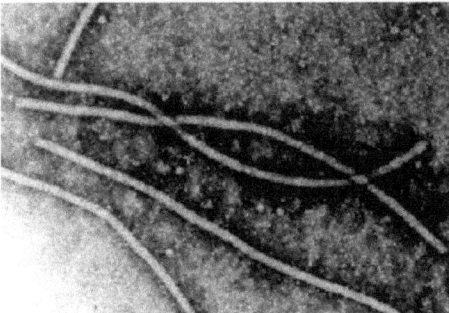

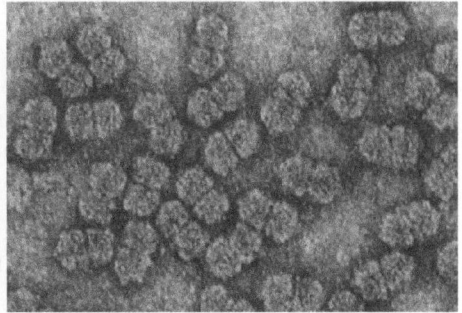

**Bacilliform:** Short round-ended rods. These come in various forms up to about 30 nm wide and 300 nm long. The example here shows *Cocoa swollen shoot virus*, genus *Badnavirus* with particles 28 x 130 nm.

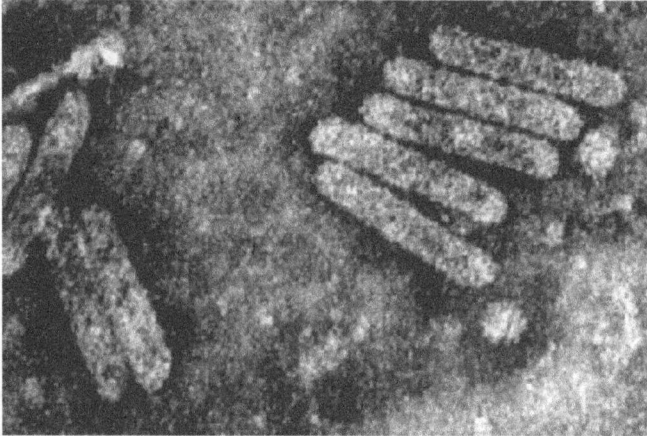

Further details can be found in the genus description pages and on the Rothamsted Electron Micrographs of Plant Viruses page.

**Genome properties:** Important features include:

❑ **Nature of the genome:** Circular (as in all known plant DNA viruses) or linear.

❑ **Number of genome components:** This varies from a single component (e.g., in the genera *Potyvirus* and *Tobamovirus*) to 11 (in some members of the genus *Nanovirus*). Individual components vary in size from about 1kb (*Nanovirus* components) to about 20 kb (in the genus *Closterovirus*).

❑ **Number of genes:** These vary considerably. Most plant viruses have at least 3 genes: 1 (or more) concerned with replication of the nucleic acid, 1 (or more) concerned with cell-to-cell movement of the virus and 1 (or more) encoding a structural protein that is assembled into the virus particle (usually called the "coat" or "capsid" protein). There may also be additional genes that have a regulatory function or which are required for transmission between plants (association with a vector).

❑ **Translation strategy:** A variety of strategies are employed to translate the genes from the genome components either directly or via mRNA intermediates and (in some cases) to permit different amounts of protein to be produced from the different genes. These are summarised for each genus in the genus description pages but 3 examples here serve to illustrate some of the variety:

❑ **Genus *Potyvirus*:** In this very large genus, there is one ssRNA component that encodes one large (*c.* 350 kDa) polyprotein. This is cleaved by 3 different proteases (all encoded by the virus itself) into 10 different mature proteins. The two proteins at the C-terminus of the polyprotein are respectively an RNA-dependent RNA polymerase (NIb, involved in replication of the virus) and the (single) coat protein (CP). Many of the proteins have multiple functions. The genome organization of a typical member is shown here, indicating the 10 mature proteins and the nine cleavage sites (arrowed).

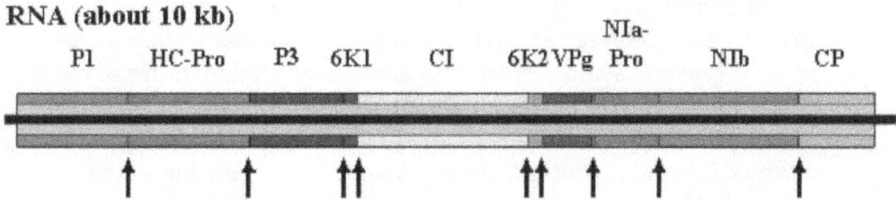

**RNA (about 10 kb)**

| P1 | HC-Pro | P3 | 6K1 | CI | 6K2 | VPg | NIa-Pro | NIb | CP |
|----|--------|----|----|----|----|----|---------|-----|-----|

↑ ↑ ↑↑ ↑↑ ↑ ↑ ↑

- ❏ **Genus *Furovirus*:** In this genus there are two ssRNA components. The 5'-proximal gene on each RNA is translated directly from the genomic RNA: on RNA1 (the larger RNA component) this gene encodes a replication protein and on RNA2 it is the coat protein. The stop codons of both of these genes are "leaky" and in a small percentage of cases, translation continues to produce a larger ("readthrough") protein. On RNA1, the replication protein is extended to include an RNA-dependent RNA polymerase (RdRp) while the readthrough region of the coat protein is probably required for particle assembly and for transmission by the plasmodiophorid vector. There is a further (3'-proximal) gene on each of the RNAs and these are translated from shorter RNA molecules transcribed from the 3'-end of the genomic RNA ("subgenomic" mRNAs). That from RNA1 is a cell-to-cell movement protein (MP) that enables the virus to move between adjacent plant cells via the plasmodesmata while the function of the product from RNA2 is uncertain but may involve suppression of the host plant defence reaction. The genome organization of a typical member is shown here.

- ❏ **Genus *Fijivirus*:** In this genus there are 10 components of dsRNA. Most of the components encode a single protein and at least 3 of these are structural proteins assembled into the complex virion.

- ❏ **Genome relatedness:** The degree of nucleotide identity (or amino acid identity in the protein sequence) between sequences is often used to examine the relationship between different viruses or isolates. For example, recent studies in the genus *Carlavirus* show that when different species are compared, they have less than 73% nucleotide identity (or 80% amino acid identity) in their coat proteins.

## BIOLOGICAL PROPERTIES

- ❏ In some families, the type of host is a useful feature for classification. For example, in the family *Reoviridae*, there are currently 3 genera with plant-infecting members (*Fijivirus, Oryzavirus, Phytoreovirus*), 1 genus of mycoviruses (*Mycoreovirus*), 1 genus containing viruses of fish and cephalopods (*Aquareovirus*), two genera that are restricted to insects (*Cypovirus* and *Entomoreovirus*) and 5 genera of vertebrate viruses that sometimes also infect insects.

❑ The mode of transmission is also a useful characteristic of some groups of plant viruses. For example in the family *Potyviridae*, members of the largest genus (*Potyvirus*) are transmitted by aphids, while viruses in the genera *Rymovirus* and *Tritimovirus* are transmitted by mites of the genus *Abacarus* or *Aceria* respectively, those in the genus *Ipomovirus* are transmitted by whiteflies and those in the genus *Bymovirus* by plasmodiphorids (root-infecting parasites once considered to be fungi but probably more closely related to protists).

Serological properties: Many viruses are good antigens (elicit strong antibody production when purified preparations are injected into a mammal) and this property has been widely exploited to produce specific antibodies that can be used for virus detection and for examining relationships between viruses. Earlier studies used agar diffusion plates but in the last 20 years these have been largely superseded by ELISA (enzyme-linked immunosorbent assay) procedures. Although serological properties are still important, their significance in taxonomy has declined to some extent now that nucleotide sequence data are available.

## ANTON DE BARY

It would not be until 1861 that Anton de Bary, who is considered the father of modern plant pathology, that the question as to the cause of the blight was finally settled. He did what would be today a rather simple experiment. He grew two groups of healthy potato plants which he subjected to the same cool and wet environmental conditions that favoured the blight fungus. To one group he applied the sporangia which he had collected from blighted plants while the other group was kept fungus free. In a matter of a few days, the group to which sporangia was applied already showed signs of the disease and eventually rotted. In the control group, disease did not occur. This convincingly demonstrated that it was indeed the fungus which caused the blight and not saturation from too much water. It was this experiment that led other scientists to critically look at, not only diseased plants, but animal and human diseases as well. This was not only the beginning of plant pathology, but a year later, in 1862, Louis Pasteur's Germ Theory would replace concept of spontaneous generation of microorganisms in dead or dying organisms. So, de Bary's work actually preceded Pasteur's Germ Theory and should probably have been credited with this theory as well.

The Potato Blight returned year after year, despite cold winters that we would assume would destroy the fungus. How then did the fungus then survive the winter to continue its devastation the following year? De Bary determined that it was the farmers themselves that perpetuated the disease. The Late Blight fungus was able to survive the winter in the potato tuber. Tubers that were not consumed that did not appear to be diseased, but in some cases were actually infected, were placed in storage bins to be planted the following year. Under such conditions, if even a *single* tuber carried the Potato Blight fungus, the rest of the tubers in the storage could become infected. When planted the following year, the disease grew upward into the stem and leaves of the potato plants and finally produced sporangia and spores that would further perpetuate the disease. Another source was the rotted tubers that were discarded in the same fields that the potatoes were planted. These tubers would produce potato plants before the potato fields were planted and became a ready source of the disease. This was the reason why entire fields of potatoes could seemingly become infected overnight. It was these practices that were responsible for the devastating famine of 1845. The previous year, the potato crop had been a particularly good one. Far more potatoes were produced than could possibly be consumed by both the Irish peasants and their livestock. The surplus potatoes were stored and many discarded. Some were infected with the Late Blight fungus which grew slowly in storage during the winter. The Blighted tubers were discarded in the spring of 1845 along with surplus healthy tubers. Thus, the Late Blight Fungus was in place before the planting of the potato plants during the summer of 1845.

Why did the Late Blight fungus suddenly destroy the potato plants of Europe in 1845? It is now known that the fungus is probably native to South America where it still causes disease on potatoes. With repeated voyages to South America, and the continued transport of potato, the Late Blight fungus was inadvertently transported with the potatoes and brought to Europe. This probably happened a number of times and during the long voyage the fungus often did not survive under the harsh conditions. However, it is thought that some faster crossings allowed the fungus to survive in the tuber which then became planted in Europe. This, together with the environmental conditions which favored the growth of the Late Blight fungus and the genetic uniformity of the potatoes throughout Europe was responsible for the sudden appearance of the disease.

By 1882, Pierre Millardet, Professor of Botany at Bordeaux University, and a student of Anton de Bary, had been carrying out research on the Downy Mildew fungus for several years and had clearly demonstrated that the infection of this fungus was much like that of the Late Blight. In October of that year, Professor Millardet was strolling through a vineyard. There had been much mildew in the locality that year, and he was surprised to see that the vines beside the roads were still leafy, while elsewhere they were bare. Examining these leaves more closely he found traces of a bluish-white deposit on them as though someone had treated them with some chemical. Millardet then went to see Mr. Ernest David, the manager of that vineyard. He learned from Mr. David that it was common practice for the vine growers to spray the vines beside the roads with a conspicuous poisonous looking substance to discourage passer-by from sampling the grapes. To do this, the grapes were sprayed with a solution composed of copper sulphate and lime. Mr. David had never really noticed that the sprayed plants remained healthy while those left untreated went the way of all other diseased vines.

For the next two years Millardet tested variations of the mixture used by Mr. David and found the copper sulphate best for not only controlling the Downy Mildew, but also the Late Blight as well. The copper sulphate is what we now call the Bordeaux mixture, named for the area of France where it was discovered. However, rumors of his copper sulphate cure had already spread far by this time. Now many others were claiming credit for the discovery of the Bordeaux mixture. Fortunately, Millardet along with Planchon were the ones to discover the Downy Mildew on grape in 1878 and since that time Millardet had continued to work on the disease. Thus, his claim to be the discoverer of this cure was the most credible. However, this did not keep many other researchers from producing variants of the Bordeaux mixture and claiming it to be their own invention.

So now a means was found by which the Downy Mildew of grapes could be controlled and the French vineyards were thought to be saved. However, this was not the end of France's problems. During the aftermath of the disease, something else happened that almost caused financial ruin to the wine industries of France and some of the Mediterranean countries, but this had nothing to do with fungi, insects or any other types of diseases of grape. Instead this came about through greed. When it appeared that the wine industry of France would collapse, some of the Mediterranean countries thought that they would be able to fill the void that France would certainly leave with the collapse of their wineries. The supply of available wine soon started to far exceed the demand and there was a glut on the market, which nearly caused financial ruin in all these countries.

**Anton de Bary:** Considered to be the father of plant pathology and the person who demonstrated that the Late Blight of Potato was caused by a fungus that he named *Phytophthora infestans*. Genus literally means plant eater.

**Bordeaux mixture:** A copper sulfate and lime solution first used to control *Plasmopara viticola*, the cause of Downy Mildew of Grapes, but also found to be effective against *Phytophthora infestans*, the cause of Late Blight of Potato. Mixture named for the university in which it was developed.

## DETECTION OF PLANT VIRUSES AND VIROIDS BY MOLECULAR HYBRIDIZATION

A viral particle is composed of nucleic acids (ribonucleic acid = RNA or deoxyribonucleic acid = DNA) and a capsid made up of several dozen to a thousand copies of coat protein subunit. In some cases the virus possesses an envelope composed of viral proteins integrated in membranes deriving from the host cell. Serological techniques detect the virus by specific recognition of the coat protein by specific antibodies developed in animals against this protein. Molecular hybridization techniques detect viral nucleic acids by specific recognition of their nucleotide sequence.

Nucleic acids are long, linear polymers of nucleotide molecules. Each nucleotide is in turn composed of several elements: a nitrogen containing base linked to a phosphate group and a sugar molecular (ribose for RNA and deoxyribose for DNA). DNA contains four different bases: Adenine (A), Guanine (G), Cytosine (C) and Thymine (T). In the case of RNA, Thymine is replaced by Uracil (U), the three other bases being the same.

DNA is usually found in a double-stranded configuration, i.e., two chains of DNA associate through specific base pairing (A pairs with T and C pairs with G). Base pairing is extremely specific and creates non-covalent hydrogen bonds that unite the molecules associated in this way. RNA is most commonly found in a single-stranded configuration but, like DNA, it possesses the capacity to form double-stranded structures through A-U and G-C pairing.

The specific pairing of the bases composing nucleic acids constitutes the basis for the formation of hybrids (double-stranded structure) between complementary molecules and, thus, for the use of molecular hybridization as a diagnostic technique.

Nucleic acid molecules differ from one another in the order and sequence of alignment of their nucleotides (= nucleotide sequence). The presence of two molecules of complementary sequences will lead to the formation of double stranded hybrids under suitable conditions. For example, TCGGCGTAT will pair with AGCCGCATA to make a DNADNA hybrid.

A probe used for virus detection in molecular hybridization experiments is a single-stranded nucleic acid molecule prepared from a viral nucleic acid, single-stranded with a nucleotide sequence complementary to that of the target viral RNA molecule.

Thus a DNA probe with the sequence: TCGGCGTAT will specifically detect RNA and DNA molecules with the respective sequences AGCCGCAUA and AGCCGCATA. An RNA probe with the same specificity would be: UCGGCGUAU.

The molecular hybridization detection system presented here is based on solid support hybridization, the samples being permanently immobilized on a nitrocellulose membrane we will describe the technique using the two most frequently used types of probe:

- Complementary DNA probes cloned in a plasmid vector;

- *In vitro* transcribed complementary RNA probes prepared from complementary DNA cloned into special purpose transcription plasmid vectors.

The probes can be labeled either radioactively or by incorporation of a non-radioactive marker such as biotin.

## MOLECULAR DIAGNOSTICS FOR FUNGAL PLANT PATHOGENS

Accurate identification of fungal phytopathogens is essential for virtually all aspects of plant pathology, from fundamental research on the biology of pathogens to the control of the diseases they cause. Although molecular methods, such as polymerase chain reaction (PCR), are routinely used in the diagnosis of human diseases, they are not yet widely used to detect and identify plant pathogens. Here we review some of the diagnostic tools currently used for fungal plant pathogens and describe some novel applications. Technological advances in PCR-based methods, such as real-time PCR, allow fast, accurate detection and quantification of plant pathogens and are now being applied to practical problems. Molecular methods have been used to detect several pathogens simultaneously in wheat, and to study the development of fungicide resistance in wheat pathogens. Information resulting from such work could be used to improve disease control by allowing more rational decisions to be made about the choice and use of fungicides and resistant cultivars. Molecular methods have also been applied to the study of variation in plant pathogen populations, for example detection of different mating types or virulence types. PCR-based methods can provide new tools to monitor the exposure of a crop to pathogen inoculums that are more reliable and faster than conventional methods. This information can be used to improve disease control decision making. The development and application of molecular diagnostic methods in the future is discussed and we expect that new developments will increase the adoption of these new technologies for the diagnosis and study of plant disease.

## REAL-TIME PCR

It differs from classical PCR by the measurement of the amplified PCR product at each cycle throughout the PCR reaction. In practice, a video camera records the light emitted by a fluorochrome incorporated into the newly synthesized PCR product. Thus, real-time PCR allows the amplification to be followed in real-time during the exponential phase of the run, and thus allows the amount of starting material to be determined precisely. Contrary to end-point PCR techniques, the result is independent from the plateau corresponding to the saturation of the reaction, the latter leading to inaccurate quantification. The main principles of the real-time PCR process, while the details and requirements necessary to obtain reliable data have been reviewed by several scientists. Besides being an alternative to some well-established laboratory techniques, real-time PCR has a number of features which makes it the choice for several types of study. Compared with the other techniques presently available, it allows the detection of a given nucleic acid target in a rapid, specific and very sensitive way. In addition, it affords the absolute quantification of the initial target. To date, the reliability of real-time PCR has never been questioned.

Real-time quantitative PCR was first developed to meet specific technical requirements, such as a high sensitivity and specificity, which were not easily achieved with other classical techniques. It is now becoming a routine tool, and it is believed that, thanks to its experienced reliability, its applications will proliferate in the forthcoming years. Thanks to its rapidity, it should even replace some widely used protocols, like Southern blotting for transgenic plant analysis. Most evidently, real-time PCR development is still limited by the high costs of the machine and reagents, but hopefully, future will make this technology economically more widely accessible.

## RELEVANT FEATURES OF REAL-TIME PCR

### Rapidity

Compared with classical PCR, one of the main advantages of real-time PCR is its rapidity to provide reliable data. Typically, the time of a whole real-time PCR run ranges from 20 min to 2 h. Indeed, the time needed to shift temperature is a major limiting factor responsible for the duration of a classical PCR experiment. The LightCycler™ PCR machine (Roche) uses capillaries instead of tubes, which are heated by light instead of a heating block. As a result, the time necessary to heat the PCR mixture is considerably reduced (from 15 s to 1–2 s). In addition, recording the amplification in real-time avoids collecting samples at different steps of the PCR experiment, making the process less tedious and time-consuming. Moreover, some machines accommodate 384 well plates and can process queuing plates over 24 h non-stops, which might be a determining advantage for high throughput studies, or if rapid sample processing is required.

### Sensitivity

Real-time PCR provides a high sensitivity for the detection of DNA or RNA due to a combination of the amplification performed by the PCR step and the system of detection. It is therefore a very convenient technique for studies with a limited amount of starting material, or for assessing the expression of a high number of genes from minute quantities of RNA. The detection is based on the measurement of the fluorescence emitted by probes incorporated into the newly formed PCR product, or alternatively released into the buffer during the amplification of the PCR product. Intercalating agents and fluorogenic probes are the two main types of molecules currently used to detect PCR amplification in real-time. The first intercalating agent used was ethidium bromide, but in 1997, Wittwer *et al.* proposed replacing it with the SYBRgreen® molecule, because of its higher affinity for double-stranded DNA. As intercalating agents bind regardless of the nucleotide nature, they can be used for any type of sequence.

This is an economical advantage for a laboratory testing a large number of genes. However, a disadvantage of SYBRgreen® is that it is equally incorporated into every amplicon, and should unspecific sequences be amplified, the signal measured would correspond to both non-specific and specific products, thereby compromising the accurate quantification of the latter (see section 'Specificity' for further developments on this issue).

In order to bypass this potential problem, intercalating agents can be replaced by labeled oligonucleotides or probes, which specifically bind to the target sequence. This technology relies on the use of probes labeled with two different fluorochromes, one of which, when excited, is able to transfer its energy to the other via Fluorescence Resonance Energy Transfer (FRET). This non-radioactive energy transfer only occurs if the two molecules are in close proximity to each other (a few nanometers). Depending on the proximity of the second fluorochrome, the first one may either emit light or transfer its energy to the second, which in turn fluoresces. Thus, bringing the two fluorochromes in close proximity to each other results in the fluorescence quenching of the first one, and fluorescence emission of the second one.

As the fluorescence from the emitting fluorochrome increases proportionally with the amount of newly synthesized DNA, both effects can be recorded to follow the amplification of the target DNA. Hence, several strategies have been developed, all of which rely on placing them in the vicinity of each other (excitation) or conversely ensuring their separation (quenching) during the amplification phase. So far, FRET-mediated excitation has rarely been used in plant studies, and its use, which requires four oligonucleotides, should be limited to studies requiring a very high level of specificity. The application of quenching systems has been more common in plant studies, being initiated by the use of the TaqMan® probes, followed by the Molecular Beacons and Scorpion™ probes. As they require the design of a labeled oligonucleotide specific for each sequence, they are economically relevant only if many experiments are to be performed on the same target.

## Specificity

Surprisingly, in a study carried out on four pea thioredoxin h (TRXh) encoding genes, Montrichard *et al.* noticed that real-time PCR yielded weaker signals than expected from northern blot analyses. This observation was explained by a cross-hybridization of the probe to the RNA encoding another isoform during the northern blot procedure. Indeed, in contrast to techniques requiring the hybridization of nucleic acids several hundreds base pairs long, such as cDNA-based microarray and northern blotting, short oligonucleotide mediated real-time PCR guarantees a high specificity in the detection of the target sequence. In fact, specificity is achieved by the use of two target sequence-specific oligonucleotides, and this can be enhanced by increasing the number of oligonucleotides nested within the initial amplification product. In this respect, FRET-mediated probes seem to ensure a higher specificity than SYBRgreen®. In any case, specificity of the process can be checked after completion of the PCR run, by testing the nature of the amplified product with gel electrophoresis, melting curves, and sequencing data.

## Quantification

Most importantly, the quantification range of real-time PCR is up to seven orders of magnitude as originally illustrated and more recently in plants. These results from the capacity of this technique to calculate, for every sample within an extremely low to high concentrations range, the number of cycles necessary to reach the Ct. The absolute amount of the target is calculated from a calibration curve. Alternatively, a relative quantification can be deduced considering Ct differences between samples and standards as nicely illustrated by several scientists.

Basically, real-time quantitative PCR may be used for quantifying DNA or RNA abundance, leading to two major types of applications: foreign DNA (e.g., transgenic or contaminating micro-organisms) detection and quantification, and gene expression studies.

## DETECTION AND QUANTIFICATION OF FOREIGN DNA BY REAL TIME PCR

### Quantification of pathogenic or symbiotic micro-organisms associated with plants

Real-time PCR assays aiming at quantifying the level of plant infection by a pathogen have been increasing for the last few years. Most of them rely on the relative quantification of two specific plant and pathogen DNA sequences. They are faster, more specific and more sensitive when compared with traditional protocols based on symptom recording or on conidiophores or colony counting and most importantly, may be transposed to virtually every pathosystem. For those reasons, they are being widely used for the diagnosis of diseases in the field and for applied purposes. For instance, seed potatoes cannot be sold in the EU unless they are devoid of the potato brown rot agent *Ralstonia solanacearum*. Classical detection methods require a labour-intensive culture and pathogenicity test on tomato seedlings. However, real-time PCR has been shown to enable the quantitative detection of *R. solanacearum* in a rapid and reliable manner, thus providing an improved alternative assay that could be implemented on a large scale.

Likewise, food contamination by mycotoxins is of great concern, since many have been found to be carcinogenic and they are not easily removed during food processing. However, since toxin abundance does not correlate with fungal contamination, but is linked to the toxinogenic properties of each strain, real-time PCR detection assays targeted at genes involved in toxinogenesis have been developed for trichothecene-producing *Fusarium* and aflatoxin-producing *Aspergillus* species. Recently implemented a refinement of this technique based on the quantification of the *nor1* mRNA, which directly addresses aflatoxin biosynthesis in infected wheat. As many countries are becoming more and more concerned about food safety, the market for such applications is growing rapidly.

Real-time PCR application in fundamental studies is still lagging behind, and only a few real-time PCR-based pathogenicity assays have been reported in this field. Most of the currently used resistance tests rely on visual assessment of the symptoms and spore or colony counting. However, recently implemented real-time PCR tests to quantify a number of pathogens on *Arabidopsis* and demonstrated that they are a very interesting alternative to classical tests.

## CONTAMINATION OF PROCESSED FOOD BY FOREIGN DNA

A requirement for methods capable of accurately quantifying food contaminants has emerged, due to the introduction of stringent food safety regulations. Since most of them enforce a tolerable level of contamination, accurate quantification is of crucial importance for agribusiness companies. Indeed, on the one hand, they have to comply with those maximal authorized levels; while on the other hand, the rejection of batches falsely labeled as contaminated would lead to unnecessary costs. In this context, real-time quantitative PCR is becoming the technique of choice for assessing food contamination or adulteration. Compared with ELISA, PCR assays are easier to develop since they do not require raising specific antibodies. They provide a higher sensitivity and are better suited for the detection of unwanted food ingredients in highly processed food, notably because DNA is more thermostable than proteins. For example, the absence of gluten in baby food can be controlled by an amplification test of cereal genes by real-time PCR. Likewise, real-time quantitative PCR has been shown to be an ideal tool for assessing common wheat (*Triticum aestivum*) adulteration in durum wheat pasta, as Spanish, Italian, and French regulations enforce a 3 percent maximal level of common wheat contamination in pasta and semolina.

Another field of applications was born with the launching of genetically modified organisms on the market. Indeed, the European Community Council recently enacted a new regulation (EC Regulation no. 1829/2003) enforcing the compulsory labeling of food containing more than 0.9% GMOs. However, GMO detection is not trivial and current assays present a number of worrying limitations. Briefly, protocols aimed at detecting transgenic DNA contamination were developed mainly for transgenic soybean and maize and more rarely for other species such as rapeseed. A major limitation of PCR-based detection assays is that a new set of oligonucleotides has to be designed for each transgenic line, except when they are targeted to common DNA regions such as the CaMV35S promoter. Alternatively, event-specific protocols have been developed for unique lines, such as Starlink and Bt11 transgenic maize. In order to make the detection specific for only one given and identifiable event, scientists have cloned the borders separating the transgene from the rest of the host genome, and used them as specific markers of the given event.

## GENETICS OF TRANSGENIC

Transgenic plants are easily amenable to genetic analyses when the transgene is inserted as a single copy within the host genome. Some scientists showed that duplex real-time PCR can be used to determine transgene copy number in transformed plants. They found a strict correlation between their results and Southern blot analyses, except for two lines (out of 37) in which the discrepancy could be ascribed to multiple insertions at a single locus. Overall, they demonstrated that real-time PCR provided a fast and robust method for this application, which could easily be automated and applied to a large number of samples. In lines containing a single insertion, multiplexed PCR could even discriminate between homozygous and hemizygous plants. If previously genotyped lines are not available to calibrate the assay, proposed using hemizygous $T_0$ plants as a standard. In this way, homozygous plants can be selected before they produce seeds, thereby avoiding labourious segregation analyses.

## QUANTIFICATION OF SPECIFIC TRANSCRIPTS

### Integrated expression analyses

Integrative developmental biology requires the parallel analyses of genes involved in the same physiological process. The study of their expression profile may be useful in understanding the cellular function of the encoded proteins. Indeed, in yeast, groups of genes displaying the same expression profile are enriched in genes encoding proteins interacting physically. However, comparisons between numerous gene expression profiles require a sensitive and reliable technique avoiding errors inherent in independent RNA preparations. Because of its sensitivity, real-time PCR is able to meet this technical requirement, as the expression of numerous genes may be tested on the same RNA preparation. Taking advantage of this, used real-time PCR to investigate the expression of eight genes involved in brassinosteroid (BR) biosynthesis, degradation or signal transduction. They were able to correlate the expression of the BR synthesizing enzymes with the presence of BR *in planta*, thus providing a more complete view of BR metabolism at the whole plant level. Likewise, the expression of genes involved in phenylpropanoid biosynthesis in pine, following phenylalanine feeding. Their study provided very interesting insights about the coordinate regulation of seven key genes (among them phenylalanine-ammonia-lyase and coumarate-4-hydroxylase) involved in lignin precursors metabolism.

## GENE FAMILIES

Mutagenic families are a distinctive feature of plant genomes, as opposed to animals. For instance, 66% of *A. thaliana* genes belong to families, with half of them containing at least five members (The *Arabidopsis* Genome Initiative, 2000), and information available from ESTs programs on other plant species shows that this particularity is not restricted to *Arabidopsis*. Analyzing all the members of a gene family is necessary in order to obtain an accurate view of its overall function. Often, studies describing the expression profile of multigene family members have been performed in *A. thaliana* because of the possibility of an exhaustive survey. However, other plants with a high number of publicly available ESTs, such as wheat, maize, tomato, rice, potato, and the moss *Physcomitrella patens*, could be used for similar analyses. However, the amplification of several highly similar sequences cannot be excluded in those organisms, and data interpretation should be performed very cautiously.

In this context, several requirements have led plant scientists to use real-time RT-PCR methods rather than northern blotting. First, the analysis of more than ten genes by northern blotting is a fairly tedious and repetitive work. Secondly, genes expressed at a very low level remain difficult to detect by northern blot techniques. In *A. thaliana*, the ARIADNE gene family is composed of 16 genes that have been identified as putative E3-ligases, based on their homology with *Drosophila* and mouse genes involved in the ubiquitinmediated protein degradation pathway. Their expression was studied by real-time RT-PCR. The transcripts of seven genes were undetectable or close to the background level, while the remaining genes were expressed, but with different absolute mRNA levels that varied up to 25-fold. The absolute transcript level remained very low, with fewer than 50 copies per nanoram of total RNA for eight of the nine *AtARI* genes analyzed. The sensitivity of real-time RT-PCR showed that, while the expression profile was quasi-constitutive for nearly all of the *AtARI* genes, two of the very low expressed genes displayed a specific expression pattern.

Thirdly, closely related genes that are very similar at the sequence level may cross-hybridize during northern blot procedures, and therefore, it may be difficult to determine the RNA level of a specific member of a gene family. This problem is resolved by the high specificity of real-time RT-PCR guaranteed by the use of at least two specific oligonucleotides. The GSK3/Shaggy kinase family is composed of ten genes in *A. thaliana* (*AtSK*). Their expression level is rather low (2-fold lower than actin; and difficult to study by northern blotting, although possible. However, some of these genes display a very high level of nucleotide sequence identity up to 98.2%., and finding probes specific for each of the ten genes requires them to be designed in the 5' or 3' UTR regions. In order to overcome these difficulties, real-time RT-PCR was used to study the expression profile of the entire gene family, both in several organs of the plant and in response to a number of abiotic stresses. Relative analyses showed that, while most of the *AtSK* genes were expressed constitutively, two of them displayed either some organ specificity or responded to osmotic stress.

Knowing the exhaustive expression pattern of a gene family opens up the additional investigation branch of molecular and functional evolution. Indeed, several studies have shown that the expression profile of gene family members was not strictly related to the position of the corresponding proteins within a phylogenetic tree. This suggested that the regulation of gene expression had undergone different molecular evolutionary mechanisms compared with those influencing protein sequences.

To date, only a limited number of additional studies have described gene family expression profiles using real-time RT-PCR. Alternative methods to address the transcriptional behaviour of members of large gene families are available, such as microarray hybridization, as illustrated for 142 members of the *Arabidopsis* cytochrome $P_{450}$ gene family. However, due to their specific limitations, an exhaustive picture can only be obtained by the parallel use of those methods.

## CONFIRMATION OF MICROARRAY EXPERIMENTS

One of the fastest expanding applications of real-time RT-PCR is the confirmation of data obtained from microarray studies. Indeed, the reliability of microarray experiments may sometimes be questioned. Since plants display a high number of multigene families, cross-hybridization between cDNA representatives of members of gene families on cDNA-based chips may lead to false interpretations. On the other hand, microarray experiments can analyze thousands of genes in one step, whereas real-time PCR is often limited to far fewer genes. Real-time PCR requires the design of specific oligonucleotides for each gene to be analyzed, and because of the limited number of both fluorophores and light spectra detected by real-time PCR machines, this allows the detection of fewer than five genes per multiplex PCR run. However, a maximum of two genes are analyzed routinely in the same tube. Therefore, a widely used strategy is to point out a handful of potentially interesting genes with microarray experiments and to confirm those candidates by real-time RT-PCR analysis. As an illustration randomly picked 12 candidates among a pool of 185 genes previously identified by Affimetrix™ microarray experiments as being up-regulated in the seeds of the *Arabidopsis pickel* mutant compared with wild-type seeds. They confirmed the up-regulation for 10 of those genes by real-time RT-PCR experiments.

While some studies are purely confirmatory, others have used real-time RT-PCR to analyze the expression pattern of the candidate genes further, either to determine fine-tuned kinetics or in conditions where the available material was sparse, thereby taking advantage of the technique's sensitivity. The hybridization of cDNA microarrays with RNA of *Arabidopsis* seedlings infected with the incompatible fungus *Alternaria brassicicola* led to the identification of functional groups of genes involved in systemic acquired resistance. Then, 23 genes, each representative of one of these groups, were chosen to perform a time-course study in response to *A. brassicicola* infection in local and distal leaf tissues by real-time RT-PCR.

Whereas numerous studies have obtained similar results by real-time PCR and microarray experiments, with a linear correlation of up to five orders of magnitude several other articles report that the *n*-fold variation measured by real-time PCR is generally lower (up to 10 times) than that measured using cDNA microarray, where cross-hybridization may occur for highly similar, yet distinct gene sequences.

Even with the Affimetrix™ chips, which are oligonucleotide-based and therefore more specific than PCR-fragment-based microarray slides, discrepancies between microarray data and real-time PCR data have been noticed. Some scientists showed that real-time PCR data could display higher induction ratios compared with microarrays, yet conserving a good correlation between the two techniques. However, it is noticed that the magnitude and the kinetics of the response of several genes to phosphate starvation differed between the two techniques, including opposite responses for some of them. These conflicting results were explained by the fact that the two experiments were performed with two different biological samples and different oligonucleotides.

In all cases, real-time RT-PCR is considered to be the most reliable technique and discrepancies can most often be ascribed to the normalization or background subtraction methods used in microarray analysis. Hence, real-time RT-PCR has even been used as a reference to compare different methods of microarray analyses.

The application range of real-time quantitative PCR is much broader than what can possibly be exposed in the framework of this review. For example, in the domain of DNA quantification exploited real-time PCR to quantify the proportion of restricted DNA corresponding to a transgene, after methylation as a result of post-transcriptional gene silencing. While the rapidity and the reliability of real-time RT-PCR has allowed the description of the expression pattern of numerous individual genes in wild-type plants and transgenic plants as well as the description of their molecular phenotypes.

# LIMITATIONS AND FUTURE DEVELOPMENTS OF REAL – TIME PCR

Despite numerous advantages, real-time PCR has some limitations. Since it is performed on small DNA fragments, real-time quantitative PCR might fail to detect biologically relevant processes like alternative splicing or partial transcript degradation occurring during post-transcriptional gene silencing events. Performing northern blots in parallel with real-time PCR should help to overcome this difficulty of the NADPH/NADP-thioredoxin gene expression. Once specific splicing sites are known, appropriate intron-hybridizing primers can be designed to monitor the accumulation of a specific transcript.

For convenience, most real-time PCR analyses are currently performed at the organ level, but further studies may take advantage of the technique's unequalled sensitivity, and address gene expression at the cellular level.

# REFERENCES

Agrios, G.N. (1997). *Plant pathology,* 4th edition, Academic Press, San Diego, U.S.A.

Alexopolvs, C.J. and Mims, C.N. 1973. *Introductory Mycology.* Wiley Eastern Ltd., New Delhi.

Anderws, J.H. and Tommemp, I.C. 1993. *Advances in Plant Pathology.* Academic Press, London.

Atlas, R.M. 1984. *Fundamentals of Microbiology.* MacMillan Publishing Co., New Delhi.

Garrow, J.S. and James, W.P. 1998. *Human Nutrition and Dietetics.* Churchill Living Stone, Edinburgh.

Kamat, N.M. 1967. *Introductory Plant Pathology.* Prakash Publishing House Pune.

Kamat, N.M. 1967. *Practical Plant Pathology.* Prakash Publishing House Pune.

Pelezar, M.J. and Reid, R.D. 1986. *Microbiology.* Mc Graw Hill Book Co., New York.

Rangaswami, G. 1979. *Diseases of Crop Plants in India.* Prentice Hall of India Pvt. Ltd., New Delhi.

Sendhilvel, V., Kavitha, K., Nakkeeran, S., Raguchander, T. and Marimuthu, T. (2004). *Glimpses of Plant Pathology.* A.E. Publications. N.10, Sundaram street-1, P. N. Pudur, Coimbatore – 641 041.

Stainer, R.Y. and Aldebelberg, 1985. *General Microbiology.* Tata Mc Graw-Hill Publishing Co., New Delhi.

Singh, R.S. 1975. *Introduction to Principles of Plant Pathology.* Oxford and IBH Publishing Co., New Delhi.

Singh, R.S. 1975. *Plant Diseases.* Oxford and IBH Publishing Co., New Delhi.

Singh S.J. (2003). *Virus and phytoplasma Diseases of papaya,* passion fruit and pineapple. Kalyani Publishers New Delhi.

Verma, H.N. (2003) *Basis of Plant virology,* Oxford and IBH Publishing Co. Pvt. Ltd, New Delhi.

Walkey, D.G.A. (1985). *Applied Plant Virology,* Heinemann. London.

Walkey, D.G.A. (1991). *Applied Plant Virology* 2nd edn. Chapman and Hall, London.

West, E.S., Todd, W.R. Mason, H.S. and Bruggan, T.V. 1968. *Textbook of Biochemistry.* Macmillan Company, New York.

www.ingramcontent.com/pod-product-compliance
Lightning Source LLC
Chambersburg PA
CBHW022045210326
41458CB00071B/308